CHANGING THE
GL

Pers

CHANGING THE GLOBAL ENVIRONMENT
Perspectives on Human Involvement

Edited by

Daniel B. Botkin
Environmental Studies Program and
Department of Biological Sciences
University of California
Santa Barbara, California

Margriet F. Caswell
Environmental Studies Program and
Department of Economics
University of California
Santa Barbara, California

John E. Estes
Department of Geography
Remote Sensing Unit
University of California
Santa Barbara, California

Angelo A. Orio
Department of Environmental Science
University of Venice
Venice, Italy

ACADEMIC PRESS, INC.
Harcourt Brace Jovanovich, Publishers
Boston   San Diego   New York
Berkeley   London   Sydney
Tokyo   Toronto

ACADEMIC PRESS, INC.
1250 Sixth Avenue, San Diego, CA 92101

United Kingdom Edition published by
ACADEMIC PRESS INC. (LONDON) LTD.
24-28 Oval Road, London NW1 7DX

Library of Congress Cataloging-in-Publication Data
Changing the global environment : perspectives on human involvement /
    edited by Daniel B. Botkin . . . [et al.].
      p. cm.
    Bibliography: p.
    Includes index.
    ISBN 0-12-118730-6   ISBN 0-12-118731-4(pbk)
    1. Man—Influence on nature.   2. Environmental policy.
3. Pollution.   I. Botkin, Daniel B.
GF75.C47   1988
363.7′05—dc19   88-14588
                                                                CIP

Printed in the United States of America
89 90 91 92   9 8 7 6 5 4 3 2 1

CONTENTS

Contributors   ix

Preface   xiii

Acknowledgments   xix

I. Perspectives on the Biosphere   **1**

  1. Science and the Global Environment   3
     D.B. Botkin

  2. Man's Role in Managing the Global Environment   15
     L.M. Talbot

  3. The Impact of Life on the Planet Earth:
    Some General Considerations   35
     P. Westbroek

  4. From Planetary Atmospheres to Microbial Communities:
    A Stroll Through Space and Time   49
     L. Margulis and R. Guerrero

  5. Sustainable Use of the Global Ocean   69
     G. Carleton Ray

  6. Deforestation and Extinction of Species   89
     T.E. Lovejoy

7. Large-Scale Alteration of Biological Productivity Due to
Transported Pollutants 99
O.L. Loucks

8. Food Problems in the Next Decades 117
S.H. Wittwer

9. Soil Degradation and Conversion of Tropical Rainforests 135
R. Lal

10. Nuclear Winter, Current Understanding 155
G. Carrier

**II. Some of the Ways We Can Learn About Our Global
Environment** **165**

11. Modern Chemical Technologies for Assessment and
Solution of Environmental Problems 167
A.A. Orio

12. The Search for Nonrenewable Resources in the Next
Twenty Years 185
J.V. Taranik

13. Remote Sensing of Environmental Change in the
Developing World 203
C.K. Paul, M.L. Imhoff, D.G. Moore, and A.M. Sellman

14. The Application of Remote Sensing in South America:
Perspectives for the Future Based on Recent Experiences 213
W.G. Brooner

15. Geographic Information Systems and Natural Resource
Issues at the State Level 231
P.G. Risser and L.R. Iverson

16. The Global Environment Monitoring System and the
Need for a Global Resource Data Base 241
M.D. Gwynne and D.W. Mooneyhan

17. The Use of Remote Sensing to Assess Environmental
Consequences of Nuclear Facilities 257
L.R. Tinney and J.G. Lackey

18. Linking Ecological Networks and Models to Remote
Sensing Programs 269
M.I. Dyer and D.A. Crossley, Jr.

19. Observing the Earth in the Next Decades 283
A.J. Tuyahov, J.L. Star, and J.E. Estes

III. Social and Economic Policy Issues   **305**

  20. Sustainable Futures: Some Economic Issues   309
     D. Pearce

  21. Ethical and Economic Systems for Managing the Global
     Commons   325
     R.C. d'Arge

  22. Managing Natural Resources: The Local Level   339
     D. Brokensha and B.W. Riley

  23. Social Values and Environmental Quality   367
     T.H. Bingham

  24. Technology and the Environment   383
     G. Bugliarello

  25. Management of High Technology: A Cure or Cause of
     Global Environmental Changes?   403
     B.W. Mar

  26. Better Resource Management Through the Adoption of
     New Technologies   419
     M.F. Caswell

  27. The Search for Effective Pollution Control Policies   437
     A. Endres

Index   *455*

CONTRIBUTORS

Numbers in parentheses indicate the pages on which the authors' contributions begin.

Tayler H. Bingham (367), *Center for Economics Research, Research Triangle Institute, Research Triangle Park, North Carolina 27709*

Daniel B. Botkin (3), *Environmental Studies Program and Department of Biology, University of California, Santa Barbara, California 93106*

David Brokensha (339), *Environmental Studies Program and Department of Anthropology, University of California, Santa Barbara, California 93106*

William G. Brooner (213), *Earth Satellite Corporation, 7222 47th Street, Chevy Chase, Maryland 20815*

George Bugliarello (383), *Polytechnic University, 333 Jay Street, Brooklyn, New York 11201*

George F. Carrier (155), *Division of Applied Sciences, Harvard University, Cambridge, Massachusetts 02138*

Margriet F. Caswell (419), *Environmental Studies Program and Department of Economics, University of California, Santa Barbara, California 93106*

D. A. Crossley, Jr. (269), *Institute of Ecology, University of Georgia, Athens, Georgia 30602*

Ralph C. d'Arge (325), *Department of Economics, University of Wyoming, Laramie, Wyoming 82071*

Melvin I. Dyer (269), *Institute of Ecology, University of Georgia, Athens, Georgia 30602*

Alfred Endres (437), *Technische Universitaet, SEKR WW15, Uhlandstrasse 4-5, 1000 Berlin 12, West Germany*

John E. Estes (283), *Remote Sensing Unit, Department of Geography, University of California, Santa Barbara, California 93106*

Ricardo Guerrero (49), *Autonomous University of Barcelona, Spain*

Michael D. Gwynne (241), *United Nations Environment Program—GEMS, 6 rue de la Gabelle, CH-1227 Carouge, Switzerland*

M. L. Imhoff (203), *Goddard Space Flight Center, NASA, Greenbelt, Maryland 20771*

L. R. Iverson (231), *Illinois Natural History Survey, Champaign, Illinois*

J. G. Lackey (257), *EG&G Energy Measurements, Inc., PO Box 1912, Las Vegas, Nevada 89125*

R. Lal (135), *Department of Agronomy, Ohio State University, Columbus, Ohio 43210*

Orie L. Loucks (99), *Butler University, Holcomb Research Institute, 4600 Sunset Avenue, Indianapolis, Indiana 46208*

Thomas E. Lovejoy (89), *Smithsonian Institution, Washington, DC 20560*

Brian W. Mar (403), *Department of Civil Engineering and Science, University of Washington, Seattle, Washington 98195*

Lynn Margulis (49), *Department of Botany, University of Massachusetts, Amherst, Massachusetts 01003*

D. Wayne Mooneyhan (241), *United Nations Environment Program—GRID, 6 rue de la Gabelle, 1227 Carouge, CH-1211 Geneva 10, Switzerland*

Donald G. Moore (203), *EROS Data Center, U.S. Geological Survey, Sioux Falls, South Dakota 57198*

Angelo A. Orio (167), *Department of Environmental Science, University of Venice, Calle Largia, S. Marta 2137, I-30123 Venice, Italy*

Charles K. Paul (203), *Office of Forestry, Environment, and Natural Resources, Bureau for Science and Technology, U.S. Agency for International Development, Washington, DC 20523*

David Pearce (309), *Department of Economics, University College London, Gower Street, London WC1E 6BT, United Kingdom*

G. Carleton Ray (69), *Department of Environmental Sciences, University of Virginia, Charlottesville, Virginia 22903*

Bernard W. Riley (339), *Environmental Studies Program, University of California, Santa Barbara, California 93106*

Paul G. Risser (231), *Scholes Hall, University of New Mexico, Albuquerque, New Mexico 87131*

A. N. Sellman (203), *Environmental Research Institute of Michigan, PO Box 8618, Ann Arbor, Michigan 48107*

Jeffrey L. Star (283), *Remote Sensing Unit, Department of Geography, University of California, Santa Barbara, California 93106*

Lee M. Talbot (15), *World Resources Institute, 1735 New York Avenue, NW, Washington, DC 20006*

James V. Taranik (185), *Desert Research Institute, University of Nevada, Reno, Nevada 89557*

Larry R. Tinney (257), *EG&G Energy Measurements, Inc., PO Box 1912, Las Vegas, Nevada 89125*

Alexander J. Tuyahov (283), *National Aeronautic and Space Administration Headquarters, 600 Independence Avenue, Washington, DC 20546*

Peter Westbroek (35), *Department of Biochemistry, University of Leiden, Wassenaarseweg 64, 2333 Al Leiden, The Netherlands*

Sylvan H. Wittwer (117), *Agriculture Experiment Station, Michigan State University, East Lansing, Michigan 44824*

PREFACE

The extent of global environmental changes and the extent to which technology can be employed to improve our global environment is the theme of this book. But having said that, the question arises, Why write another book about the environment? After all, the environment became a household topic in the 1960s and 1970s; there has been a flood of books on the subject. We are all now well aware of the problems that our modern civilization poses for our environment and the risks that alteration of the environment pose for us. There are three reasons for this new book. First, modern environmentalism, in order to sound the cry of warning and catch public attention, has emphasized the negative. Its supporters believed that a concern with the environment was not only morally correct, but necessary to save civilization. Opponents, in turn, have tended to view environmentalism simply as a negative movement, opposed to progress, economic development, and the advance of civilization. But we have begun to move past such confrontation; it is becoming a more widely accepted idea that a concern with the environment is simply good economics and planning. With this in mind, it is useful to examine a wide variety of environmental issues from this changed, more constructive perspective.

Second, our knowledge has grown rapidly in the past generation and there are new attempts to integrate knowledge among disciplines. Ecology, the scientific study of the relationships between living things and their environment, has made major advances; so, too, have the many other disciplines in the physical, social, and natural sciences that concern the environment. New technologies have great potential to help us solve our environmental problems. Environmental economics has begun to develop and provide useful insights into workable policies. It is time to consider the implications of the new knowledge from each traditional discipline and the integration of these individual sets of knowledge in our approach to environmental issues.

Third, our perspective has become global. We have learned that the remarkable feature of our planet is life and its persistence over a long time. In the last human generation, we have learned that life has existed on the Earth for more than three billion years and that life has profoundly changed our planet. We are also being rudely awakened to the fact that our impact on the environment is affecting our entire planet. Satellites give us a planetary perspective on life; new discoveries in biology and geology point out the global effects of life on our planet.

The more we understand our planet's life-support system, the more likely it is that we can take positive, constructive actions to provide for sustainable uses of natural resources and for the enhancement of environmental quality. Without this understanding, our range of viable alternatives may be unduly limited. Such understanding is absolutely essential to the design of policies and for taking actions necessary in the future. And so it is time for our generation to take a retrospective view, to consider what we have learned, and what we might be able to know and do in the future. The purpose of *Changing the Global Environment* is to provide such a retrospective as we near the end of the twentieth century.

Changing the Global Environment is dedicated to two previous landmark books of the modern industrial age, each of which examined the human environmental dilemma of its time: *Man and Nature,* published in 1864 by George Perkins Marsh,[1] and *Man's Role in Changing the Face of the Earth,* a symposium volume edited by W.L. Thomas, Jr.,[2] published in 1955. The latter was the result of a meeting of a renowned group of scientists and scholars in Princeton, New Jersey. The book that resulted provided a wealth of information about what our species had done to the Earth during its tenure on the planet. It covered the use of fire as a landscape-shaping tool, the environmental effects of peasant agriculture and herding of domestic animals, the cutting of much of the world's forests, the history of the environmental effects of our disposal of wastes to that point, and the effects of our cities on the environment. That book provided an invaluable summary of the history of human alterations of the environment. It represents a summary that is still of great use today.

Man's Role in Changing the Face of the Earth was dedicated to George Perkins Marsh, whose book *Man and Nature* began the modern discussion of environmental issues. A native of Vermont, Marsh became ambassador to Egypt and Italy. He was struck by the differences in the landscape of Italy and Egypt that resulted from the long use of natural resources by human beings in comparison to the wilderness of his native New England. Marsh eloquently recounted the history of human impacts on the environment and the possible effects that these impacts have had on civilization. "The ravages committed by man subvert the relations and destroy the balance that nature had established between her organic and her inorganic creations," he wrote. "She avenges herself upon the intruder, by letting loose upon her defaced provinces destructive energies hitherto kept in check by organic forces."[3] The decline of the Roman empire, Marsh suggested, was hastened by the destruction of forests and soils. Speaking for his time, Marsh made use of scientific information available in the mid-nineteenth century. His theme is a warning of the negative consequences of these changes. But his book was pragmatic and progressive, seeking to tame nature to our benefit, not to abandon all alterations. The "arrangements of nature," he wrote, "are highly desirable to maintain, when such regions become the seat of organized commonwealths. In developing an area, it is of primary importance that the transforming operation should be so conducted as not unnecessarily to derange and destroy what, in too many cases, it is beyond the powers of man to rectify or restore."[4]

Both books looked back into human history and described the environmental dilemma that faced our species. Unlike so many subsequent works, neither of these books could be characterized as negative or anti-progress. Marsh was pragmatic; the Princeton book represented many points of view and was in part an appreciation of the new technologies, such as the airplane, that permitted a different perspective on the environment. Soon after the publication of *Man's Role in Changing the Face of the Earth,* with the printing of Rachel Carson's *Silent Spring* in 1962, environmentalism became a popular movement. Around the world, laws were passed to conserve endangered species and require environmental assessments of new activities. The passage of some legislation conflicted with development and commercial goals, thus making environmentalism controversial.

Today, much is changing in our perception and concern with the environment. The technologies discussed in the chapters that follow either did not exist or were in a comparatively primitive stage 30 years ago. In 1955, W. L. Thomas, Jr., editor of *Man's Role in Changing the Face of the Earth,* wrote that "the most striking symbol of the new scale in time and space that has been brought into being since the time of Marsh is the airplane." The newer techniques of digital remote sensing had yet to be invented; the first artificial satellite was yet to be put into orbit. Satellite and aircraft remote sensing now provide synoptic views that radically change our perception of our planet. New chemical methods discussed in this book let us measure the concentration of chemicals rapidly and on location in parts per billion, at accuracies and speeds not dreamed of in 1955.

Looking ahead today to the twenty-first century, we see technology again radically changing the relationship between people and nature, but now on a global scale and with great potential for constructive management for achieving sustainable uses of natural resources and for the enhancement of environmental quality. This book is intended to lead a change in emphasis from negative to positive aspects of our abilities to manage our environment and from a local and regional to a global perspective. In the last 30 years—one human generation—we have learned a great deal about our environment, and we have a wealth of new technologies and methods that we are applying to understand and mitigate our effects on our environment. It is time to look at these developments and summarize how well we are doing. This is a task useful to each generation.

To accomplish this task, an international conference was held in Venice, Italy, in October 1985. At this conference, a unique group of experts from 70 nations was assembled: experts on biological conservation and the science of ecology; experts on some of the new technologies that are being used to deal with environmentalism issues; and economists and other social scientists who were interested in assessing the approaches that could be taken to integrate a concern with the environment into economic development and planning. The results of the conference were divided into two parts: technical papers have been published in two special issues of *The Science of the Total Environment,*[5] while a selection of the broader papers of more general interest are published here. They provide an overview of the role of human beings in changing the global environment: what we have done, are doing, and could accomplish in the future. *Changing the Global Environment* is divided into three sections: I. Perspectives on the Biosphere; II. Some of the Ways We Can Learn

About Our Global Environment; and III. Social and Economic Policy Issues. The first section sets forth the broad perspective of the book and focuses on specific conceptual issues of central concern. The second discusses technologies of remote sensing, computer-based data systems, and advanced chemical analytical techniques, which together radically change our perception of our environment and our ability to analyze complex issues of environmental concern. These technologies also improve our ability to mitigate the effects of our actions on the environment. The third section discusses the interactions among social, environmental, and economic goals, and the role that technological advances might play in attaining these goals.

It is our hope that this book will be helpful in the attempt to understand how our new technologies are changing our perception of our environment and to further integrate a concern with the environment into all our activities.

Daniel B. Botkin, Margriet F. Caswell, John E. Estes, and Angelo A. Orio

Notes

[1] G.P. Marsh, 1964, *Man and Nature; or, Physical Geography as Modified by Human Action*, Charles Scribner, New York, reprinted in 1965 and edited by D. Lowenthal, Belknap Press, Cambridge, Massachusetts.

[2] W.L. Thomas, 1956, *Man's Role in Changing the Face of the Earth*, University of Chicago Press, Chicago, Illinois, 1193 pp.

[3] Marsh, 1964, *op. cit.*, p. 42.

[4] *Ibid.*, p. 35.

[5] *The Science of the Total Environment*, 1986, Vols. 55 and 56, Elsevier Science Publishers, Amsterdam.

ACKNOWLEDGMENTS

Many organizations and individuals have made this book possible. In particular, the editors would like to thank the following: the Woodrow Wilson International Center for Scholars, Washington, D.C., for providing a year-long fellowship to D. B. Botkin during which the ideas that laid the foundation for this book developed; the Rockefeller Bellagio Study and Conference Center, Bellagio, Italy, for providing a month-long fellowship to D. B. B. during which the initial structuring and editing of the book began; the Office of University Affairs of the U.S. National Aeronautics and Space Administration, for providing some support of activities in the preparation of the book; the University of California, Santa Barbara, and the University of Venice for sponsoring an international conference titled "Man's Role in Changing the Global Environment," held in Venice, Italy, October 21–26, 1985; the Veneto Region of Italy and the City of Venice for additional help with the conference. At that conference scholars gathered from more than seventy nations, and a small fraction of these were asked to contribute to this book. We would also like to thank the University of Venice for providing funds for color plates. Carol Johnson, Leslie Campbell, Corey Elias, and Marylee Prince provided invaluable assistance in typing and communication with the authors.

SECTION I
PERSPECTIVES
ON THE BIOSPHERE

Our perception of our role in the global environment is changing in two fundamental ways: 1) we perceive that environment as dynamic and greatly affected by life over a very long time; and 2) we recognize that we are altering the environment at a global level. This section discusses the importance of this new perception (Chapters 1 and 2); introduces some of the key concepts concerning the dynamic nature of the global environment and the impact of life on this environment (Chapters 1 through 5); and provides examples of key human effects (Chapters 5 through 10). Of the many global environmental issues, a selection has been chosen for presentation here. The chapters included were selected to cover a representative range of topics from food production (Chapter 8) and sustained use of the ocean (Chapter 5) to certain impacts of deforestation (Chapter 6); from large-scale pollution (Chapter 7) to soil degradation (Chapter 9). Each chapter illustrates changes and advances that have occurred in the past 30 years. The chapters in this section represent our current view of global environmental issues; they provide an introduction to the second and third sections of the book.

1 SCIENCE AND THE GLOBAL ENVIRONMENT

DANIEL B. BOTKIN
University of California
Santa Barbara, California

INTRODUCTION

⌈A very great change is taking place in our perception of our affect on the environment. In the last three decades, there has been an explosion of concern with the adverse effects of technology on local and regional environments.⌋Looking to the twenty-first century, we see technology again radically changing the relationship between people and nature, but now on a global scale and with great potential for achieving sustainable uses of natural resources and for the enhancement of environmental quality. Solving global environmental problems requires a new perspective, a perspective that involves science. Yet this perspective goes beyond science; it pertains as well to the way that everyone perceives the world.

In part the change in perception of environmental issues is from local to global. In addition, the change is from negative (pointing out problems) to constructive (developing solutions). With the last three decades of experience, we now have a rich background in conservation, science, and technology, which together provide an opportunity to take positive approaches to global environmental issues.

The concern with the environment is older than three decades. In the conservation of natural resources, our modern discussions can be traced back to George Perkins Marsh's seminal 1864 book *Man and Nature.*

As the American ambassador to Italy, Marsh was struck by the effects of two thousand years of human settlement on soil and vegetation in comparison to his native Vermont, and he warned of the danger and foolhardiness of ignoring nature in the process of developing civilization.[1]

These concerns were echoed and strengthened in our century by people such as Paul Sears, who, in his book *Deserts on the March,* pointed out that we must understand nature's rules as we continue to modify our landscapes.[2] Such foundations were being laid in the concern with our environment throughout the twentieth century. Meanwhile, scientific discussions about our planetary life support system, called the biosphere, have also been developing in many fields, such as ecology, geography, geology, oceanography, climatology, and atmospheric sciences.

Concomitantly, advances in technology have begun to provide unprecedented means to observe and measure the environment at every scale. Advances in computers, in remote sensing, in chemical analysis, and in many other techniques can provide us with the basis for a positive approach to our environmental problems. Remote sensing can be traced back to nineteenth-century photographs taken from balloons and to the first aerial movies taken by the Wright brothers from an airplane in the first years of this century.[3] The constructive use of remote sensing techniques for environmental problems began just over a decade ago when the technique was used to track the rapid spread of a corn blight in North America. In the last three decades remote sensing has developed into the use of satellites and aircraft to study the Earth's surface for improvement of our environment.

Remote sensing and computer-based data systems allow global studies that were not possible before. Recent advances in ecological scientific research—advances in information, knowledge, techniques, and concepts—allow new approaches to understanding nature. Recent developments in the study of the biosphere can move us toward the required understanding to deal with global environmental issues. However, to achieve constructive management, sustainable uses, and an enhanced environment, or even to sustain a merely livable environment, we must undergo several major changes in perspective. Some of these changes are in our scientific perspective, but others are deeper and involve our fundamental perspective about life on the Earth, including our assumptions and preconceptions with which we begin any analysis. It is the purpose of this chapter to discuss aspects of this new perspective that underly the discussions that follow in this book.

THE BIOSPHERE: ITS NATURE AND
DYNAMIC PROPERTIES

The changes that must take place in our perspective are twofold: the recognition of (1) the global, planetary view of life on the Earth; and (2) the dynamic rather than static properties of the Earth and its life-supporting system. What is meant by a global view of life on the Earth? The word "global" is used in two ways in this context: first, some actions that occur in a single place or a few places affect the entire biosphere. For example, point source burning of fossil fuels may lead to global changes in climate.

Second, there are local activities and local changes that are repeated around the world at so many locations that they are global in quantity. For example, in many parts of the world forests are being rapidly logged. In many cases, the decision to cut is local, but the total amount of forests that is being cut around the world may lead to significant changes in the global carbon cycle. It has been suggested that this in turn might change the climate. Thus the repetition of local actions can, in their summation, have a global impact.

What is there that is special about a global perspective on the environment, except the scale? The term "biosphere" is not new; it can be traced back one hundred years when it was coined by Edward Suess,[4] who used it to refer to the total amount of organic matter on the Earth. Since then the term has evolved to mean the planetary scale system that sustains and includes life. The term "biosphere" is used with this meaning in our book. The biosphere includes the biota and the atmosphere, oceans, and sediments that are in active exchange with the biota.

In the last 20 years, it has become common knowledge that we live on a special planet. The observation is not new that life has evolved to suit the planet. Henderson recognized the special qualities of our planet more than 70 years ago when he wrote *The Fitness of the Environment* in which he pointed out remarkable qualities of the Earth's environment that support life.[5]

Two things in our perspective our new: the growing understanding of the extent to which life has influenced the biosphere over the Earth's history, and the growing understanding that the biosphere is a complex and unusual system requiring new approaches to its analysis.

For example, we are accustomed to thinking of life as a characteristic of individual organisms. Individuals are alive, but it is only a system with living and nonliving components that makes possible a flow of energy and cycling of chemical elements that sustain life.[6] We can imagine, and some ecologists repeatedly attempt to create, simple, closed systems that sustain life—a small glass vial with water, air, a little sediment, and two or three species. These generally persist a short while—a year or so. The longest-lived closed systems of this kind are those sealed about twenty years ago by Clair Folsome of the University of Hawaii. However, the biosphere has persisted for more than three billion years, in spite of, *or perhaps because of*, its immensely greater complexity. The biosphere is estimated variously to have somewhere between three and ten million species collected in tens of thousands of local ecological communities and ecosystems. These ecosystems can be grouped into 20 to 30 major kinds, which are called "biomes," such as tropical rainforests, coral reefs, and grasslands. The biosphere is complex. An example is the abundance and distribution of the life changes in time and space at different scales, from seasonal fluctuations of insects in a single field to transitions in entire forests over thousands of years. There are complex pathways that connect these subunits of the biosphere. Different pathways allow different rates of flux of energy, material, and information. In the aggregate, then, the collections of species represent a complex patchwork of subsystems at different stages and states of development spread across the Earth's surface.

I would like to consider an example to illustrate the complexity of the biosphere, an example originally discussed by G.E. Hutchinson and repeated here because it nicely illustrates several characteristics of the biosphere.[7] Phosphorus is one of the major elements used by all living things. It becomes available to plants from the soil. Phosphorous does not occur as a gas in our atmosphere, where it is a trivial constituent, existing in the atmosphere only in dust particles.

Within the biosphere the general pathway of phosphorus is "downhill." When it occurs in rocks on the continents, phosphorus is slowly eroded, used temporarily within terrestrial ecosystems, then slowly washed to the oceans via streams and rivers. In the oceans, it is temporarily used by marine organisms, but is eventually deposited in the deep oceans or marine sediments.

There is no short-term significant nonbiological return of phosphorus to the surface. Without life, the only return of phosphorus is long term, through geological uplift.

Phosphorus is, however, a major constituent of life. It is an important agricultural fertilizer, often in low supply. For many years, commercial sources of phosphorus for fertilizers were obtained from deposits on small islands off the coast of South America and Africa.[8] These deposits were formed by the accumulation of bird guano, which resulted from the nesting of thousands of birds on the islands over thousands of years.

The presence of birds on these islands depends on a peculiar and complex interplay among the living and nonliving components of the biosphere. Phosphorus is deposited in bird excrement on the islands. The birds obtain the phosphorus when they feed on fish. The fish feed on planktonic animals, and the animals on algae. The algae grow abundantly along the coasts where there are oceanic upwelling currents. These upwellings bring phosphorus from deep waters to the surface, where the algae use it.

The upwellings are driven by winds, which affect surface currents. The winds require the presence of a continent and an ocean, as well as climatic conditions where the winds are predominantly from the continents to the oceans. The winds push the surface waters away from the continents, and the deep waters move upward to replace the surface waters, thus carrying phosphorus back to the ocean surface.

Guano deposits can only remain for long time periods in dry climates, otherwise they would be dissolved in water and washed back to the seas. Guano accumulates in significant amounts only when there are many birds nesting in large colonies. Such birds appear to prefer islands, perhaps because the islands are free from predators.[9]

Thus, a major source of phosphorus fertilizer requires a special climate, a special geography, and a special set of species. The relationships among these components sometimes change. Occasionally climate changes, the winds fail, the upwelling stops, the fish and birds die in huge numbers. The combination of these events is part of El Niño, which has obtained notoriety in recent years, when, as happens periodically, the upwelling failed, greatly affecting commercial fish harvests in the western Pacific.

This example illustrates the complexities and interrelations that influence the cycling of chemical elements in the biosphere. It reinforces the need for a new interdisciplinary scientific basis as a guide for any applied biospheric issues and thus the need for the development of a new science of the biosphere. Understanding the phosphorus cycle requires the knowledge of many disciplines—biology, climatology, geography, geology, oceanography. Understanding all other biogeochemical cycles and other ways that life and its environment function together at a global level requires this new science. This science would bring together even astronomy and zoology, as it requires that we view the Earth as a planet with life on it and as a planet strongly influenced by life. Yet these disciplines must be integrated in new ways to produce a unified approach to understanding the biosphere. This is what I mean by developing a new science.

A CONTEMPORARY VIEW OF THE CONCEPT OF HARMONY BETWEEN HUMAN BEINGS AND NATURE

The biosphere is what people have referred to as "nature" taken in its largest sense. I now want to compare current knowledge about the biosphere with current beliefs about nature. As the historian Glacken has pointed out, throughout the history of Western civilization there have been several major unanswered questions concerning the relationship between people and nature. What is the character of nature undisturbed? What is the influence of nature on people? What is the influence of people on nature?[10] There have also been several major beliefs about nature: 1) nature achieves constancy, 2) when disturbed, nature returns to that constant condition, and 3) that constant condition is desirable and good.

These ideas were well expressed by George Perkins Marsh in *Man and Nature*. He wrote:

> "Nature, left undisturbed, so fashions her territory as to give it almost unchanging permanence of form, outline, and proportion, except when shattered by geologic convulsions; and in these comparatively rare cases of derangement, she sets herself at once to repair the superficial damage, and to restore, as nearly as practicable, the former aspect of her dominion . . . In countries untrodden by man, the proportions and relative positions of land and water, the atmospheric precipitation and evaporation, the thermometric mean, and the distribution of vegetable and animal life, are subject to change only from geological influences so slow in their operation that the geographical conditions may be regarded as constant and immutable."[11]

Marsh had remarkable insight for his time. His statements are eloquent assertions of some of our dilemmas and also serve to clarify the difference between his nineteenth-century perspective and ours. Marsh stated the need for a concern about the effects of human actions on nature and reiterated the belief in the constancy of nature as a norm and a good. Both of these assertions have formed the predominant perspective during the last 20 years with the resurgence of contemporary environmentalism. Marsh's view of nature is the one generally espoused in textbooks on ecology and in popular environmental literature. It has formed the basis of twentieth century scientific theory about populations and ecosystems. More important, it is the basis of most national laws and international agreements that control our use of wild lands and wild creatures. It has been an essential part of the 1960's and 1970's mythology about conservation, environment, and nature.

Although the belief about the constancy and stability of nature as a norm and a good is a theme found in the writings of the classical Greeks and Romans, and one that runs throughout Western history, the evidence available to us today contradicts this belief. Until the last few years, the predominant theories in ecology either presumed or had as a necessary consequence a strict concept of a highly structured, ordered, regulated, *steady-state* ecological system.[12] Individual populations were supposed to grow according to the S-shaped growth curve to a constant abundance and remain at that abundance indefinitely unless disturbed. Ecosystems were supposed to attain a climax condition and remain there indefinitely. More recently, these same assumptions have been applied to the biosphere. For example, in the analysis of the global carbon cycle the biosphere is generally assumed to be a steady-state system, one with exactly the kind of stability that Marsh attributed to a single forest.

Ecologists know now that this view is wrong at local and regional levels—at the levels of populations and ecosystems. Change is intrinsic and natural at many scales of time and space in the biosphere. Nature changes over essentially all time scales and in at least some cases these changes are necessary for the persistence of life, because life has adapted to them and depends on them.

There are many examples that illustrate the way that some species are adapted to environmental change. The most familiar examples to ecologists are plants adapted to disturbed conditions. For example, following a fire or windstorm in a forest, the first vegetation to re-grow are of species adapted to the bright light conditions, the disturbed soil, and so forth. The adaptation to periodic change is illustrated by a famous case in the conservation of an endangered species, the Kirtland's warbler. This warbler lives in jack pine woodlands in Michigan. The jack pine is a fire-adapted species, with seed cones that open after they are heated by a fire. The pine grows well in the bright light conditions of a forest opening but cannot reproduce and grow in the dense shade of a mature forest. When fires were suppressed in Michigan during the first half of the twentieth century, the jack pine stands became uncommon and the warbler became endangered. When the necessity for disturbance was recognized, fires were allowed to burn, the jack pine stands reestablished, and the habitat of the Kirtland's warbler became more common. We are just beginning to understand the way that larger ecological systems—entire ecosystems and landscape units—are dependent on change over time.[13]

The idea that change is natural has created problems in natural resource management at local and regional levels. How do you manage something that is always changing? A classic, well-known example of this is a large, legally established wilderness area shared by Canada and the

United States, the Boundary Waters Canoe Area in Minnesota and Ontario. The goal of management of this million acre wilderness is to keep the forest in its natural state. But what is *the* natural state? Imagine a manager whose job it is to maintain that forest in its natural condition. What does he or she choose as a goal? The obvious answer appears to be to choose a forest similar to the natural, presettlement forest of that region—the one that the American Indians saw. George Perkins Marsh would have believed this to be a forest that had existed from time immemorial. But we know now that this forest has changed periodically since before the end of the last ice age. About every thousand years a completely different landscape could be found, from arctic tundra 10,000 years ago to the dense forests of today.[14]

Which of these types represents the natural state of the United States-Canadian border where the Boundary Waters wilderness has been legally established? Is it tundra? Is it the modern forest? Each appears equally natural in the sense that each dominated the landscape for a long period. The range of choice is great, representing kinds of vegetation now distributed over thousands of miles. There is no single equilibrium condition to be attached to this wilderness. These changes reflect fluctuations in climatic periods of cooling and warming, and the differential rate of migration of the species as they returned north following the melting of the ice.

In discussing the Boundary Waters Canoe Area forest, I have mentioned only the last 10,000 to 100,000 years. As one goes back in time, one encounters successive changes in life, climate, ice volume, sea level, the atmosphere, and the Earth's crust. Perhaps even more noteworthy, during Earth history there have been a series of biological innovations that have changed the biosphere unidirectionally. These historically unique changes must have required adjustments in many components of the biosphere.[15]

As a familiar example, the evolution of photosynthesis in the ancient biosphere led to the biological production of oxygen, which altered geochemical cycles by changing the rates and sites of mineral weathering and organic decomposition. It appears, for example, that the early introduction of oxygen from photosynthesis led to the formation of the economically important iron ore deposits of the Earth.[16] This is discussed further by Westbroek and by Margulis and Guerrero.

Since the origin of life on the Earth, there have been a number of major events in biological evolution that have changed the rest of the biosphere—altered the atmosphere, oceans, soils, and rocks, of which the evolution of photosynthesis is one striking example. These are

TABLE 1-1 SOME MAJOR UNIDIRECTIONAL
CHANGES IN THE HISTORY OF THE BIOSPHERE[a]

Biological innovations
  1. Origin of life
  2. Origin of photosynthesis
  3. Origin of aerobic photosynthesis
  4. Origin of aerobic respiration
  5. Origins of other biogeochemically important metabolisms
  6. Origin of eukaryotic organisms
  7. Origin of calcium-containing skeletons
  8. Origin and expansion of bioturbating organisms
  9. The colonization of land by plants and animals
  10. The evolution of angiosperms
  11. The evolution of humans

[a]Table 3.1 National Academy of Sciences, 1986, Remote Sensing of the Biosphere (D.B. Botkin, ed.) National Academy Press, Washington, D.C. (p. 43).

one-way events—once they have taken place, the biosphere cannot move backwards from them to previous conditions. In this way, the biosphere has a history—a one-way change over time. (See table 1-1.) The steps in each one of these major events are: 1) biological breakthrough in evolution—a new opportunity is opened up by the evolution of a new group of species and these evolve rapidly, "taking advantage" of the new opportunities; 2) new kinds of life change the biosphere; 3) other life forms evolve to adjust to the new environmental conditions and evolution takes place within this new environment with its new set of problems and opportunities.

The evolution of diatoms provides an example of this process.[17] Diatoms are single-celled algae that have a hard shell made of silicon. When they evolved they were the first major group of organisms to make use of the dissolved silicon in the oceans. Their hard silicon shells provided protection against enemies and in this way were a major evolutionary advance. The diatoms made use of a previously unused resource and gained an evolutionary advantage from it. There were two results: 1) the biological result was the evolution of many kinds of diatoms; 2) the biospheric result appears to be a change in the cycling of carbon. Diatoms live at the surface of the ocean where there is enough light for photosynthesis, but when they die the cells sink to the ocean floor. This led to a new major storage of silicon and carbon in the ocean floor, creating diatomaceous earth, which we know as chalk.

The evolution that led to calcareous shells and internal skeletons

containing calcium also led to new opportunities for animals because the
shells provided greater protection against predators and other advantages. This biological use of calcium led to an increase in the production
of limestones and in the storage of major amounts of carbon in
limestones, which indirectly allowed an additional buildup in the atmosphere of a large amount of free oxygen. Just as in the case of the
evolution of diatoms, the evolution of calcium-based shells and skeletons
led to a new wave of biological evolution and to a major change in the
biosphere.

So far, I have emphasized the importance of natural change as a
feature of the biosphere, but there are other qualities about our
planetary life-support system that can only be described as peculiar, as
least peculiar in comparison to the engineering systems in cars, air-
planes, and radios that we have become used to building and analyzing
since the start of the modern scientific and technological era. I will
mention just a few.

1. *Life affects the biosphere out of proportion to its mass.* For example, the
   total mass of all the bacteria that fix nitrogen is small compared to the
   mass of all other life, yet this small amount of bacterial material has a
   great effect on the atmosphere and on all life. As another example,
   the mass of ozone is a small fraction of the mass of the atmosphere,
   and yet the amount of ozone has a large effect on life and atmospheric chemistry.
2. Another peculiar property of the biosphere is that it has what can be
   called *mutual causality*—for example, climate affects the distribution
   of land vegetation, but the distribution of land vegetation can affect
   climate.
3. *A third peculiar property is that the importance of an event is often inversely
   related to the frequency of its occurrence*—a hurricane is more important
   than common rain showers; in most areas, most of the soil erosion
   occurs during only the few heaviest of storms, which represent only a
   small percentage of precipitation events.
4. Another important quality is the *heterogeneity in time and space* at every
   level. Some parts of the biosphere are more important than other
   parts in regard to specific dynamics. For example, upwellings and
   coastal marshes have an importance in biogeochemistry out of
   proportion to the relative size of these biomes.

The biosphere is made up of a hierarchy of systems. At each level a
system consists of some biological phenomena and an environment.[18] At

the highest level of aggregation, the biosphere is made up of the lower atmosphere, all of the oceans and bodies of fresh water, all of life, all soils, and those sediments that are in active interchange with other parts of the biota. The fundamental unit of the biosphere is the ecosystem, which is the smallest unit that has the characteristics necessary to sustain life. These characteristics are the capability to maintain a flow of usable energy and the cycling of chemical elements necessary for life.

SUMMARY

We are living in a time of transition in the relationship between science and our global environment. There is a need to develop a new science, a science of the biosphere. This science requires that we view the Earth as a planet with life on it and as a planet strongly influenced by life. It must integrate disciplines in new ways to produce a unified approach to understanding the biosphere.

In discussing the characteristics of the biosphere, I have considered what has been thought of, throughout the history of civilization, as nature in its largest aspect. At this moment in history, our science and technology give us a new perspective on this ancient idea of nature. But we have not yet achieved this new perspective. Our analyses of ecological systems, including the biosphere, are tied as yet to an older mechanical perspective. We use the mathematics derived for machines, and in our concepts we still hold onto the idea of the constancy of nature, stated eloquently more than a century ago by George Perkins Marsh. From that older perspective, the managerial goal was steady-state operation.

This then is the heart of the matter that confronts us. Modern technology has changed our view of nature. Remote sensing has changed our perceptions of our planet. Computers have changed our view of life. We can no longer rely on nineteenth-century models of analysis for twenty-first century problems. More than any other issue, confronting and recognizing these deep-seated assumptions is a major challenge to us in dealing with our global environmental issues.

When we do recognize, confront, and change these assumptions, we will be able to achieve a more comfortable relationship between ourselves and nature. When we do this, we can then proceed more rapidly to develop a constructive approach to the management of our global environment and to decide what research should form the first steps toward the solutions to global environmental problems.

NOTES

[1] G.P. Marsh, 1864, *Man and Nature* (reprinted 1965 and edited by D. Lowenthal), Belknap Press, Cambridge, Massachusetts.

[2] P. Sears, 1935, *Deserts on the March*, University of Oklahoma Press, Norman.

[3] D.B. Botkin, J.E. Estes, R.M. MacDonald, and M.V. Wilson, 1984, "Studying the Earth's Vegetation from Space," *BioScience*, Vol. 34, pp. 508–514.

[4] E. Suess, 1875, *Die Entstehun Der Alpen*, pp. 158–160.

[5] L.J. Henderson, 1913, *The Fitness of the Environment*, Macmillan, New York.

[6] H.J. Morowitz, 1979, *Energy Flow in Biology*, Oxbow Press, Woodbridge, Connecticut.

[7] G.E. Hutchinson, 1950, "Survey of Contemporary Knowledge of Biogeochemistry 3: The Biogeochemistry of Vertebrate Excretion," *Bull. Amer. Museum Natural History*, Vol. 96.

[8] Hutchinson, 1950, *op. cit.*

[9] *Ibid.*

[10] C.L. Glacken, 1967, *Traces on the Rhodian Shore: Nature and Culture in Western Thought from Ancient Times to the End of the Eighteenth Century*, University of California Press, Berkeley.

[11] Marsh, 1864, *op. cit.*, pp. 29–30.

[12] D.B. Botkin, and M.J. Sobel, 1975, "Stability in Time-Varying Ecosystems," *Amer. Naturalist*, Vol. 109, pp. 624–646.

[13] More discussion of the importance of change in ecological systems can be found in Botkin and Sobel, 1975, *op. cit.*; D.B. Botkin, S. Golubeck, B. Maguire, B. Moore, III, H.J. Morowitz, and L.B. Slobodkin, 1979, "Closed Regenerative Life Support Systems for Space Travel: Their Development Poses Fundamental Questions for Ecological Science," in *Life Sciences and Space Research, XVII*, R. Homquist (ed.), Pergamon Press, New York, pp. 3–12; D.B. Botkin, 1980, "A Grandfather Clock Down the Staircase: Stability and Disturbance in Natural Ecosystems," in *Forests: Fresh Perspectives from Ecosystem Analysis*, R.H. Waring (ed.), Proceedings of the 40th Annual Biology Colloquium, Oregon State University Press, Corvallis, Oregon, pp. 1–10.

[14] M.L. Heinselman, 1970, "Landscape Evolution, Peatland Types, and the Environment in the Lake Agassiz Peatland Natural Area, Minnesota," *Ecological Monographs*, Vol. 40, pp. 235–261.

[15] D.B. Botkin (ed.), 1986, *Remote Sensing of the Biosphere*, National Academy of Sciences, National Academy Press, Washington, DC, 135 pp.

[16] S.M. Awramik, 1981, "The Pre-phanerzoic Biosphere—Three Billion Years of Crises and Opportunities," *Biotic Crises in Ecological and Evolutionary Time*, Academic Press, New York, pp. 83–102.

[17] D.B. Botkin, 1986, *op. cit.*

[18] R.V. O'Neill, B.R. Del Angelis, J.B. Waide, and T.S.H. Allen, 1986, *A Hierarchical Concept of Ecosystems*, Princeton University Press, Princeton, New Jersey.

EDITORS' INTRODUCTION TO:
L.M. TALBOT
Man's Role in Managing the Global Environment

The second chapter, *Man's Role in Managing the Global Environment,* is by Dr. Lee M. Talbot, one of the world leaders in international conservation. Trained and well known as an expert on ecology and wildlife research, Dr. Talbot has been a member of the United States Council on for the Conservation of Nature. Along with Dr. Sidney Holt, he is the author of "New Principles for the Conservation of Wild Living Resources." He is also an author of the World Conservation Strategy, which has been adopted by many nations as a foundation for their environmental policies. In this chapter, Dr. Talbot draws on his extensive experience in environmental management in more than 100 countries during the past 30 years to discuss our present worldwide situation. By setting our global environmental concerns within an historical context and by discussing these issues from a variety of points of view, Dr. Talbot sets the stage for the rest of the book. His chapter represents both an overview and a summary of what is to follow.

2 MAN'S ROLE IN MANAGING THE GLOBAL ENVIRONMENT

LEE M. TALBOT
East-West Center
Honolulu, Hawaii
and
World Resources Institute
Washington, DC

INTRODUCTION

Throughout history there is a dominant theme of "the growing mastery of human beings over their environment. Antiphonal to this is the revenge of an outraged nature on human beings. It is possible to sketch the dynamics of human history in terms of this antithesis."[1] This is how the great geographer Carl Sauer summed up our relationship with the environment.

These words are as true today as when they were written 50 years ago. But today the impact—and the revenge—are global. The human population has doubled, then doubled again. The environmental impact of this vast population has increased by orders of magnitude, profoundly affecting the face of the Earth and the climate above it. If humankind is to survive, this historical pattern of environmental conflict—of action and revenge—cannot continue.

In 1955 an international conference on Man's Role in Changing the Face of the Earth was held by the Wenner Gren Foundation at Princeton, New Jersey. This landmark effort brought together a broad array of scientists and other academicians who compiled the most comprehensive picture ever developed of the impact of human beings on the environment. While the resulting book was largely retrospective, it did sound a warning of the dangers ahead if these trends of action and revenge continued.[2]

The three subsequent decades have been the most significant period in history in terms of the emerging changes in the nature and magnitude of human impacts on the environment and of humankind's emerging capabilities to detect and analyze those impacts. Human beings have continued to cause unintended and ever more profound changes in the environment, in the process reducing the capability of that environment to sustain human life, much less its capability to improve the quality of human life. But this period has also seen the development of emerging capabilities for the management of human actions to effect environmental changes that are *intended* and can lead to improvements in the capabilities of the environment to sustain as well as to improve the quality of that life.

The purpose of this chapter is to provide a brief review of these changes and to consider the challenges they pose in the coming years. It presents the author's perspective based on more than 30 years of experience working on environmental issues in over 100 countries.

HISTORICAL PERSPECTIVE

Much environmental concern today tends to focus on the results of modern society's impacts on the environment, such as oil spills, acid rain, pesticides, and urban sprawl. However, for hundreds and possibly thousands of millennia before the Industrial Revolution, human beings had a profound and, I believe, still inadequately recognized impact on the face of the Earth.

Human being's prehistoric activities modified vegetation, soils, waters, and wildlife of much—if not most—of the Earth. Their domestication of fire enabled them, intentionally or unintentionally, to change profoundly the vegetation, which in turn altered the soil structure and location, water regimes and wildlife, and, at least in some cases, climate.

In various publications, Sauer postulated that most of the Earth's grasslands were the result of fire, for which human beings were probably entirely responsible.[3] He also emphasized the human role in affecting plant evolution during the Pleistocene.

My ecological studies specifically on savanna ecology[4] and my other ecological studies and environmental work throughout the world have convinced me of the soundness of Dr. Sauer's basic conclusions and that, if anything, he has understated the case. With few exceptions, the present location and composition of tropical savannas and most other grasslands is largely if not totally anthropogenic, through the use of fire and, in some cases, subsequent grazing of domestic livestock.

Human beings played a similar role in many of the areas that are now desert. The vegetation composition of many of the forests and woodlands of the temperate and drier tropics also appear largely influenced, if not determined, by early humans.

How "early" is still a matter of some uncertainty. Until recently the earliest generally accepted use by humans of fire was at Chowkowtien, China, about half a million years ago. It was argued that this early use was only to keep warm and that use of fire for cooking and other activities only came about 150,000 years ago. Sauer, of course, questioned this, believing correctly that as more data appeared it would push the horizons of early humans' use of fire much farther back. My own earlier published estimate was about one million years,[5] but recent research has indicated the use and control of fire by early humans 1.4 million years ago.[6] I am convinced that with additional research we will continue to push back that date.

But the significant point is that human activities have played a major role in the evolution of the Earth's ecosystems. In a real sense, human beings have been changing the face of the Earth since their earliest times.

Early human beings exerted a major impact on wild fauna as well as flora, both by changing vegetation and through direct hunting. There is much evidence, for example, that they played an important role in the extinction of some of the large fauna in the late Pleistocene. And clearly, the more recent but still prehistoric spread of agriculture further altered the faunal scene.

For example, from my work in southeast Asia I am convinced that shifting cultivation allowed the spread of that area's rich variety of large wild animals (including many species of wild cattle, deer and deerlike animals, elephants, rhinos, and pigs) into areas that otherwise would have been closed tropical forests in which such animals cannot thrive or even survive.

It has long been my conviction that human activities that involved the clearing of forests and other vegetation resulted in local and possibly regional or global climatic changes. For many years this view has not been popular or generally accepted. Now, however, there is increasing evidence that anthropogenic environmental changes may have caused significant regional climatic changes and "that humans have made substantial contributions to global climate changes during the past several millennia, and perhaps over the past million years."[7]

The domestication of livestock added another dimension to the leverage early humans exerted over the vegetation, soils, waters, and wildlife. The results of overgrazing by goats and sheep are particularly dramatic in the Middle East and the Sahara edges, but it should be remembered that domestic livestock have a significant impact on virtually all environments, from reindeer in the arctic to carabao and pigs in the tropics, and from camels, cattle, and sheep in the lowlands to yaks in the alpine zones.

The exponential growth of human numbers has brought a corresponding exponential increase in environmental impacts. Larger forms of wildlife, being high on the food chains, are sensitive indicators of the state of the environment. Indeed, from my earliest international work it was clear that the status of a country's wildlife resource was usually a good indicator of the status of that country's natural resources as a whole.[8]

And the fate of wildlife has reflected the fate of the environment as a whole. From the time of Christ to around 1800 AD the rate of exterminations of large mammals was roughly one each 55 years. In the next 150 years this increased to roughly one a year.[9] This latter period also corresponded with a period of unprecedented exploitation of other resources involved in lands newly settled. In 1938 Sauer wrote, "In the late 18th century the progressively and rapidly cumulative destructive effects of European exploitation became marked . . . In the space of a century and a half—only two full lifetimes—more damage has been done to the productive capacity of the world than in all of human history preceding."[10]

Sauer went on to note the apparent paradox "that the lands of recent settlement were the worn and wornout parts of the world, not the lands of old civilization," and he cited the United States heading the list of exploited and dissipated land wealth, followed by South Africa, Australia, southern Russia, and Argentina. This was in 1938, before the second world war and when the world's human population was only about two billion.

Since that time, the trends he described have continued, but the pressure of a doubling population has been felt most in the developing nations of the world, and there has been consequent dramatic change to the landscape, damage to the environment, and adverse impact on human welfare in those areas.

Most people think of environmental impacts as a new phenomenon, something that has arisen since the first Earth Day in 1970. And it is true that much of the public awareness of environmental change dates from that period. But it is important to recognize that what we are now seeing is not a new process. It is a simple acceleration of the processes that have been underway for one or more million years. The primary changes are a massive increase in the number of people involved combined with a dramatic increase in the leverage over the environment that these people can exert through the development of new technologies. It is simply a continuation of Carl Sauer's historical dynamic of human being's growing mastery over their environment countered by the revenge of an outraged nature.

However, the process is accelerating at an exponential rate. As a consequence, the magnitude and global significance of the environmental changes in the past few years far surpasses those of all humankind's previous history. This situation can be illustrated by a brief review of some of the changes since the 1955 Wenner Gren Conference on Man's Role in Changing the Face of the Earth. In this brief period:

- The human population has almost doubled, from less than three billion to over five billion people.
- Half the world's tropical forests have been lost. There are 56 nations today where the situation of tropical forests is considered critical and at the present rate of destruction most of those countries will have lost all tropical forests by the year 2000.
- The processes of overgrazing and desertification have greatly increased, resulting in dramatic spread of deserts, accelerated loss of agricultural lands' and the degradation of most of the world's rangelands.
- Overfishing, pollution, and conversion of estuarine habitats have significantly altered the world's marine habitats.
- In 1955 relatively few species were known to be endangered. The problem of endangered species received little attention and when it did it was considered to be of local concern at best. Today it is recognized that loss of biological diversity is a worldwide problem. Ten to twenty percent of all species are endangered, and it is believed that if we had information we would find that a much higher proportion is in danger. Science has only named about 1.7 million kinds of animals out of a total now estimated to be over 30 million, and many scientists fear that species are being lost at a faster rate than they are being identified.
- Chemical pollution, hardly even recognized in 1955, has become all-pervasive and few species or lands escape its effects. The residues of pesticides, for example, are found in birds and mammals that live their whole lives in Antarctica, many thousands of miles from the places where the chemicals were first applied.

- Other hazardous wastes, not even mentioned in the 1955 conference, now represent a major problem worldwide.
- With increasing industrialization throughout the world, the combustion of fossil fuels has greatly increased, leading to a host of local, regional, and even global problems, including acid rain and CO_2 buildup.
- Production of CO_2 and other greenhouse gasses now appears to be leading to a global warming, threatening to alter the world climate in a significant way.
- Assisting Third World countries with economic development has become one of the world's largest industries. The annual flow of resources to developing countries is in the range of $100 billion, yet much of the development effort does not result in sustainable benefits for the Third World people involved. Today a higher percentage of the world's population is undernourished and lives in poverty than was the case in 1955. Because they did not take environmental factors into account, many development projects have had the effect of reducing, rather than enhancing, the capacity of the lands involved to support people.

During this period, as the nature and seriousness of environmental problems have been recognized, there has been a proliferation of governmental and nongovernmental actions to deal with them.

- In 1955, there were only a few dozen nongovernmental environmental organizations (NGOs), mostly in the United States. Today there are tens of thousands of such NGOs throughout the world.
- In 1955, no government had environmental institutions (that is, agencies, departments, or other governmental structures), as such. Today virtually all nations have such institutions, most of them supported by comprehensive environmental legislation developed within the past decade or so.
- In the past 15 years, the environment has become an important concern of much of the United Nations system, led by the United Nations Environment Programme, which was established as a result of the UN Conference on the Human Environment held in 1972.
- In 1955, there were few international agreements or conventions that addressed environment. Today there is a comprehensive body of such international law and a number of nations have negotiated bilateral agreements for cooperation on environmental matters.

Thus, while the basic fact that human beings are causing environmental changes is merely a continuation of the age-old process, in the last few decades the nature, extent, and significance of those changes has increased out of proportion.

EMERGING PERCEPTION OF OUR ROLE

Recognition came early that human actions that changed the environment also affected human's interests and well-being. As early as the third century B.C., the precursors of national parks had been established in India to protect parts of the environment from exploitative human activities, and laws had been promulgated to protect and regulate the exploitation of economically important wild animals, birds, and fish.[11]

In the same period Plato wrote eloquently of the Hills of Attica being "like the skeleton of a body wasted by disease."[12] Man's impact on parts of the Mediterranean environment had been profound, but it had taken place over such a long period that virtually no one had realized what was happening. In Plato's time the only evidences that remained of Attica's former forest cover, deep mantle of soil, and perennial springs were the tree trunks in the temple structures and the stone shrines at the sites of long-dry springs.

This illustrates a basic characteristic of people's perception of their relationship to their environment. When change occurs too slowly, it is not noticed, much less acted upon. It is only when change occurs quickly enough that the results can be seen and felt within one person's memory span that there will be recognition and action. We seem to need to experience a disaster before we will make significant changes in our ways.

Throughout the world the rate of anthropogenic environmental change was slow until the last century. Then, in North America, energetic European settlers made profound changes in a short time. Vast areas of forest were cut and burned; agricultural lands from coast to coast were opened, abused, and often abandoned; wildlife was decimated, the endless herds of buffalo destroyed; and the seemingly inexhaustible passenger pigeons exterminated. All this happened in a few short decades. People experienced it, they recognized what was happening, and they took action.

The seminal writings of George Perkins Marsh about the impact of man on nature[13] were soon followed by others. Unprecedented actions were taken to conserve forests, lands, wildlife, and other natural resources. For the first time concern for the environment in the form of conservation of natural resources became a priority issue for a national government.

Nearly 30 years later another incident brought the environment back to America's national attention. Misuse of farmland combined with drought conditions led to massive erosion and created the "dust bowl." The resultant clouds of dust blew as far as the nation's capital. Again, recognition—experience of the problem—led to national action.

In the following two decades, worldwide public and governmental concern was focused on World War II and its aftermath. The evidence of environmental problems—Sauer's "revenge of an outraged nature"—was mounting, but it went largely unnoticed by the public, in spite of the writings of perceptive individuals such as Fairfield Osborn and William Vogt.[14]

The 1955 Wenner Gren Conference on Man's Role in Changing the Face of the Earth brought unprecedented academic attention to the environmental impacts of humans, but in spite of the intellectual insight of the participants and the significance of the resulting publication,[15] no disaster was felt by the public and essentially no environmental action resulted. The proceedings of the conference also reflected the different realities and perceptions of environmental impacts of that era. There was not a single paper on hazardous wastes. Nor was there a paper involving remote sensing; there was little focus on the ocean.

The situation changed in the following decade. Dramatic technological developments of the post-war period brought with them wholly new dimensions of environmental impacts. They also brought dramatic developments in communications. Through the media of television millions of people throughout the world "experienced" severe pollution incidents, such as the mercury poisoning at Minimata, Japan, the Torrey Canyon oil spill in the English Channel, "killer smogs" in London and Los Angeles, and the Santa Barbara oil leak. These developments led to unprecedented levels of awareness of environmental problems and consequently to efforts to deal with them.

Since the 1955 Wenner Gren conference, a number of other international meetings on environmental change have been convened. These have included the 1968 UNESCO "Biosphere" conference; the 1972 UN Conference on the Human Environment in Stockholm; the subsequent UN "Theme Conferences" on subjects related to the environment, such as desertification, population, human settlements, and water; the 1984 "Global Possible" conference of the World Resources Institute;[16] and the 1985 Venice Conference on Man's Role in Changing the Global Environment.[17] Other international meetings have been held in connection with a variety of environmental programs including those of the International Council of Scientific Unions, regional intergovernmental organizations, and individual governments.

Each of these efforts has sought to improve the state of knowledge and awareness of the role of human beings in changing the environment, and each has contributed to scientific understanding or administrative efforts. Each has become somewhat more self-consciously aware of the *global* nature of environmental change caused by humans.

GLOBAL PERSPECTIVE

The last few years have seen remarkable developments in our capabilities to perceive, understand, and deal with environmental change. These developments have enabled us to begin to understand and deal with environmental change *on a global basis*. For example, there have been dramatic technological developments in remote sensing and computer science, economic thought and theory, and ecological understanding and capabilities. Most important, there has been rapidly growing recognition of the *global* dimension of environmental change and a growing realization that our present scientific, philosophic, and administrative framework is not adequate to deal with global issues.

Environmental issues become global in several ways:

- Some activities occur in a few places but may affect the biosphere as a whole. For example, large-scale combustion of fossil fuels may only occur in a few industrial areas but it may affect the climate as a whole. The same is true of the release of other greenhouse gasses.
- Smaller-scale local activities, when repeated around the world, also may have global effects. Estuaries near coastal cities are heavily polluted, and, while each case is of local pollution, the wide repetition of the problem and its total impact make it global. The same is true, for example, of local changes in land use that cumulatively can affect the planet's albedo and the local clearance of forests, which can have major cumulative effects on species loss and CO_2 balance.
- Species extinction is an important dimension of the problem. Although in former times the loss of species was considered to be a local problem, maintenance of biological diversity is now recognized to be of major global importance.
- All environmental problems ultimately have social and economic impacts on people. Some global issues, like climate change, can have clear impacts on humans throughout the world. But with the increasing interdependence among nations and peoples, more apparently localized environmental problems have increasingly pervasive economic and social impacts in other parts of the world. The localized loss

of a resource, such as the Peruvian anchoveta, can have an economic domino effect, causing economic dislocations in other nations across the world.

- The socioeconomic impacts of environmental problems are particularly evident in the case of development assistance to Third World nations. With the annual flow of resources to developing nations of over 100 billion U.S. dollars, much of it in the form of commercial loans, the size of the debt of many developing countries is staggering. Yet environmental degradation in many countries is reducing their capability to support themselves, much less to repay debts. This, in turn, affects other nations elsewhere in the world, developing as well as developed.

ESSENTIAL TRANSITIONS

These negative environmental changes are pervasive. Their actual and potential impact on the present and future well-being, and indeed, survival, of mankind must not be underestimated. But at the same time, the emphasis on the negative aspects of technology—and in a broader context, the negative aspects of our relationship with the environment—can obscure the courses of action that must be taken to avoid the negative consequences.

The need, then, is for a cautiously positive approach. The grave dangers of the present resource and environmental deterioration must be recognized, along with the potential for disasterous consequences if the present trends are allowed to continue. But sustainable use of resources can be achieved, and improvements in environmental quality can be obtained, if human wisdom and ability are well served by modern science and technology.

Therefore there must be—and can be—a transition from largely inadvertent *changing* of the environment to planned and intentional *management* of people's use of the environment.

To achieve this objective there must be a set of transitions,[18] all of which are interrelated and interlinked:

- a transition to stability in the world population
- a transition in the use of renewable resources, to assure that they are used sustainably and safely
- a transition to the use of energy, which is both efficient and non-threatening to the biosphere
- a transition to the development and application of high technology in the service of environmental management and improvement

- a transition to a new economics, which supports and undergirds sustainable resource management and environmental improvement
- a transition to economic development, which is sustainable and equitable
- a transition to an integrated science of the biosphere, which can provide the information on which effective environmental management must be based; a holistic science, which goes beyond the traditional fragmented disciplinary approach and which also takes into account the relevant socioeconomic considerations
- a transition to effective implementation of measures to conserve biological diversity

Accomplishing these transitions will require the dedicated efforts and cooperation of all segments of society. It will require broad public understanding and support, and—most particularly—effective political will and the administrative and managerial capabilities to carry it out.

Most of these transitions have been initiated. The need is to see that they continue to progress and that they are completed. The needed transitions in the economic, technological, and scientific areas illustrate the nature of the challenges and the possibilities.

ECONOMIC TRANSITION

The status and trends in resource and environmental conditions over the greater part of the world are determined by economic development—by the direction it takes where it is present and the effects of poverty where development is not present. Until recently, environmental concerns and development were considered to be in opposition. Now, due largely to efforts such as the 1972 Stockholm Conference, the World Conservation Strategy,[19] the Brandt Commission,[20] and, most recently, that of the United Nations World Commission on Environment and Development,[21] environment and development are increasingly recognized as being two sides of the same coin. Environmentally sound development is a prerequisite for good environmental management, and environmental considerations are a prerequisite for development that is successful and sustainable.

However, while this truth is recognized by leaders in government and development assistance institutions, it often has not filtered down to the practitioners in the field. The application of the more traditional narrow economic analysis is no longer appropriate, and what is necessary is the incorporation of natural resource and environmental factors in macro-economic analysis and planning of nations and development assistance institutions.[22]

Emphasis also should be given to ways to affect the behavior of individuals through economic policies and incentives. There is increasing recognition that human motivation plays a critical role in determining present environmental use and in effecting change. For example, government subsidies for pesticides or to the timber industry can have profound impact on the environment and on the sustainability of the resource use involved. Consequently, economists should also focus on development of systems of incentives (and the resultant policies) to bring about optimum patterns of resource and environmental use.

Lack of data has been cited as one reason why economists have not adequately taken environmental factors into account in the past. Now, however, the increasing availability of data on resources and environment provides economists with necessary current information and the capabilities to project ahead. Consequently, the prospects for economists to integrate resources into economic policy are greater than ever before.

At present there is a fundamental shift in emphasis in the thinking and practice of some economists, but economic theory still has difficulty incorporating biological and ecological understanding and models into its framework. The accomplishment of this shift will make a profound difference in the directions of public and official governmental motivation and, hence, on the status of management of the global environment. This process is essential to achieving effective global environmental management and, as such, this transition is of the highest priority.

TECHNOLOGY TRANSITION

Technology represents another area that is somewhat parallel to economics. Technology has been viewed by some as the answer to all environmental problems. The reaction to blind faith in technology has been an equally blind opposition to it on the part of some concerned with the environment. But rationally, technology per se is not at fault. Instead it is the inappropriate use of it, including its use without proper consideration of the environmental consequences.

There have been dramatic technological developments in remote sensing, automated information systems, data processing, and associated activities that provide unprecedented capabilities for study, monitoring, and analysis of environmental conditions and problems at a truly global level. These developments have also led to significant expansions in our capabilities to detect and cope with pollution and various terrestrial hazards.

There has been an almost exponential growth of technological capabilities for environmental uses. At the time of George Perkins Marsh, the available technology was limited to binoculars, horses, trains, and boats. By the time of the 1955 Wenner Gren conference, surface vehicles, aircraft, and a variety of photographic capabilities had been developed. With the coming of the space age and the recent computer revolution, the technological possibilities for environmental uses have increased by orders of magnitude. At the 1955 conference, there was one paper based on the use of technology—aircraft.[23] At the 1985 Venice conference, there were nearly 40 papers on remote sensing and information systems alone.

Modern remote sensing capabilities, coupled with existing computer capabilities for information handling, analysis, and interpretation, offer new horizons to environmental science and management.

SCIENTIFIC TRANSITION

The needs for transition are particularly clear in the field of biology and related sciences applied to the environment. There is a great creative opportunity for truly innovative science built on fundamental information that we need but that we are not now getting.[24]

In the first place, our basic information about the environment is woefully inadequate. As one example: in spite of available technologies and clearly stated needs, we still do not have such elementary information as the basic extent, status, and trends of global forest cover—or even information for many individual countries. The paucity of basic scientific information was a major finding of the 1955 conference and remains true today.[25]

In the second place, our ability to effect change outpaces our ability to recognize the impacts, much less control them. Scientific understanding of what we are doing often lags behind our doing of it. Our present approach—if and when consideration to environmental factors is given—usually is to predict what is likely to happen from a given environmental perturbation on the basis of past experience and incomplete models. This approach is clearly inadequate. Our need here is to develop better capability to understand and predict and, on the basis of better knowledge, to determine how to build more effective safety factors and midcourse correction feedback into our procedures.

In the past few years, we have gained much knowledge about the special characteristics of the biosphere and the contributions of life to all its aspects, from the rocks below the surface to the composition of the atmosphere and the climate. This knowledge is far from complete, but it provides enough of a glimpse at the complexity of our biosphere to show us that our traditional scientific and philosophical approaches to it are no longer adequate.

A major obstacle is the fragmented approach enforced by the traditional disciplinary divisions of scientists and institutions. A further obstacle is the theoretical base that these separate disciplines have constructed. Many of the current mechanistic ideas of the biosphere date from the nineteenth century and have little relevance to our present state of knowledge. If we are to deal effectively with the science of the biosphere, our approach should incorporate all the present biological sciences, plus the disciplines that claim the other aspects of the biosphere such as geology, oceanography, atmospheric chemistry, and climatology.

Our dismal record of managing marine resources illustrates vividly how unsuccessful our single-discipline approach has been, even when it has been applied to managing a single resource. The marine example also illustrates a further dimension of the problem: while the basic scientific basis for fisheries management has been inadequate,[26] another important contribution to management failure has been socioeconomic. People are an ecological dominant in most aspects of the biosphere, and resource management usually intimately involves people. Yet all too often people are not included in scientific considerations of the environment.

There should be an interdisciplinary approach to biosphere studies. Under the present disciplinary systems, cross disciplinary communication and fertilization is rare, and true interdisciplinary work is rarer. Indeed, it is effectively discouraged by the nature of most academic and administrative institutions.

A further consideration is that most global environmental research is done on the basis of temporary grants and projects. As a consequence, continuity in such studies is rare. Therefore, there is a real need for a facility specifically dedicated to global environmental research where researchers can undertake the necessary long-term as well as shorter duration research.

If such a facility were created, it could foster the development of a new scientific framework—a science of the biosphere. The most needed major areas of relevant research focus on life. Life is fundamental to the biosphere, and life is the truly unique characteristic of our planet. Examples of the major areas of research that appear appropriate for such a new facility could include:

- The current state of the biosphere, which would provide basic information for virtually any other aspects of research and which would include basic inventories of biospheric components, such as tropical forests and, more broadly, biomass
- Energy flow through the biosphere
- The cycling of matter through the biosphere (that is, biogeochemistry)
- The maintenance of life over a long period (that is, what determines the distribution and abundance of life)
- Development of biosphere theory

BIOLOGICAL DIVERSITY TRANSITION

The key to maintaining the integrity of the biosphere as we know it, and therefore to maintaining human survival, is the maintenance of biological processes and biological productivity. Stated another way, the basis for our concern for the global impact of the activities of human beings is our concern with what we are doing to the health of the biosphere. Biological diversity is a key component of that health because of the role in the biospheric processes played by individual species and their contribution to the evolution of new species. It has been suggested that one key measure of our success in global environmental management will be how well we have maintained the Earth's biological diversity.

As more of the Earth's surface is altered, more of the species that rely on the habitats involved are being lost. The most effective way to avoid further losses is to establish a global system of protected areas to maintain adequate and representative samples of the Earth's ecosystems.

However, the areas still most in need of protection lie in nations whose economic conditions render them least able to take such action. Therefore, there is an urgent need for international cooperation—which must involve both effective national and international action—to establish the protected area system necessary for conservation of biological diversity. Effective international agreement to this end probably would require some form of assistance to many of the developing nations involved to enable them to set aside and manage the necessary areas. Various proposals have been made, ranging from inclusion of biological diversity as a priority objective in development assistance programs (as it is in U.S. Agency for International Development) through linking conservation programs in developing countries to some form of adjustment of their international debts.

Maintaining biological diversity is a low priority for most developing countries and development assistance agencies. Consequently, there is need to incorporate biological diversity concerns more effectively in their basic policies and practices. In addition, more effective means of environmental assessment are required, along with improved information about the global distribution and status of species.

SUMMARY

The history of humankind is the history of actions that have, intentionally and otherwise, changed the environment. These changes often have resulted in serious negative impacts on human welfare (that is, Carl Sauer's "revenge of an outraged nature"). With exponentially increasing population coupled with the even more rapid expansion of technology, these environmental impacts have become pervasive and truly global.

The present patterns of environmental and resource degradation cannot continue without grave risk to human welfare and even human survival. However, if human wisdom and ability are well served by science and technology, it will be possible to make the transition from inadvertent *changing* of the environment to planned and intentional *management* of the use of the global environment by human beings.

NOTES

[1] C.O. Sauer, 1938, "Theme of Plant and Animal Destruction in Economic History," *Journal of Farm Economics*, Vol. 20, pp. 756–775.

[2] W.L. Thomas (ed.), 1956, *Man's Role in Changing the Face of the Earth*, University of Chicago Press, 1193 pp.

[3] Examples include: C.O. Sauer, 1947, "Early Relations of Man to Plants," *Geographical Review*, Vol. 37, pp. 1–25; C.O. Sauer, 1950, "Grassland Climax, Fire and Man," *Journal of Range Management*, Vol. 3, pp. 16–21; C.O. Sauer, 1962, "Fire and Early Man," *Paideuma*, Vol. 7, pp. 399–407.

[4] L.M. Talbot and M.H. Talbot, 1963, "The Wildebeest in Western Masailand, East Africa," *Wildlife Monographs*, No. 12, 88 pp; L.M. Talbot, 1964, "The Biological Productivity of the tropical Savanna Ecosystem," in *Ecology of Man in the Tropical Environment*, IUCN, New Series No. 4, Morges, Switzerland, pp. 88–97; G.D. Anderson and L.M. Talbot, 1965, "Soil Factors Affecting the Distribution of the Grassland Types and Their Utilization by Wild Animals on the Serengeti Plains, Tanganyika," *Journal of Ecology*, Vol. 53, pp. 33–56; L.M. Talbot and R.N. Kesel, 1975, "The Tropical Savanna Ecosystem," *Geoscience and Man*, Vol. 10, pp. 15–26.

[5] L.M. Talbot, 1980, "A World Conservation Strategy," *Journal of the Royal Society of Arts*, Vol. 27, pp. 493–504.

[6] J.A.J. Gowlett, U.W.K. Harris, J.D. Walton, and B.A. Wood, 1981, "Early Archaeological Sites, Hominid Remains and Traces of Fire from Chesowanja, Kenya," *Nature*, Vol. 294, pp. 125–129.

[7] C. Sagan, O.B. Toon, and J.B. Pollach, 1979, "Anthropogenic Albedo Changes and the Earth's Climate," *Science*, Vol. 206, pp. 1363–1368.

[8] L.M. Talbot, 1957, "The Lions of Gir: Wildlife Management Problems of Asia," *Trans. North American Wildlife Conference*, Vol. 22, pp. 570–579.

[9] F. Harper, 1945, "Extinct and Vanishing Mammals of the Old World," American Committee for International Wildlife Protection, Special Publ. No. 12, 850 pp.

[10] Sauer, 1938, *op. cit.*

[11] L.M. Talbot, 1964, "The International Role of Parks in Preserving Endangered Species," in *First World Conference on National Parks*, U.S. Department of Interior, National Park Service, Washington, DC, pp. 295–304.

[12] Plato, 1929, *Timaeus and Critias, Translation from the Greek* by A.E. Taylor, Metheun and Co., London, 136. pp.

[13] G.P. Marsh, 1864, *Man and Nature*, Scribners, New York, reprinted 1965, Belknap Press of Harvard University Press, 472 pp.

[14] F. Osborn, 1948, *Our Plundered Planet*, Little, Brown and Co., Boston; and W. Vogt, 1948, *Road to Survival*, Sloane, New York.

[15] Thomas, 1956, *op. cit.*

[16] World Resources Institute, 1984, *The Global Possible: Resources, Development, and the New Century*, World Resources Institute, Washingon, DC, 39 pp.; and R. Repetto (ed.), 1985, *The Global Possible*, Yale University press, 538 pp.

[17] A.A. Orio and D.B. Botkin, 1986, "Man's Role in Changing the Global Environment," *The Science of the Total Environment* (Elsevier), Vol. 55, pp. 1–400, and Vol. 56, pp. 1–416.

[18] For a more complete discussion of the needed transitions, see World Resources Institite, 1984, *op. cit.*; Repetto, 1985, *op. cit.*; R. Repetto, 1986, *World Enough and Time*, Yale University Press, 147 pp.

[19] IUCN, UNEP, and WWF, 1980, *World Conservation Strategy: Living Resource Conservation for Sustainable Development*, IUCN, Gland' Switzerland, 72 pp.

[20] Independent Commission on International Development Issues, 1980, "North-South: A Programme for Survival," Pan Books, London and Sidney, 304 pp.

[21] Independent Commission on Environment and Development, 1987, *Our Common Future*, Oxford University Press, 383 pp.

[22] Probably the most outstanding work in this field applied to development is that of the Environment and Policy Institute of the East-West Center, Honolulu, Hawaii. See for example, M.M. Hufschmidt and E.L. Hyman (eds.), 1982, *Economic Approaches to Natural Resource and Environmental Quality Analysis*, Tycooly International, Dublin, 333 pp.; M.M. Hufschmidt, et al., 1983, *Environment, Natural Systems and Development: An Economic Valuation Guide*, Johns Hopkins University Press, Baltimore; J. Dixon et al., 1986, *Economic Analysis of the Environmental Impacts of Development Projects*, Asian Development Bank, Manila, 100 pp.

[23] W.L. Thomas, 1956, *op. cit.*

[24] D.B. Botkin (ed.), 1986, *Remote Sensing of the Biosphere*, National Academy of Sciences Press, Washington, DC, 135 pp.

[25] M.W. Holdgate, M. Kassas, and G.F. White, 1982, *The World Environment 1972–1982: A Report by the United Nations Environment Programme*, Tycooly, Dublin, 637 pp.

[26] S. Holt and L.M. Talbot, 1978, "New Principles for the Conservation of Wild Living Resources," *Wildlife Monographs*, No. 59, 33 pp.

EDITORS' INTRODUCTION TO:
P. WESTBROEK
The Impact of Life on the Planet Earth: Some General Considerations

The two preceding chapters argued that there is a new global perspective on the environment and that an understanding of this perspective is necessary to deal with major environmental problems. In the next chapter, Dr. Peter Westbroek, a geologist and biochemist from the University of Leiden, the Netherlands, tells us about some of the more remarkable changes in our understanding of the global environment. From this chapter we learn that the Earth is a dynamic planet, that the rocks on which we stand that *seem* unlifelike and unconnected to the effects of life are much affected by life. Life plays an important role in geological processes, affecting the supplies of raw materials that living organisms require and playing a cleansing role by removing toxic chemicals from the environment.

In addition to introducing this new geological perception on life's role in the global environment, Dr. Westbroek discusses two approaches to the development of scientific theory appropriate to this new perception. This new perception is the product of scientific research of the last two decades; the ideas have developed in our generation and were not available at the time of the 1955 conference, *Man's Role in Changing the Face of the Earth*. In this important way we have made a positive step in our generation in the understanding of our planetary surface and the way in which it has supported life for more than three and one-half billion years.

3 THE IMPACT OF LIFE ON THE PLANET EARTH: SOME GENERAL CONSIDERATIONS

PETER WESTBROEK*
University of Leiden
(The Netherlands)

LIFE IN A GEOLOGICAL PERSPECTIVE

Developments in the geological sciences over the last decades have led to an integrated outlook on the evolution of our planet. Geotectonics became established in the sixties; it provides a physical understanding of the structure and dynamism of the Earth. From the early 1970s, geochemical models have been proposed representing the cycling of the elements at a global scale. The historical analysis of the geologic record has advanced to a point where major events that have affected the Earth as a whole may be distinguished from more localized developments. Finally, the role of the biota—all of Earth's life—in the global dynamism is receiving renewed attention. There is overwhelming evidence indicating that the biosphere—the thin film of living things at the interface between the atmosphere, the hydrosphere (oceans, lakes, rivers, etc.), and the rocky, solid Earth has profoundly influenced the history of our planet.

The picture that we now have of our planet may be epitomized as follows: In the deep Earth, convection currents in ever-changing pat-

* I gratefully acknowledge Professor Daniel Botkin for valuable criticisms and a thorough revision of the manuscript.

37

terns carry the materials of the outer parts of the planet down into the interior and then may bring them back toward the surface. These slow "endogenic" fluxes are energized by radioactive decay. A complementary system of "exogenic" material streams is generated by solar radiation in the peripheral parts of the globe. Here the movement may be faster by many orders of magnitude. The exogenic forces immediately affect the outermost layer of the crust, the atmosphere, and the hydrosphere. Each of the chemical elements is channelled along a characteristic maze of cyclic routes through this global system and may travel at different rates in the various subsystems. The geochemical cycles of the different elements are intertwined to form a network of interconnected routes.

In geochemical terms, life is a localized pattern of circuitous detours and transformations maintained in this circulatory system. In the biosphere the fluxes of materials and energy are vastly accelerated; they are knitted together into transient webs of extreme complexity. Life is a geochemical process, a self-perpetuating organization of geochemical cycles that emerged from the abiotic fluxes in the early Earth. From this perspective, biochemistry can be seen as a specialized subdiscipline of geochemistry. More specifically, the cycling of chemical elements between life and the rest of the Earth, through complex pathways, is called biogeochemistry. Life and Earth together form an integrated whole and must be studied in conjunction.

STRATEGIES OF LIFE

A biological system could only have been generated by evolution if a sufficient supply of nutrients had been provided by the environment over an adequate period of time. This simple principle has important consequences.

1. Out of the chemical elements present on Earth, approximately 20 are utilized by the biota. These elements have been selected by evolution because a) they can perform together the full spectrum of functions needed for the maintenance and proliferation of biological organization and b) they are abundant on the outer Earth. On the other hand, elements occurring in trace amounts in the outer Earth may only be used at extremely low concentrations for specialized functions or they may not have any function. Many of them are toxic.[1] An example of a functional trace element is molybdenum, which is an essential component of nitrogenase, the enzyme responsible for the fixation of nitrogen from the air.

2. Selection has favored the emergence of powerful biological "mining" mechanisms whereby nutrients are scavenged efficiently from rocks, water, and air. It should be kept in mind that biological systems are intimately involved in the weathering of rocks and in the formation and maintenance of soils.

3. Natural selection has favored the arrangement of biologically catalyzed transformations of nutrients in cyclic patterns, so that extensive re-utilization of these materials is warranted, even if their supply by the geochemical fluxes is minimal. An example is the biologically catalyzed cycle of sulfur. The element plays an important role in metabolism and, in addition, it may serve alternately as an electron donor or an electron acceptor in energy-retrieving reactions. It may be used over and over again, but when there is an abundant supply, the operation of the cycle may be less tight and sulfur may be exchanged with the external environment. Such "biogeochemical" cycles exist for all the major nutrients. They are intimately linked with one another, with the rate of cycling of one affecting the rate of cycling of the others. The nature of the interactions differs vastly in soils, forests, grasslands, deserts, rivers, coasts, and oceans.

A biological system can only have been generated by evolution if the concentrations of elements and chemical compounds remained below toxic threshold levels during sufficient periods of time. This constraint had several important consequences.

1. Natural selection has favored the emergence of powerful detoxifying mechanisms at the level of individual organisms. Toxic elements tend to be removed from the cellular interior of living things by specific biochemical pumps in the surrounding plasma membrane or else they are immobilized by specialized intracellular macromolecules such as metallothioneins, by cell wall constituents, in storage tissues, or in skeletons.[2]

2. Natural selection has favored the emergence of biological catalytic systems whereby toxic substances, such as hydrogen sulphide or nitrite, are converted into less toxic or even useful products (sulphate or nitrate). These catalytic mechanisms tend to be arranged as integrated components of biogeochemical cycles, so that the resulting supply of useful products can be fully exploited. Alternatively, new biochemical pathways have evolved whereby the toxic substances

could be utilized. An example is molecular oxygen, a waste product of photosynthesis and highly toxic to the early biota. At present, anaerobic life is limited to restricted environments; natural selection has led to the widespread use of oxygen as a resource, allowing very high energy reactions.

3. Natural selection has favored biologically catalyzed mechanisms whereby toxic substances are removed from the biosphere. Such mechanisms will be arranged in linear arrays rather than in cycles; they tend to channel their products away from biogeochemical circulation into the exterior "abiotic" fluxes. For example, the microbial methylation of mercury or the reduction of ions of that metal to the metallic form results in volatile products that are readily released from an ecosystem into the external environment.[3] As another example, many heavy metals in the oceans are removed from the upper (and most densely populated) water layers by association with sinking cell debris and skeletons.[4]

4. Biological catalysis of geochemical reactions tends to suppress the accumulation of toxic intermediates in the environment. For instance, hydrogen sulphide spontaneously reacts at *millimolar* concentrations with atmospheric oxygen to form sulphate. In addition, many intermediary products are formed, and some of these tend to accumulate. In contrast, bacterially catalyzed oxidation proceeds at a rapid rate at *micromolar* concentrations of hydrogen sulphide. Under these conditions, intermediary products tend to have short half-lives and their further oxidation is efficiently catalyzed by specific organisms. In the absence of life, both hydrogen sulphide and the intermediary products would rapidly accumulate in the hydrosphere and atmosphere, even if the present oxygen tension in these media would be maintained.[5]

One can interpret the overall effect of these properties of living systems to say that the outer Earth is cleansed by the biota so that suitable conditions for biological activity are promoted and maintained. Supplies for raw materials, offered by the geochemical fluxes, are efficiently exploited, nutrients are kept in circulation, and toxic substances are removed.

LIFE AND THE "ABIOTIC" CYCLES

Biogeochemical cycles are leaky. Nutrients are not recycled indefinitely in any local area, but they are eventually dispersed into the sedimentary environment, the hydrosphere, or the atmosphere. The wastes are absorbed, and fresh nutrient supplies are delivered by the overall dynamism of the Earth. A delicate balance is maintained between the global geochemical cycling and the cleansing activities of life. The ultimate limiting factor for the biological activity on Earth is the sluggish endogenic fluxes. They serve as ultimate sinks for refuse, and it is in the deep Earth that the waste is re-assembled into extractable raw materials for the next cycle. It would follow that the activity of the inner Earth is essential for the maintenance of life. It is unlikely that extensive biological activity can be supported by planets where the internal dynamism has come to a standstill.

At first sight, it may appear as if the biological elaboration of the geochemical cycles only represents a local alteration within the global system. But the impact of life reaches far beyond the limitations of the actual biosphere, owing to the intimate association between the biota and plate tectonics. This is exemplified by Figure 3–1, which represents the relation between metabolic and geologic cycling of carbon and oxygen. In photosynthesis, the energy of solar radiation is transformed into a redox potential gradient: organic carbon (CH_2O) and oxygen are produced from CO_2 and water. The elimination of this gradient is the driving force for respiration. In principle, it could be a closed cycle, but in the natural environment this is not the case. Less than 0.1 percent of the produced organic carbon is trapped in sediments and buried in the crust. This reduced material is then carried through the geological cycle until it becomes exposed at the surface, is weathered, and so undergoes a delayed reaction with atmospheric oxygen.

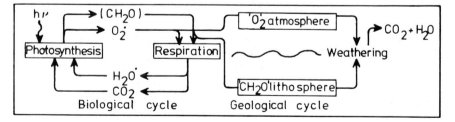

FIGURE 3–1 Relation between metabolic and geologic cycling of organic carbon and oxygen. (For explanation, see text.)

Plate tectonics is a sluggish process: the mean residence time of organic carbon in the crust is estimated to be approximately 400 million years. So, in spite of the fact that a minute fraction of the organic carbon escapes the photosynthesis-respiration cycle, a lithospheric reservoir as big as 10^{21} moles exists. It is generally agreed that the atmospheric reservoir of oxygen has originated in response to this accumulation of CH_2O in the crust.

But atmospheric oxygen is not the only component where the oxidation power, released in response to this massive burial of organic carbon, has been dumped. Examples of other major sinks of the oxidation power that were (and still are) released are the crustal reservoirs of gypsum ($CaSO_4 \cdot 2H_2O$), ferric oxides, and nitrate. The reducing power of organic carbon is also transferred in part to other substances. So, ferric iron, sulphate, and nitrate can be converted into ferrous iron, sulphide, and molecular nitrogen, respectively. The redox potential gradient is manifested in a multitude of chemical forms on Earth and turns the outer domains of the planet into an energized and highly reactive medium that is far removed from thermodynamic equilibrium.

The reactivity of this system is well illustrated by the fact that the residence time of free oxygen in the atmosphere is approximately 2,000 years. The energized condition of the outer Earth is maintained through the combined operation of photosynthesis and the endogenic cycle; it is essential for the rapid flux of biogeochemical reactions.

CONTINUITY OF LIFE

The continuity of life on Earth over 3.5 billion years is remarkable, especially because the dynamism of the planet has undergone dramatic changes during that period. The internal heat production and hence the endogenic activity has strongly decreased, whereas there is evidence suggesting that the intensity of solar radiation may have increased by as much as 25 percent. Stable isotope ratios in sedimentary rocks indicate massive transfers of sulphur, carbon, and other elements from their oxidized to their reduced reservoirs and vice versa during the Phanerozoic (the last 600 million years). Yet the effects of these changes on the conditions in the biosphere have never exceeded the limits of tolerance for life.

For an explanation of this apparent contradiction, two approaches are possible. In one, which I call the "constructionist" approach, model representations of the Earth's dynamism are built using geochemical evidence concerning the operation of individual global processes and on the chemical and energy fluxes between these processes. Quantitative estimates of the size of major global chemical reservoirs and fluxes of chemicals among these reservoirs, along with assumptions as to an appropriate set of mass balance equations, form the basic ingredients of these models. The stability of the biosphere is explained in the constructionist approach as arising from nonbiologically regulated feedback loops. Such representations can be expressed in a mathematical form, and this allows the behavior of these constructionist systems to be studied in computer simulations.

The alternative, which I call the "conceptualistic" approach, takes a general hypothetical concept of the global dynamism as its starting point. The stability of the biosphere is explained as arising from biologically regulated feedback loops. Here, too, computer simulations are used. These computations are implemented in order to study the implications of the conceived principle for planetary behavior. In order to verify the general applicability of both the constructionist representations and the hypothesis underlying the conceptualistic models, the results of the simulations are compared with independent evidence on the behavior of the Earth (for instance, the continued capability of the outer Earth to support life).

CONSTRUCTIONIST MODELS

A simple model studied by Garrels et al.[6] may serve here to illustrate the principle of the constructionist approach. It relates the following processes: 1) the production of biomass and release of oxygen into the atmosphere and the removal of carbon dioxide from the atmosphere by photosynthesis; 2) removal of atmospheric oxygen and biomass, and release of carbon dioxide in the atmosphere by oxygen-consuming respiration; 3) the utilization of biomass and sulphate in anaerobic respiration, leading to the accumulation of pyrite (FeS_2) in the lithosphere; 4) the accumulation of residual organic carbon in the lithosphere; 5) weathering of the lithosphere, leading to the oxidation of organic carbon and pyrite, removal of oxygen from the atmosphere, and the formation of carbon dioxide and sulphate; and 6) dissolution and formation of calcium carbonate. All the reactions are catalyzed by living systems.

In this model, the system is stabilized by several nonbiological feedbacks. For example, increased erosion exposes more pyrite and old organic carbon to oxidation, resulting in a net drain on atmospheric oxygen. As oxygen slowly declines, the fraction of the biomass reaching the seafloor increases and the net production of oxygen from photosynthesis increases as a result.

In addition, the reduction of sulphate to sulphide, and hence the production of pyrite, is stimulated by the increased accumulation of organic material in the oceanic sediment. This negative feedback would counteract the decrease in size of the pyrite reservoir as a result of the accelerated rate of erosion.

Initial conditions for the model were chosen to represent the biosphere prior to human influences. The dynamic relations between the reservoirs and fluxes were defined by an appropriate set of mass balance equations. The effects on the entire system of a change in the rate of any of the reactions could be calculated. A stepwise procedure was followed whereby the condition of the system was calculated over small, successive increments of geological time.

According to this model, tripling the rate of erosion would lead to a moderate increase of the carbon dioxide reservoir (from 0.055 to 0.140 $\times 10^{18}$ moles) over a period of ten million years and to minor changes in the rest of the system. While a doubling in the rate of photosynthesis would not lead to dramatic changes, the pivotal role of life is demonstrated by an experiment whereby photosynthesis is brought to a standstill. The oxygen reservoir is depleted in 12 million years and, over that same period, atmospheric CO_2 is increased from 0.055 to 3.88 $\times 10^{18}$ moles.

It should be borne in mind that this model is a modest representation of the real world and that it only explores a limited aspect of the global oxygen-carbon-dioxide system. As a general rule, the strength of constructionist models available to the present day is their potential to reveal trends and unexpected geochemical relationships, rather than the actual numerical outcome of the calculations.

A much more ambitious model, studied by Lasaga et al.,[7] explores the relation between the carbonate-silicate geochemical cycle and the redox cycle. This model was used to reconstruct the geochemical evolution of the outer Earth over the last 100 million years. Preliminary investigations designed to test the robustness of this model were carried out in our laboratory. It was found that the system has only moderate stability and that it is especially sensitive to changes in rates of weathering and seafloor spreading. This outcome raises doubts as to the capability of the constructionist approaching with current knowledge of geochemistry to explain the long-lasting suitableness of the outer Earth to support life.

The geochemical influence of life is widely recognized among investigators concerned with constructionist modeling. However, the implicit consensus appears to be that, as a rule, stabilizing feedback mechanisms are related to nonbiological, chemical, and physical constraints, rather than to growth characteristics of living systems. This latter possibility is explored by the conceptualistic models of Lovelock.

THE CONCEPTUALISTIC APPROACH

According to Lovelock,[8] the continuity of life on Earth and the precise adaptation of the biota to the conditions on our planet are insufficiently explained by the outcome of the constructionist models. His Gaia hypothesis maintains that the outer Earth and the biota together must be considered as a single self-regulatory system, where the biota actively create optimum conditions for their own propagation. The postulate of Gaia, which may be considered as an extrapolation of the biological concepts of homeostasis and organization, serves as a guide in the search for biologically regulated stability and optimization of the global environment. Examples of his conceptualistic approach are models for biological modulation of atmospheric carbon dioxide concentrations[9] and of global albedo (daisy world model),[10] both leading automatically to optimum temperatures for life. The idea of biologically catalyzed "cleansing" of the biosphere, developed earlier, may be considered a contribution to the Gaia hypothesis.

The daisy world model is an excellent illustration of the principle involved. A planet is illuminated by a star with increasing luminosity. Seeds of daisies with flowers of dark, intermediate, and light shades occur on the planetary surface. All daisies have the same growth characteristics: they flourish between 5 and 40°C, and the optimum growth temperature is 20°C. The dark daisies retain the stellar radiation while the light daisies reflect the radiation. The behavior of the daisies and the resulting surface temperature of the planet is calculated throughout the simulation. As soon as a temperature of 5°C is reached, the dark daisies come up and rapidly cover the entire surface of the planet. As a result, the temperature shoots up to 20°C. This optimum temperature is maintained, while increasingly lighter daisies colonize the planetary surface until the intensity of stellar radiation has reached a point whereby all daisies die off. The surface temperature shoots up as a result, and from there on it is determined only by the planetary albedo and radiation intensity. The system may be perturbed in many ways, but, as long as the daisies are in charge, its stability is high: the temperature automatically returns to the optimum value.

This remarkable stability is brought about by the combination of three factors: 1) the capacity of the organisms to influence sufficiently the effects of a changing parameter in the environment; 2) the fact that different organisms exert different effects; and 3) a growth curve that is nonlinear, and bell-shaped, with respect to the varying parameter.

Daisy world provides a general model for homeostasis in biological systems, be it at the biochemical, cellular, organismal, or ecological level. In these situations, the three conditions for automatic optimalization and stabilization are met. The question arises whether this simple conceptual model can be modified and elaborated to describe adequately the behavior of the planet Earth and explain the stability of the biosphere. The answer seems to depend on whether life's influence on our planet is of a magnitude comparable to the size of the forces that tend to destabilize the global environment. This problem cannot be answered offhand.

SUMMARY

A revolution in our thinking about life has developed during the past two decades, beginning with the acceptance of the theory of plate tectonics and an improvement in our understanding of global chemical cycles. We now understand that life and the geological processes are intimately connected, with each affecting the other. We are in a stage in our science where we are attempting to better understand these relationships and to learn how life has persisted for more than 3.56 billion years. Two approaches have been developed to try to explain the functioning of the biosphere: the constructionist and conceptualistic approaches. The constructionist works from the bottom up and builds a model of life-environment interactions based on knowledge of chemical and physical fluxes and of specific mechanisms. The conceptualistic approach begins with the "big picture"—assumptions as to how a global life-support system might function and achieve stability. Each approach has its advantages and dangers. The constructionist approach can explain details but may not readily lead to a direct understanding of the stability of the entire system. The conceptualistic approach gives insight into possible kinds of stability, but may not readily yield a precise model of the entire system.

At present, constructionism and the Gaia hypothesis lead to contradictory pronouncements regarding the role of the biota in the maintenance of suitable conditions for life on Earth. But this does not imply that the two approaches are incompatible. On the contrary, certain

Gaian elements can be recognized in recent constructionist models, and it is well possible that with the ongoing sophistication of constructionist modeling the explicit integration of the Gaian principle will become unavoidable. Independent evidence on the history of our planet is essential for testing the explanatory value of emergent models, and the further development of such critical knowledge deserves special attention.

On the whole, the Gaia hypothesis has not been widely accepted among the scientific community. Much of the criticism is based on ideological rather than scientific arguments. The concept would be teleological (an argument that has been rebutted by the daisy world model), and it might lead to a dangerous complacency with regard to the environmental crisis. Intuitively, many scientists may feel that the overwhelming explanatory potential of this concept will swamp a critical evaluation of the Earth's condition. On the other hand, Gaia is emphatically embraced by individuals of various ideological denominations, including Christians, spiritualists, and new wave enthusiasts.

It should be realized that the meaning of words is different within and outside science. In the context of ideology the word "Gaia" may have a connotation of stability and safety, it may be the subject of hatred, love, or belief and refer to the unity of the individual with the cosmos. In a scientific context, however, Gaia represents a hypothesis about the geologic role of life. It opens the possibility to study optimization of the global environment by and for the biota. Within science, Gaia may or may not turn out to be an operational concept, but it never can be the subject of belief. It is essential that we remove ideological connotations from the scientific enterprise; science can only thrive in relative independence. Ideology follows: It bridges the gap between the new and uncertain conditions to which we are continually exposed by the ongoing scientific and societal development and our inner drive to be meaningfully integrated in our environment.

NOTES

[1] J.J.R. Fausto da Silva and R.J.P. Williams, 1978, in *New Trends in Bioinorganic Chemistry*, R.J.P. Williams and J.J.R. Fausto da Silva (eds.), Academic Press, London, pp. 67–121.

[2] P. Westbroek, 1983, in *Biomineralization and Biological Metal Accumulation, Biological and Geological Perspectives*, P. Westbroek and E.W. de Jong (eds.), Reidel, Dordrecht, pp. 1–11.

[3] A.O. Summers and S. Silver, 1978, *Ann. Rev. Microbiol.*, Vol. 32, pp. 637–672.

[4] M. Whitfield and A.J. Watson, 1983, in *Biomineralization and Biological Metal Accumulation, Biological and Geological Perspectives*, P. Westbroek and E.W. de Jong (eds.), Reidel, Dordrecht, pp. 57–72; W.S. Broecker and T.-H. Peng, 1982, *Tracers in the Sea*, Columbia University Press, Palisades.

48

[5] J.G. Kuenen, personal communication.

[6] R.M. Garrels, A. Lerman, and F.T. MacKenzie, 1976, *Amer. Sci.*, Vol. 64, pp. 306–315.

[7] A.C. Lasaga, R.A. Berner, and R.M. Garrels, 1985, *An Improved Geochemical Model of Atmospheric CO_2: Natural Variations Archaean to Present*, pp. 397–411.

[8] J.E. Lovelock, 1979, *Gaia*, Oxford University Press, Oxford.

[9] J.E. Lovelock and M. Whitfield, 1982, *Nature*, Vol. 296, p. 561.

[10] J.E. Lovelock, 1983, in *Biomineralization and Biological Metal Accumulation, Biological and Geological Perspectives*, P. Westbroek and E.W. de Jong (eds.), Reidel, Dordrecht, pp. 15–25; J.E. Lovelock, 1986, *New Scientist*.

EDITORS' INTRODUCTION TO:

L. MARGULIS AND R. GUERRERO
From Planetary Atmospheres to Microbial Communities:
A Stroll Through Time and Space

The previous chapter by Dr. Peter Westbroek showed us the great impact that life has had on geological processes during the Earth's history. In the next chapter by Drs. Lynn Margulis and Ricardo Guerrero, we learn that life has greatly changed the atmosphere of our planet. But the major actors in this change are revealed as bacteria and their descendants, the chloroplasts in the cells of green plants. The effects of human beings simply as biological organisms—without our agricultural and industrial capabilities—on the global environment is seen as extremely small and unimportant, especially when viewed from a perspective of the long history of life on Earth. Contrary to the usual way we have portrayed ourselves in relation to our environment, this chapter suggests that our species is a minor and short-lived phenomenon from a planetary perspective. Research in the last two decades has shown that microbial communities have existed for several billion years and have not only greatly changed the Earth, but have also continued to persist. From this perspective, human beings appear as a geologically short-lived "weed" species, possibly exerting only a short (although large) effect on the Earth.

Drs. Margulis and Guerrero are microbiologists who study life's impact on the biosphere; Dr. Margulis has written numerous books on the early history of life and evolution of symbiosis and is, along with Dr. James Lovelock, the originator of the Gaia hypothesis.

4 FROM PLANETARY ATMOSPHERES TO MICROBIAL COMMUNITIES: A STROLL THROUGH SPACE AND TIME

LYNN MARGULIS
University of Massachusetts
Amherst, Massachusetts

RICARDO GUERRERO*
University of Barcelona
Barcelona, Spain

Nothing in biology makes sense except in the light of evolution.

Theodosius Dobzhansky[1]

INTRODUCTION

The United States in 1975 launched the Viking spacecrafts—two orbiters and two landers—to search for life on Mars. Even in earlier telescope images the planet Mars appeared to be extremely different from Earth. Many discoveries previously made by ground-based astronomy were verified by the Viking landers, which for years, beginning in 1976, continued to send back data from Mars to the Earth.

PLANETARY ATMOSPHERES

The atmosphere of the Martian planet, composed primarily of carbon dioxide, is thin, so thin that it exerts a pressure of only 6 millibars. Mars is drier than any place on Earth; the Martian air contains no more than 3 percent nitrogen and only trace amounts of oxygen.

One of the long series of Soviet spacecraft, Venera 9, approached and landed on our other neighboring planet Venus in 1976. Data returned

* We are grateful to Gail Fleischaker and John Kearney for manuscript preparation. This work was supported by NASA-NGR-004-052, the Lounsbery Foundation, the Boston University Graduate School (to L.M.), and the CAICYT and FIS, Spain (to R.G.).

confirmed that, in many essential ways, Venus is quite like Mars. Both our neighbors are extraordinarily dry. The atmosphere of Venus is also abundant in carbon dioxide. The air on Venus contains about 98 percent carbon dioxide and 2 percent nitrogen. As we compare the surface of these flanking planets with Earth, we are deeply impressed with three observations: our striking quantity of gaseous oxygen and of liquid water (the well-known hydrosphere) and our virtual absence of carbon dioxide (only 0.03 percent in the Earth's atmosphere). The atmospheres of Mars and Venus seem peculiar to us, yet from a solar-system perspective, it is the Earth's atmosphere that is strange. Measurements of the composition of the atmosphere on Earth reveal, to our surprise as chemists, gas mixtures that are incredibly discrepant. In its atmosphere, the Earth has a huge quantity (over 20 percent) of the reactive gas oxygen, whereas it contains almost no carbon dioxide. The concentration of carbon dioxide in the Earth's atmosphere is far less than what is expected for a "typical inner planet" situated between Mars and Venus in the solar system. Further, the Earth's atmosphere contains small amounts of large, peculiar organic compounds such as the epoxide disparlure (moth sex pheromone) and dimethyl sulfoxide. The Earth's atmosphere also contains enormous amounts of particulate matter. For example, from a chemical cycling point of view, the atmosphere has in it motile phosphorus particulates that measure up to a meter across (some call these locomotive particles gulls or albatrosses). These atmospheric components are absolutely inexplicable from the point of view of chemistry or physics alone. There is no physical or chemical science to foresee the presence of thousands of these carbon-hydrogen-rich compounds and complexes such as dimethyl sulfide derivatives, methyl

TABLE 4–1   ATMOSPHERIC COMPOSITION OF THE INNER PLANETS

	Venus	Earth	Mars
Gas			
CO_2	98	0.03	95
N_2	1.7	78	2.7
O_2	trace	21	<0.1
Pressure			
(approximate, in atmospheres or bars)	91	1	0.006
Temperature			
(° C, range)	447 ± 10	17 ± 50	-53 ± 40

mercaptan, camphor, catnip, myoporum, and DNA-rich propagules that have actually been discovered in the Earth's atmosphere.

From these and other observations, we are driven to conclude that the entire atmosphere of the Earth has been modified by the phenomenon of life. The composition of the atmospheres of Venus, Earth, and Mars is compared in Table 4–1. The atmospheres of Venus and Mars are virtually identical to each other, having 95 percent to 98 percent carbon dioxide, over 2 percent nitrogen, and trace quantities of other gases. The impressive differences between Mars and Venus are related to surface temperature—a factor easily explained by proximity to the sun.

The atmospheric mixtures of both these nearby planets can be considered "spent gases," that is, gases already consumed by chemical reaction. However, the atmosphere of the Earth forms a highly reactive mixture: nitrogen, hydrogen, and methane in the presence of 20 percent oxygen. The many trace gases on the Earth, comprising nearly 1 percent of the weight of the atmosphere, include compounds like methyl iodide, dimethyl sulfide, ammonia, hydrogen and methane, all trace gases that react strongly and easily with oxygen. The extent of the chemical imbalance of the Earth's atmosphere is illustrated in Table 4–2, where it is seen that many gases are out of chemical equilibrium by factors greater than 10^{35}! Chemists know, of course, that nitrogen (our major atmospheric gas) as well as hydrogen and methane and the other trace gases react strongly and easily with oxygen.

Not only is the solid Earth blanketed by a reactive mixture of atmospheric gases today, but from paleontological observations, we infer that for millions of years such reactive mixtures of atmospheric gases have persisted on Earth. What explains this anomaly? The process that

TABLE 4–2   THE ATMOSPHERE PROBLEM (ASSUMES 20% OXYGEN)*

Gas	Abundance	Expected equilibrium concentration	Discrepancy	Residence time (yrs)	Output (10^6 tons/yr)
Nitrogen	0.8	10^{-10}	10^9	3×10^6	1,000
Methane	1.5×10^{-6}	$<10^{-35}$	10^{29}	7	2,000
Nitrous oxide	3×10^{-7}	10^{-20}	10^{13}	10	600
Ammonia	1×10^{-8}	$<10^{-35}$	10^{27}	.01	1,500
Methyl iodide	1×10^{-12}	$<10^{-35}$	10^{23}	.001	30
Hydrogen	5×10^{-7}	$<10^{-35}$	10^{23}	2	20

* Calculations of J.E. Lovelock, see Lovelock 1988.

removes carbon dioxide and produces oxygen is easily recognized: it is oxygenic photosynthesis. This mode of metabolism characterizes some bacteria (cyanobacteria), many protoctists (algae), and nearly all green plants. Although the major oxygenic photosynthesizers today are land plants, before their appearance 450 million years ago, oxygenic photosynthesis was exclusively carried out by microorganisms.[2] Further, most of the contributing chemical reactions that have made the Earth's atmosphere so different from that of Mars and Venus are exclusively carried out by bacteria. Huge populations of bacteria are generated daily in the Earth's soils, fresh water, and oceans. From the planetary perspective of Mars or Venus, we become aware of the crucial role played by the bacteria in maintaining the reactive mixture of the gases of the Earth's atmosphere and the surface sediments. By comparison, the effect of human beings simply as biological organisms on the chemistry of the Earth is extremely small. Metabolically, people have no unique or necessary metabolic modes in relation to our other planet-mates; we are oxygen-using and carbon dioxide-releasing organisms like many bacteria, many protoctists, all fungi, and all animals.

MICROBIAL MATS

One of the major changes in our perspective of the Earth as a planet in the last 30 years is the recognition of the continuing importance of nonhuman life, especially the subvisible life of the microbes, in the geochemistry of the Earth. From this point of view, man appears to be an incredible egotist with a compelling self-centeredness. We have anthropocentric and unwarranted concepts of our power and importance on this Earth. Our planet functioned as a living system long before human beings. Communities of organisms maintain the surface sediments and atmosphere of this planet; this maintenance has been in continuous existence for at least three billion years. The Earth will continue to function in its own way long after the plausible extinction of our species.

How do some communities of organisms maintain a dynamic ecological stability that so far has eluded humankind? The study of the complex layered soil communities of microorganisms known as microbial mats may provide some clues. Microbial mats probably have covered all coastlines and emerged continents for more than three billion years. They most likely have influenced changes in Earth's lithosphere and atmosphere. We are only just beginning to understand the organization of such communities. Such compact living communities still dominate the landscape in a number of places in the present world: for example, Abu Dhabi in the Persian Gulf; Shark Bay in western Australia; Alicante in Mediterranean Spain; Mellum Island in the North Sea of Germany; Laguna Figueroa in Baja California, México; and throughout the Caribbean.

Most people living in areas where microbial mats occur consider these barren coasts, evaporite flatlands, or sandy intertidal zones to be totally unproductive. Life remains unnoticed. At closer view, slices of sediments display striped layers extending only a few centimeters deep. These layers are composed of highly organized communities of microorganisms, which in fact are very productive. As photosynthetic communities, they make and sediment large quantities of organic matter. Today's mats are the descendants of bacterial communities that probably have persisted for over 3,000 million years.

A microbial mat is a multilayered benthic community. The external layers of the mat are formed by photosynthetic and oxygen-producing primary producers, mostly a complex association of different cyanobacteria (blue-green bacteria), both multicellular and unicellular. Their colors are dominated by different shades of green. These bacteria remove carbon dioxide from the atmosphere, precipitating and binding it into calcium carbonates that may become limestone. Carbon dioxide is also converted into food in prodigious quantities in the form of organic matter: sugars, carbohydrates, nucleic acids, proteins and the like. Simultaneously, some of these same bacteria produce oxygen.

Under the surface layers, other layers can be observed comprised of organisms that do not live in environments with oxygen—anoxygenic phototrophic bacteria belonging to the following families: purple sulfur bacteria (Chromatiaceae) and green sulfur bacteria (Chlorobiaceae). The pigmented bacteria give these intermediate layers their characteristic coloration. Finally, at the bottom, a black layer of varying thickness formed by sulfate-reducing *Desulfovibrio*-like bacteria is found. Sulfate-reducing bacteria use dissolved sulfate ion as an electron acceptor and produce hydrogen sulfide (H_2S) as waste. In iron-rich water, as H_2S bubbles through, iron sulfide precipitates. The black color of this layer is primarily due to the formation of an iron sulfide mineral called pyrite ("fool's gold"). Gaseous sulfide is also used by the overlying layers of purple and green sulfur bacteria as the electron donor for their anoxygenic photosynthesis.

Analysis of the striped soil reveals many kinds of bacterial communities. In detail, bacterial mats are organized "tissues" like those of animals. As documented by the fossil record, these kinds of microbial communities have persisted throughout time. During the past 2,500 million years, these bacteria have removed enormous quantities of carbon dioxide from the atmosphere and simultaneously produced oxygen on such a scale that they have permanently changed the Earth's atmosphere.

In a very few cases, complex communities of bacteria have left extremely conspicuous remnants, as in the Kuuvik formation of northwest Canada (Figure 4–1). These limestone mountains, estimated to have been laid down some 2,000 million years ago, were constructed by communities of carbonate-precipitating bacteria much like some still living in restricted zones today (Figure 4–2).

Year after year at seaside locations, in subtropical climates, such bacterial communities removed carbon dioxide from the atmosphere, precipitating, layer by layer, these columns of rock. Comparable to fossil coral reefs, these mountains are composed of biogenic rocks—rocks made by life. Such continual precipitation of carbon dioxide over millions of years has virtually changed the entire planet Earth. Later limestone deposits were produced by shells of animals and algae, but early limestone deposits were made by bacteria. Throughout Earth's history, bacteria have produced much of the early limestone; bacteria also have other powers.

BACTERIA AND BANDED IRON FORMATIONS

Another ancient biological product on the Earth are iron-rich mineral deposits of commercial interest called banded iron formations (BIF). Although the formation of BIFs probably required a peculiar combination of geological circumstances, it is doubtful that these rich iron deposits can form at all on a planet devoid of life, and the extensive mineable iron deposits would not have been produced on Earth in the absence of life. Further, the genesis of BIFs is related directly to bacterial growth and evolution. They are the product of a sequence of events that occurred in the history of the Earth between approximately 2,200 million years and 1,800 million years ago. Nearly 90 percent of the world's mineable iron comes from these deposits. Iron used for the manufacture of automobiles and machines, the iron upon which our culture depends, is mined from such deposits. The accumulation of iron on the external layers of the solid Earth is a phenomenon uniquely

FIGURE 4–1 Stromatolite reef mound in the Kuuvik Formation, Kilohigok Basin in northwestern Canada. (Photo: Geological Survey of Canada, Dr. F. Campbell.)

FIGURE 4–2 Living stromatolite from Hamelin Pool, Shark Bay, western Australia.
(Photo: Professor Paul Strother.)

associated with life, and, therefore, we predict, not to be found on Mars
or Venus.

In most places, as in the iron deposits of northern Michigan, northern
Minnesota, and southeast Ontario, the rocks have been highly metamor-
phosed (greatly altered by heat and pressure) and evidence of life, if it
ever existed in association with the rocks, has been obliterated. But in
certain fortunate locations, for example, at Schreiber Beach and Frus-
tration Bay, Ontario, some of these iron-bearing rocks escaped meta-
morphosis. From these favorable localities, thin sections of the flinty
rocks can be cut with a diamond knife. After polishing, they are viewed
under an optical microscope. Dark areas richer in iron alternate with
light iron-poor/silica-rich regions. Light and dark, such a layered
appearance is reminiscent of the view of the structure of the microbial
mats of Baja California or those from modern Abu Dhabi under the
microscope. Higher magnification of this gunflint rock, which is a black,
smooth cryptocrystalline quartz rock called chert, reveals structures

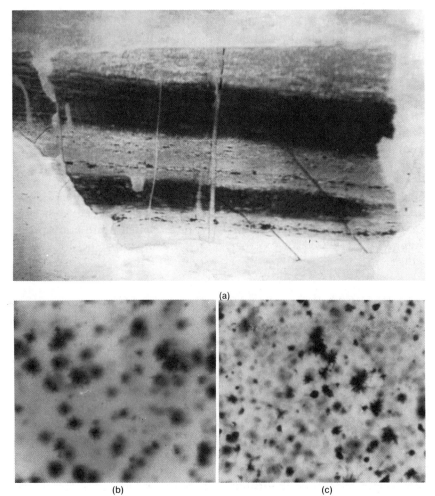

FIGURE 4–3   (a) Hand sample of laminated Gunflint chert (b) *Eoastrion* or "dawn star" (c) *Metallogenium* iron-precipitating microbes which may be modern analogues of *Eoastrion*.

called "*Eoastrion*" (or "dawn star") (See figure 4–3b). The rock hardly looks like a biological sample (Figure 4–3a). What are these tiny "stars"? No one is positive, yet most of us believe that the "dawn stars" are fossil counterparts of the structures found in extant microbial mats (Figure 4–3c). In 1963, two scientists, Barghoorn and Tyler,[3] first recognized these to be microfossils. Probably they are remains of bacteria that accumulate iron.

We argue that BIFs are due in two ways to bacterial processes: iron-bacteria affected the accumulation of metal required for the metal precipitation, and oxygen was a waste product of cyanobacterial photosynthesis. The formation of massive iron-oxide deposits precipitated by bacteria, including oxygenic ones, is thought to be utterly unique on the Earth. BIFs are likely to be 2,200 to 1,800 million year old products of biogeochemical processes, including bacterial growth, metabolism, and evolution, on a planetary scale. This busy movement of matter by life occurred on an iron-, oxygen-, and water-rich planet.

How long has life, as bacteria, been modifying the environment? Probably for at least 3,000 million years. Direct evidence for life on Earth in the form of familiar large organisms, plants and animals, exists for only about 570 million years. By contrast, we have direct evidence for communities of microbes that modified the sediments and atmosphere from as long as 3,000 million years ago. Long, of course, before human beings.

GEOCHEMICAL ROLE OF BACTERIA

Bacteria have had great impact on our planet. We need to become more aware of the dynamic, complex gas exchange and sediment production that involves these organisms. Microbial mat communities of Baja California Norte, México along the Pacific coast can be recognized in Skylab photographs taken from orbit in 1978. Such mats are rather limited in their distribution today, primarily because of the influx of people, but also because of high salt concentrations and competition with salt marsh and other vegetation.

A sample cut from one of these mats reveals layers of sediment topped by a greenish layer of the cyanobacterium *Microcoleus*. As we noted, these *Microcoleus* communities are thought to resemble those that grew much like this, covering the marine coastal regions of the Earth throughout most of its history. We have discovered that the Baja California microbial mats can survive desiccation; they can dry out for months or years. Re-wetting revives them within a few minutes. Somehow, these bacterial communities have solved the ecological problems of self-maintenance; they are enormously persistent.[4] The *Microcoleus* mat community, similar to that studied by Golubic[5] on the Persian Gulf, always forms in the intertidal zone. For *Microcoleus* communities to form, certain conditions must be met. *Microcoleus* mat communities do not form in the open ocean where they would be subject to strong wave action. A barrier to the ocean must be present. These communities form landward of the barrier. Tides from the ocean must be present, as well as dunes or other barriers that block the access of direct ocean water to the mat. Rocky rubble from volcanoes or mangrove trees—in the absence of dunes—may form the barrier, but in all cases, a barrier must protect the mats from the open ocean. The productive cyanobacterium *Microcoleus* moves toward sunlight. As it glides, it traps pieces of sand, crystals of salt, and calcium carbonate on the surface of its sheath. *Microcoleus* filaments glide over each other; they glide to the mat surface and construct these communities (Figure 4–4a) whose stripes are much like the ones observed in the ancient rock samples. Electron micrographs of *Microcoleus* filaments reveal radially aligned membranes, the site where carbon dioxide is removed from the atmosphere and oxygen is simultaneously produced (Figure 4–4b).

In microbial mats, there are millions of organisms per cubic centimeter that comprise what looks like dirty soil. Occasionally, animals are present in microbial mats; but, in fact, few types of animals are able to penetrate these oxygenless bacterial communities. Certain nematodes can manage to live inside bacterial mats. Some swimming organisms associated with bacterial mats, for example, *Metopus contortus,* a protist, move quickly in the absence of oxygen.[6] There also are organisms that eat bacteria that precipitate metals such as iron or manganese. Apparently the bacteria-eaters can re-dissolve the iron and manganese that coats their bacterial food. Most of the details of the relationships between metals and microbes are not known, but metal accumulation and precipitation is common in the bacterial world.

We have cataloged approximately 150 different kinds of microorganisms that live in the top 2 centimeters of the *Microcoleus* mats, and we are nowhere near listing the total diversity. There is an incredible biological diversity in a small volume of these mats, a diversity unknown before the 1960s and comparable to the well-known diversity of a forest. To try to describe the organisms and their activities that make up any of these systems meaningfully is an enormous task. At levels from microbial mats to tropical forests, we share an immense ignorance towards nature and the interaction of life on Earth, but most especially towards microbial life.

Remember that everything discussed here exists only in the top 2 centimeters of a *Microcoleus* microbial community. The density of microbes is so thick and complex that we still have no complete description of most of them. Many of these bacteria fix nitrogen: they take nitrogen as gas from the air and convert it into organic matter—the kind of solid nitrogen compound that can be used as food. Nitrogen

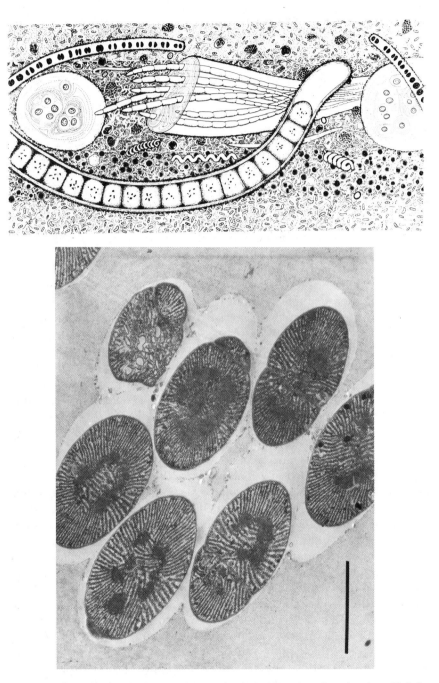

FIGURE 4–4   (a) Laminated microbial mat dominated by *Microcoleus* (drawing: Christie Lyons from Margulis et al., 1986.). (b) Electron micrograph of *Microcoleus* from the laminated microbial mat at Laguna Figueroa, Baja California Norte, Mexico. (Micrograph: Dr. John F. Stolz.)

fixation is entirely a bacterial process. No plant or animal can take nitrogen as gas directly from the air and use it as food. People and animals would not exist if the microbial ability to fix nitrogen into organic matter did not also exist. From these examples, we see that microbes are not just curious little germs that are occasionally dangerous, but that they are essential to our survival. Their communities are diverse and their adaptations often amazing.

A further example of the diversity and adaptability of bacteria includes another mat organism, a flagellated purple phototrophic bacterium called *Chromatium,* which have wheel-like structures. Other bacteria are also capable of such rotation. Although we speak loosely as if they were single-celled, not all bacteria are. Many bacteria (e.g., myxobacteria, cyanobacteria) form large multicellular structures.[7]

NEED TO STUDY THE ECOLOGY OF MICROBES

A forest or desert, pond, or other large ecological system is always underladen, infiltrated, covered, penetrated, and embedded in some kind of microbial system. Yet, rarely is a microbial community studied by ecologists.[8]

In Baja California, in 1979 and 1980, immense spring floods covered our field site. By 1981, the floodwaters receded. Mud from these floods remained, blanketing the microbial mats. The *Microcoleus* community responsible for oxygen production disappeared. Another complex bacterial community replaced it. The new community was composed of interacting bacterial forms that produce food and sulfur globules but did not produce oxygen. Using the electron microscopic techniques employed by physicians, histologists, and other biologists to study livers, kidneys, lungs, and other tissues, John Stolz studied the newly forming sediment.[9] He showed that what looks like a bit of mud (Figure 4–5a) is a wall-to-wall extension of diverse interacting microorganisms (Figure 4–5b).

Our point is that we systematically ignore the contribution to ecology of the microbial communities that sustain us. These communities ultimately produce our food and oxygen. In nature, these microbes are packed one next to another, as can be seen in pictures of purple photosynthetic bacteria taken directly from "mud" samples (Figure 4–5a). Even if we choose to ignore them, these bacteria are productive beings. They and their descendants, the chloroplasts in green plants, make the food for the rest of the biota. The appearance of land plants and agricultural grasses—from 400 million years ago to the present—is

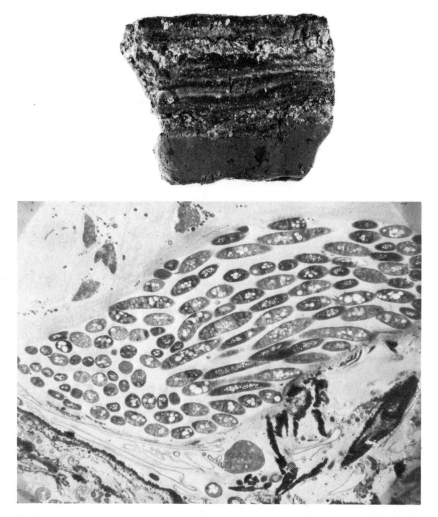

FIGURE 4–5   (a) Mud sample from flooded microbial mat. Sample is about 6-mm high.
(b) Electron micrograph of purple photosynthetic bacteria. (Photo: Dr. John F. Stolz.)

an extremely recent phenomenon. Plants, including our food grasses,
essentially depend on the productivity of the chloroplasts inside their
cells. It is well documented that these chloroplasts are of direct bacterial
origin—indeed, cyanobacterial origin.[10]

The *Microcoleus* mat of North Pond, San Quintín, México, has been studied since 1975.[11] It was flooded and covered with mud in 1980, after which the weather in the area returned to usual dry and windy conditions. New *Microcoleus* mat grew out from the channels and a re-colonization occurred. New laminae formed. Remarkably, we have analogs of these laminae in the fossil record. The mud preserved some of the structures of these microbes. Organisms preserved in the mud, similar to those preserved microfossils that date back 3,000 million years, are further examples of communities existing long before human beings.

GAIA

This story ends with the idea of the hand of Gaia, of "Mother Earth" (Figure 4–6). What does the design tell us? Gaia, the hypothesis developed by J.E. Lovelock,[12] states that Earth's lower atmosphere is maintained by the sum of the life on the surface (biota). It is hypothesized that the reactive gases, acidity-alkalinity, oxidation-reduction state, and temperature of the Earth's surface are maintained by the biota. Figure 4–6 portrays an Earth that is an extraordinary planet not just because it has water, but because it has living processes on its surface. From a planetary point of view, the major effect of the living process is the removal of carbon dioxide into the compounds of food and calcium carbonate rocks (sometimes the size of mountains), oil, coal, and other carbon-containing materials, and the simultaneous production of oxygen. Such living processes do not occur on Mars, Venus, or the moon. Bacteria of diverse types change carbon dioxide and nitrogen into usable carbon compounds with the simultaneous production of oxygen or sulfur. Carbon dioxide, nitrogen, sulfur oxides, and oxygen, and their removal and production, are crucial for the maintenance of the ecosystems upon which human beings depend.

Homo sapiens are a recent example of one of thirty million extant species of life on the planet. From this perspective, humankind appears as a mammalian weed, which like all weeds, grows rapidly and is destructive of its environment.[13] Fast-growing organisms, given the opportunity, grow until their immediate environment becomes unlivable. Like many other rapidly growing organisms in the history of the Earth, humans simply accelerate their extinction. Meanwhile, these microbial mat communities, which began more than 3,000 million years ago, have been in continuous existence. Although they flourished a thousand-million years ago and they have declined somewhat in extent in the last 500 million years, they persist. In North America, for

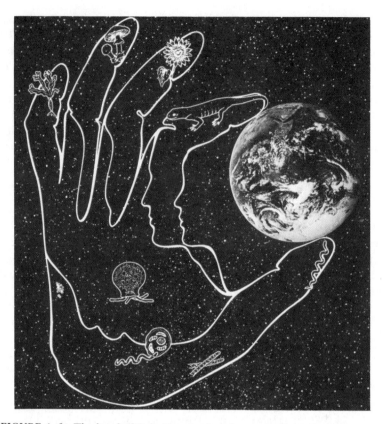

FIGURE 4–6   The hand of Gaia (drawing based on a sketch by Dorion Sagan).

example, *Microcoleus* mats still extend down the Atlantic Coast from
Nova Scotia to North Carolina. These bacterial communities, which live
in an ecological balance of gas exchange and food production, certainly
will persist long after the disappearance of humans. The major mechanisms of Gaian environmental control are the metabolism, growth, death,
and extinction of the populations comprising Gaia. Most species ever to
have lived are extinct. It is solipsistic nonsense to expect any fate other
than extinction for *Homo sapiens*. What, perhaps, we can control are the
qualitative details and the timing of our inevitable demise.

We must learn from the microbial communities. We must be aware
that our well-being and health is entirely embedded in that of our
planet-mates. The concept of independence from the biosphere is
illusory and dangerous. People do not dominate nature, nor have they

ever done so in spite of ancient and weighty prose arguing the contrary. Our attempts to wrest control and independence from the rest of the biota will simply assure us of unpleasant death and hasten the extinction of our species.

NOTES

[1] T. Dobzhansky, 1973, "Nothing in Biology Makes Sense Except in the Light of Evolution," *American Biology Teacher,* Vol. 35, pp. 125–129.

[2] T.D. Brock, D.W. Smith, and M.T. Madigan, 1984, *Biology of Microorganisms,* Prentice-Hall, Inc., Englewood Cliffs, N.J.

[3] E.S. Barghoorn, and S.A. Tyler, 1965, "Microorganisms from the Gunflint Chert," *Science,* Vol. 14, pp. 563–577.

[4] S. Brown, L. Margulis, S. Ibarra, and D. Siqueiros, 1985, "Desiccation Resistance and Contamination as Mechanisms of Gaia," *BioSystems,* Vol. 17, pp. 337–360.

[5] S. Golubic, 1973, "The Relationship Between Blue-green Algae and Carbonate Deposits," in *The Biology of Blue-Green Algae,* N.G. Carr and B.A. Whitton (eds.), Blackwell Scientific Publications, Oxford, England pp. 434–472.

[6] B. Dyer, 1984, "Protists from Microbial Mats, Baja California, Mexico," Doctoral dissertation, Boston University.

[7] L. Margulis, and K.V. Schwartz, 1987, *Five Kingdoms—An Illustrated Guide to the Phyla of Life on Earth,* 2nd ed., W.H. Freeman & Co., New York.

[8] L. Margulis, L. Lopez-Baluja, S.M. Awramik, and D. Sagan, 1986, "Community Living Long Before Man," in *The Science of the Total Environment,* Vol. 56, pp. 379–397.

[9] J.F. Stolz, 1984, "Fine Structure of the Stratified Microbial Community at Laguna Figueroa, Baja, California, Mexico. II. Transmission Electron Microscopy as a Diagnostic Tool in Studying Microbial Communities *in situ,*" in *Microbial Mats: Stromatolites,* Y. Cohen, R.W. Castenholz and H.O. Halvorson (eds.), Alan Liss Co., New York pp. 23–38.

[10] L. Margulis, 1981, *Symbiosis in Cell Evolution,* W.H. Freeman, San Francisco, California.

[11] R.J. Horodyski, and S.J. Von der Haar, 1975, "Recent Calcareous Stromatolites from Laguna Marmona (Baja California) Mexico," *J. Sediment. Pet.,* Vol. 45, pp. 894–906.

[12] J.E. Lovelock, 1988, *Ages of Gaia,* W.W. Norton Pub. Co., New York. Oxford, England.

[13] L. Margulis, and D. Sagan, 1986, *Microcosmos—Four Billion Years of Evolution from our Microbial Ancestors,* Summit Books, New York.

EDITORS' INTRODUCTION TO:
G. CARLETON RAY
Sustainable Use of the Global Ocean

In the past 30 years much has changed in our understanding and perception of the oceans. As Professor G. Carleton Ray, a marine ecologist who is an expert on marine mammals and a faculty member of the University of Virginia explains, we are undergoing a "marine revolution." Only a generation ago, experts still believed that the oceans were so large that we had not and could not exert any noticeable effects on them. Today, he writes, pollution of the ocean is "obvious," with petroleum, the largest single source, contributing millions of metric tons per year. But beyond our pollution of the oceans, the role of the oceans in global processes is just beginning to be understood. Although considerable advances have been made in recent decades in our knowledge of the oceans, our scientific perception of them remains primitive. How we should categorize the ocean and divide it into meaningful units is not clear. The role of the oceans in the global carbon budget has many unknown and poorly estimated numbers.

However, some portions of the oceans stand out as especially crucial to all life on the Earth. Among these, Dr. Ray suggests, are the coastal zones, near which most of us live and within which many of our important impacts on the oceans are occurring. His central message to us is that only by correcting our landward bias and integrating our view of the environment to include the oceans fully will we be able to understand and come to terms with our global environment.

5 SUSTAINABLE USE OF THE GLOBAL OCEAN

G. CARLETON RAY*
University of Virginia
Charlottesville, Virginia

INTRODUCTION

Thirty-two years have passed since the publication of *Man's Role in Changing the Face of the Earth.*[1] That book contained minimal consideration of the oceans and coasts, but both the omissions and statements are revealing. Overall, there is little recognition of the oceans as an integral part of planet Earth or that human culture could much affect them. Michael Graham[2] ended his chapter on harvests of the seas with "It seems that the effect of man on the ocean has been small, that there remain relatively untouched sources of wealth, and that even if these are greatly exploited in the future, the ocean will remain much as it is and has been during the human epoch . . . it would, indeed, seem that here at the beginning and the end is the great matrix that man can hardly sully and cannot appreciably despoil." John C. Bugher[3] did forecast the future, however, when he recognized the possible effects of fission material and concluded that "we need to divert some of our scientific

* I extend thanks to M. Geraldine McCormick-Ray and C. Richard Robins for reviewing earlier versions of this chapter, and especially to Daniel B. Botkin for suffering through several iterations. Max J. Dunbar contributed to a reinterpretation of classification of environments (Figure 5–1). Judith Peatross also deserves credit for converting my scribbling into Figures 5–2 and 5–3.

preoccupation with land-locked problems and to apply a steadily increasing effort to the advancement of knowledge of marine biology."

In the 1950s, the intensive use of ocean resources had hardly begun. Surely, certain places, particularly near shore, and certain depleted fisheries bore signs of the heavy hand of humankind, but large areas were still largely pristine and the bulk of species—even the whales—were still abundant. The world population was only half of the present 5 billion; now, an equivalent number of the entire 1950s population inhabits just the coastal zone. Today, the shortage of global resources and space forces humans to develop new technology for exploration and exploitation of the sea. Concurrently, a new sense of urgency is evolving from the recognition of widespread coastal and marine pollution, habitat degradation, and depleted living resources. Most significantly, perhaps, technology now permits human beings to explore beneath the surface and to see the beauty that still exists, side by side with the despoliation that has occurred. The oceans have, for the first time in history, become a personal matter to a vast number of people.

These fundamental shifts, taken together, have resulted in what I have called the "Marine Revolution."[4] This is in stark contrast to the fact that humans remain predominantly preagricultural hunter-gatherers of oceanic resources and are largely at the mercy of oceanic processes for harvest of the sea's resources and processing of humanity's wastes. That this situation is highly unstable and commands heightened awareness and preventative action is self-evident.

This brief essay attempts to illustrate some of the changes that have occurred over the past three decades and to present a context for some of the challenges that lie before us. The task is not simple. We might ask: What might be the global impact of humanity's burgeoning billions and ever-increasing per capita demands on the global ocean? What impacts of humankind have been detected and which among those not yet detected are insidiously at work? What are the implications for management and institutions? Finally, how can diverse societies, linked on a global scale by ecologic, economic, and social factors, arrive at the sustainable use of the oceans? These are comprehensive questions, each the subject of many books. At the heart of the matter lies the question: What are we doing to the life of the oceans? It is on this question that I will focus, first by considering the oceans on the global scale and second by examining their biological and ecological diversity. Third, the coastal zone will be defined—that Africa-and-a-half-sized part of Earth along which most of the human population now lives and, therefore, an area which is of most direct concern to us. Finally, I will depict some activities of humankind in order to make several statements about sustainable use.

THALASSOGAIA—THE GLOBAL OCEAN

A view of the Earth from the South Pole reveals a contiguous body of water, interrupted only by continents. Subdivisions, which we call "oceans" with their appended "seas," are bounded by the land. Looking down into the water's depths, we see that these oceans and seas are strongly three-dimensional; that is, not only are they subdivided over their surfaces, but they are also layered into water masses, each carrying distinct biotic assemblages. The thin skin of terrestrial life is obviously three-dimensional too, but it could be swallowed in this 135 million-cubic-kilometer volume of living ocean about 100 times.[5]

Recognition of the ocean's great expanse, volume, continuity, and importance was emphasized by James Lovelock[6] when he hypothesized on Gaia that "The biosphere is a self-regulating entity with the capacity to keep our planet healthy by controlling the chemical and physical environment"; it is "a complex entity" involving the Earth's biota, atmosphere, oceans, and soil, "the totality constituting a feed back or cybernetic system which seeks an optimal physical and chemical environment for life on this planet." Lovelock regarded planet Earth as an "operating system" and derived his view largely on the basis of the Earth's unique atmosphere. The implication is that the Earth—a somewhat unfortunate name for a planet dominated by water—is an integrated functional unit in which the ocean, and especially its life, is a critical part.

In the absence of the oceans, there would be no life as we know it on this planet. This is made clear by virtue of the ocean's role as the storehouse of most of the Earth's heat and water. This has been recognized for decades, perhaps centuries, according to which human culture is used for reference. Western thought has seen the world differently from some other cultures that also had Gaian thoughts about an Earth Mother, in that Western reductionist science has led to the view of life driven by its physical surroundings. But planet Earth is not merely a physical system on which life depends. Otto Solbrig put it this way: "It is widely recognized that the geosphere and the biosphere are intimately connected. In fact, the sharp distinction between them is mainly semantic."[7]

Examples of ocean life's role in global processes abound. Carbon, nitrogen, phosphorus, sulfur, and other elements are now known to be cycled on a global scale, in large part through the agency of living organisms. Carbon, in particular, is of interest, due to the effect of increasing carbon dioxide on climate. Unfortunately, scientists have not yet been able to balance the global carbon budget; the life of the oceans may hold a key.[8] A "biological pump" for carbon that involves uptake by primary producers, transport from the surface to deep water, and sedimentation has resulted in an estimated 99 percent of both organic and inorganic carbon being sequestered in marine sediments. How the carbon enters the sediments and how it gets out and is recycled is an extremely complex topic that is by no means clearly understood, but key questions seem to involve patterns and processes of oceanic productivity, biogeochemistry, mixing, consumption, and regeneration—truly a whole-Earth question, with ocean life at the epicenter!

Trace gases also illustrate the global importance of the oceans and their life. The science of atmospheric chemistry is only about two decades old, but we now know that a large group of compounds is involved. One is dimethyl sulfide that is produced by phytoplankton and which is responsible for the odor of sea air. Dimethyl sulfide may help prevent Earth from being overheated. Its concentration in the atmosphere over ocean space is low, but because the ocean is so immense, about 60 million tons of it are produced a year. It exists as minute aerosols on which cloud droplets form; these are responsible for reflecting and re-radiating much of the heat of solar radiation. Meinrat Andreae[9] has pointed out that such compounds are "produced and consumed by a variety of marine organisms, each with different rates and possibly different pathways." This implies that, if the patterns and compositions of marine communities vary, so will the patterns and production of such globally important compounds as dimethyl sulfide and many others.

John McGowan and Patricia Walker[10] have pointed out that oceanic planktonic communities, far from shore, are predictable in species composition and that "the regulatory forces are strong and almost certainly biological, rather than physical." Evidently, the activities of humans have not yet significantly altered such communities beyond their resiliency, but it may be only a matter of time, in my estimation, before the impact will be felt. Whether we will be able to recognize the alterations is problematic. Whether we will be able to do anything about it is dubious at best, for, as McGowan and Walker also point out, the regulatory forces are unknown.

Lovelock's Gaia hypothesis remains controversial, but there seems to be no question that global biological processes help shape the character of Earth in a way unimagined three decades ago and that the life of the oceans is responsible for a large part of the shaping process.

BIOLOGIC AND ECOLOGIC DIVERSITY

The diversity of life, habitats, and ecosystems of the oceans must be understood so that conservation of resources may be based on ecological principles. Diversity is a widely misunderstood term. One view, held by the majority of the public and successfully promulgated by Norman Myers[11] and others, presents a yardstick based on species richness, that is the number of species in a given area. This viewpoint has made a useful impression on decision makers, but is an oversimplification. It neglects a vast range of interactions from genetics to differences among landscapes and ecological processes that may have little to do with richness *per se.*

Therefore, we must also consider interactions within ecosystems. It may be best to approach diversity simultaneously in two fundamental ways: 1) to consider such matters as which species are dominant, which may be considered "key" or "indicators," and what the effect of depletion or of introduced species might be, and 2) to be more concerned with redundancy among species, spatial and temporal hierarchies of communities, and biogeographic distributions of biological assemblages. In both respects, it is especially important to determine, if possible, how spatial and temporal scales relate to environmental variables.[12] Both viewpoints may be constructively used to examine how diversity may affect such global processes as, for example, the production of biogenic substances or the recycling of organic materials discussed earlier.

An impediment to interpretation of oceanic diversity in either sense is that we are still forced to resort to theory in order to guess its dimensions. For example, largely due to the relative ease of terrestrial over marine logistics, it has been assumed that the land is about four times as species rich as the oceans, though only about one third as big in surface area. That is, the oceans are supposed to hold only about 20 percent of planet Earth's species. Actually, this figure is probably seriously in error, as new marine species are being discovered at a rapid pace. Even if species turn out not to be as numerous in the oceans as on land, species accounting does not necessarily yield a proper interpretation of the variety of its life. Terrestrial species diversity is dominated by such numerous groups as insects and orchids, which are absent from the seas. However, roughly twice the number of plant and animal phyla are represented in the oceans as on land and in fresh water.[14] From a different point of view—that of ecology—the oceans may be interpreted as being more diverse than the land. This is indicated by several factors, including the fact that the size range of marine fauna is much greater than for land fauna. One important life-style is present in aquatic systems that is lacking in terrestrial ones—filter-feeding. Finally, the strong three-dimensionality of the oceans has the result that marine food chains are longer and food webs more complex than are those of most terrestrial systems.[14]

The biogeographic description of the diversity of ecosystems adds still another dimension. Intuitively, the thought of placing boundaries about ecosystems is attractive. Nevertheless, it remains hypothetical, due largely to the difficulty of delineating the time-space hierarchies of ecosystems[15] within the three-dimensional living space of the oceans. Kenneth Norris and I[16] have characterized the ocean as "a bouillabaisse of animals and plants, of uncountable microorganisms, of nutrients, of degradation products of life, of inorganic contributions from land, from chemical precipitation, and of dust from the atmosphere. Its 'winds,' which are the ocean currents, move at all levels from the surface to the deepest sea where water generally creeps northward from the Antarctic Convergence." That is, the sea has texture, but its boundaries are ever mobile and difficult to recognize, even with sophisticated instruments and complex computer models.

Nevertheless, we do recognize coastal landscapes and marine sea-scapes that can be interpreted as having ecosystem properties. Several means have emerged since the birth of oceanography in the nineteenth century to define oceanic units. Biogeography is one way; the determination of physical units, such as water masses, is another. Figure 5–1 represents one rather simplistic interpretation of our present state of knowledge, in an attempt to synthesize these two approaches. The figure does not reflect the three-dimensional nature of the hydrosphere, nor does it take into account a variety of data, both biotic and benthic, too numerous to mention here, but it does illustrate the fact that there exists no consistent coastal-oceanic environmental classification. Biotic "provinces" have been defined largely by the distributional patterns of selected endemic or dominant species; this approach serves to emphasize the uniqueness of each province. The inevitable result is that insufficient attention is given to the similarities and relationships among systems. This state-of-the-art marine environmental classification has

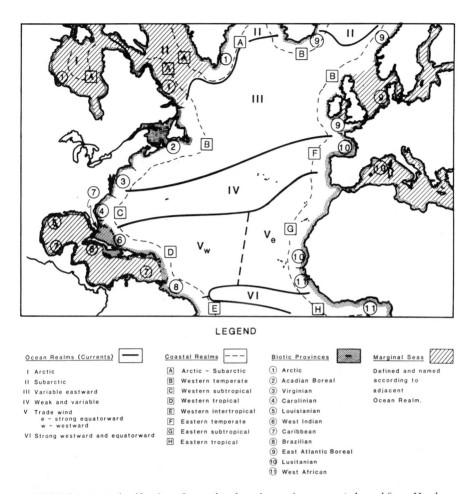

LEGEND

Ocean Realms (Currents) ▬	Coastal Realms ---	Biotic Provinces ▓	Marginal Seas ⧄
I Arctic	Ⓐ Arctic – Subarctic	① Arctic	Defined and named
II Subarctic	Ⓑ Western temperate	② Acadian Boreal	according to
III Variable eastward	Ⓒ Western subtropical	③ Virginian	adjacent
IV Weak and variable	Ⓓ Western tropical	④ Carolinian	Ocean Realm.
V Trade wind	Ⓔ Western intertropical	⑤ Louisianian	
e – strong equatorward	Ⓕ Eastern temperate	⑥ West Indian	
w – westward	Ⓖ Eastern subtropical	⑦ Caribbean	
VI Strong westward and equatorward	Ⓗ Eastern tropical	⑧ Brazilian	
		⑨ East Atlantic Boreal	
		⑩ Lusitanian	
		⑪ West African	

FIGURE 5–1   A classification of coastal and marine environments (adapted from Hayden et. al., 1984). This representation is symbolic and not drawn to scale for coastal areas. Ocean realms are for surface waters only. Coastal realms are highly variable, especially for temperate areas, which contain attributes of both subarctic and subtropical coastal waters.

shortcomings from the ecological point of view, even though it has pragmatic applications.[17]

The purpose of concentrating on diversity is to focus attention on units of management and conservation, as well as science, whether it is the species or the ecosystem that is of concern. Even given the state of the art of biogeography, for example, attempts to describe marine ecosystem boundaries on a global scale can and should be intensified. The effort should lead us, most importantly, to look at oceanic, ecosystemic units in relational, functional terms, within which characteristic biota of each unit may also be identified. In short, it remains obscure to me how global marine management can evolve in the absence of a realistic environmental taxonomy, within which time-space scale attributes are made clear.

INTERACTIONS WITHIN THE COASTAL ZONE

The segment of planet Earth called the *coastal zone* is especially important in the contexts of diversity and human interactions. This is by far the most populated and urbanized portion of our planet, as well as the richest and among the most perturbed. It is a bit surprising that it is also not the most familiar in concept and dimension, in view of the fact that so many people live within it or near it.

An examination of coastal biological or physical processes reveals that we live in a tripartite world, the parts of which we might describe as upland, open ocean, and coastal zone. Terrestrial continental plains are geologically contiguous with the oceans' continental shelves.[18] Sea level rise and fall over recent geologic time has almost completely covered it at times, left it almost completely dry at others. Its area comprises about 8 percent of the globe, or one and a half times the size of Africa. Its extent for the eastern United States is shown in Figure 5–2. Throughout history, the apparent sharp divisions between terrestrial and aquatic systems at the water's edge have deceived most human societies. Presently, the land-sea discontinuity is encodified into law and custom so strongly that there is little hope of unification in the way that nature might indicate and that might most benefit management (see Figure 5–3). Policies now revolve around such questions as How may the coastal zone best be conceived? Is it an ecotone (transitional between land and sea) or an ecosystem in its own right? Distinguishing between these is not trivial; ecotones and ecosystems exhibit different boundary conditions that strongly influence the way they must be conserved and managed.

The ecotone versus ecosystem viewpoint may be illustrated by consideration of how one may wish to manage waterfront property versus the conservation of living resources, or beyond that to the biosphere at large. *Coastal zone management,* as most commonly defined, is concerned with near-shore processes such as conservation of beaches and wetlands. But this narrow perspective will not suffice for the greater bulk of

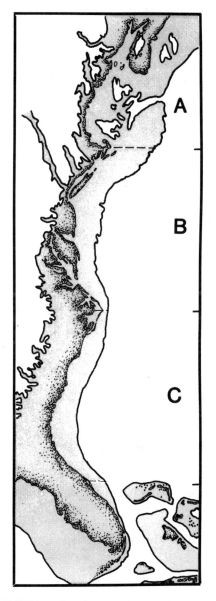

A. ACADIAN - BOREAL

1. Ocean-dominated
2. Dissected shelf with basins and canyons
3. Strong tidal mixing
4. Rocky shores; glacial history
5. Narrow coastal plain
6. Small barriers; many islands
7. Few long rivers
8. Small bays, coves, pocket beaches
9. Cool temperatures
10. Influenced by Labrador Current

B. VIRGINIAN - MID ATLANTIC

1. Coastal/ocean-integrated
2. Wide, flat shelf with canyons
3. Mixing of coastal and offshore waters
4. Extensive wetlands and lagoons
5. Rolling coastal plain
6. Extensive barrier beaches
7. Extensive river drainages
8. Very large estuaries
9. Variable seasonal temperatures
10. Influenced by Gulf and Labrador currents

C. CAROLINIAN - SOUTH ATLANTIC

1. Terrestrial-dominated
2. Wide, flat shelf; no canyons
3. Coastal waters variable and turbid
4. Poorly drained tidal marshes
5. Broad, flat coastal plain
6. Extensive barrier beaches
7. Long, silt-laden rivers
8. Small lagoons and estuaries
9. Warm temperatures
10. Influenced by Gulf Stream waters

FIGURE 5–2   The coastal zone of the eastern United States. The biotic provinces are delineated and their characteristics are listed in the legend. (Figure drawn by Judith Peatross.)

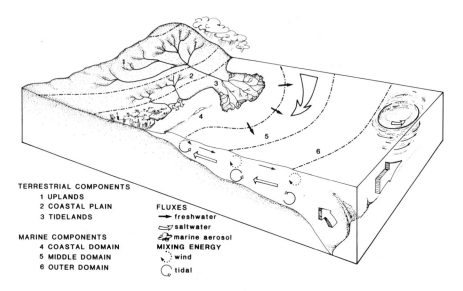

TERRESTRIAL COMPONENTS
  1 UPLANDS
  2 COASTAL PLAIN
  3 TIDELANDS

MARINE COMPONENTS
  4 COASTAL DOMAIN
  5 MIDDLE DOMAIN
  6 OUTER DOMAIN

FLUXES
  → freshwater
  saltwater
  marine aerosol
MIXING ENERGY
  wind
  tidal

FIGURE 5–3   The coastal zone is shown to consist of terrestrial and marine components tied together by various processes and fluxes. Biotic processes (not shown) are also important, for example, the movements of organisms to and from continental shelf waters, estuaries, and rivers. Three types of coastal units are illustrated: a) a tidal unit consisting of the coastal and near-shore domain and tidelands, b) a river-estuarine unit in which interactions with both the coastal and middle domains are possible, and c) a large watershed-deltaic unit with interchanges mostly with middle to outer marine-shelf domains. All are frought with problems of multiple jurisdictions among agencies. (Figure drawn by Judith Peatross.)

management issues. Even a cursory review of terrestrial and marine life reveals the dependency of many plants, invertebrates, fishes, reptiles, birds, and mammals on coastal habitats from coastal plains to rivers and wetlands, to estuaries, to the great variety of habitats of the continental shelves—for example, coral reefs, sea grass beds, mud flats, and offshore banks and shoals. We also have known for some time that the coastal zone exhibits among the highest rates of productivity of our planet. Clearly, the conservation and management of these resources requires a broad, regional perspective. Further, it has become known that this zone is especially important with respect to global ecological processes. For example, scientists concerned with global carbon cycling have proposed that an important "sink" for carbon occurs on the continental shelves and slopes.[19] Many questions remain about how organic matter is deposited, utilized, and recycled there,[20] but one thing is certain—the coastal zone is important on a global scale out of proportion to its size.

I have made a crude estimate by multiplying mean productivity by area that indicates that the 8 percent of the planet Earth that is the coastal zone may equal the 65 percent of open ocean (beyond the continental shelves) or the remaining 27 percent that is upland (above the continental plains) in total production and recycling of organic matter. This serves to emphasize that how we manage ourselves within the coastal zone is of critical importance on a global scale. For example, marine "farming" or *mariculture* is rapidly expanding as a result of strong market and social pressures, and is generally considered in a positive light. However, the effect of extensive mariculture may not prove to be all to the good. Lovelock warns: "We should not, however, assume that the sea, and especially the arable regions of the continental shelves, can be farmed with the same impunity. Indeed, no one knows what risks are run when we disturb this key area of the biosphere."[21]

If the coastal zone is so vital—out of proportion to its size—an obvious question concerns the role of specific segments of its biota. Natural history provides many examples of biotic interactions among species and habitats that might be extrapolated upscale, admittedly at some risk. For example, the territorial behavior of damselfishes (family Pomacentridae) discourages other herbivores, thereby leading to the development of an "algal lawn," which, in turn, has the effect of altering reef communities[22] and their diversity.[23] Other studies of reef-dwelling fishes have shown that a significant portion of coral reef nutrition derives from fish (especially grunts, family Haemulidae) that feed on sea grass beds at night and excrete nutrients over the reefs by day.[24]

Can such interactions be carried to the global level? Logic would tell us that the compounded interactions of the many species that alter community structure should have a combined global effect. Some species clearly have a significant regional impact. Within the Bering and Chukchi seas, where lies the largest continental shelf area on Earth, walruses (*Odobenus rosmarus*) (Figure 5–4) and gray whales (*Eschrichtius robustus*) overturn massive amounts of benthic sediments during their feeding— approximately 100 thousand square kilometers per year.[25] The hypothesis is inescapable; these large organisms may exert strong controls on both community structure and nutrient flux. An interesting speculation is that these species may be "cultivating their own garden" on a seawide scale, as the little pomacentrid fish do on a more local scale. That is, the Bering and Chukchi seas may exhibit greater biotic productivity due to the feeding of walruses and gray whales than they would in the absence of these species and are richer as a result.

Such examples suggest important implications for interactions among biota and the physical environment. Nevertheless, the roles of oceanic

FIGURE 5–4   A herd of walruses, *Odobenus rosmarus,* on the sea ice of the Chukchi Sea.

and coastal biota are poorly understood, particularly the roles of large organisms in ocean systems. We are presently forced to construct models composed of bits and pieces of information in the midst of an explosion of knowledge and a revolution in technology in other fields. Slowly, this revolution is reaching ocean science. It is bound to alter permanently how we view global interactions among oceans and lands, which are most intensive in the coastal zone. Hopefully, we will learn to manage ourselves accordingly.

SUSTAINABLE USE

Intuitively, *sustainable use* appears to be an oxymoron. How can use of coastal and ocean resources possibly be made sustainable given present population levels and per capita demands? One thing is clear. The answer depends on the point of view, training, and physical location of the respondent, as well as on the time-space scales involved. Sustainable? For how long? By whom? Where?

There is no doubt that humans have had significant impact on the coastal zone and the global ocean, but conclusive evidence is often hard to find. Pollution is, of course, obvious, but pollutants number in the hundreds of thousands, and the case for each may be different. The pollutant entering the world ocean in greatest quantity is probably petroleum; estimates range from 1.7 to 8.8 million metric tons per annum. Although petroleum occurs naturally, it now represents a far more sudden and significant input than previously. Nevertheless, a U.S. National Academy of Sciences' report concluded that "there has been no evident irrevocable damage to marine resources on a broad oceanic scale by either chronic inputs or occasional major oil spills."[26] Consequently, research rather than management was the major emphasis of that report. Similarly, the United Nations Environment Programme[27] stated: "In the open sea we have not detected significant effects on the ecosystem. Trends have indeed been observed, some up, some down, but these are not reflected in environmental deterioration." This report also noted, however, the "effects can be seen in semi-enclosed seas, shelf seas and coastal zones." Present assessments of the situation are not much different than when those reports were issued. This form of pollution, as well as others, is most observed in coastal areas where human-environment interactions are strongest and where concern is greatest. This does not signify, however, the lack of dangerous and chronic effects elsewhere. Chronic oil pollution may or may not be immediately observable in organisms and may have an effect, on reproduction for example, a year or more after the event.[28]

Pollution is, of course, not the only way that we disturb the ocean environment. Resource exploitation has been conducted for centuries with the results of altered biological community structure and losses of production. Fisheries have been depleted one after the other—the situation for whaling being the most dramatic and perhaps best known among the public. The whaling situation is a strong example of the way that the perturbations that human beings impose on the marine environment are insinuated into the realm of subjective interpretation. The great whales have been tragically and selfishly depleted in the pursuit of short-sighted greed. Will their populations return to abundance? There is good evidence that some populations are recovering, but are there larger and more dangerous matters at stake? At what point can one say that deleterious activities are threatening the "health" of the global ocean? One reason that this is such a difficult question is that there is no clearly defined norm against which to make a judgment and that is neutral enough to be applicable to all users of the environment. All too often, an environment is judged "good" when it produces societal values and "poor" when it no longer produces those values. In a real sense, this places additional burdens on science to "prove" cause and effect; it also places a burden on the economist to show whether effects are, in the long term, deleterious or not. This strongly suggests that new approaches for both science and societal values are needed.

Against the rubrics of global pollution, intense fisheries, waste disposal at sea, coastal habitat alteration, growing mariculture, and intense human occupation of coastal habitats, I cannot conceive that coastal-oceanic processes have not been altered on a global scale. A long litany of environmental effects can be cited, that combine to indicate the widespread nature of coastal and marine perturbations. Nevertheless, until cause and effect on an ecosystem level can be established, the significance of these alterations cannot be appraised. Meanwhile, a wait-and-see attitude prevails on the parts of most people and their governments. A question is: Given current trends, is it not merely a matter of time before effects are noted for the oceans as a whole? Following from this are more serious questions: What effects on global processes will result? and What can we do about these?

Unfortunately, the threats that humankind pose to coastal and marine ecosystems are often obscure and related to the still poorly understood processes that drive ecosystems. Just how ecosystems may be described, what makes them work, and how species' populations have evolved within them are of intense scientific interest today. These matters are also obviously of intense conservation and management interest. Finding the nidus of a perturbation is at the core of environmental decision, and if scientists cannot come up with *the* solution, decisions are most often put aside. It is not often clear how many data are necessary for particular decisions to be made. Nor has sufficient attention been paid to addressing bureaucratic inactions and inconsistencies that render many agencies and governments unable to address problems on a holistic basis, no matter how extensive or inclusive the data may be. Pollution and threats to the resources we value most clearly exist. Straightforward cures are difficult, particularly as societies become more complex and demanding.

Sustainable use (or *sustainable development,* as it is also called) is a recent byword—nearly a buzzword—in conservation and management communities. It represents an attempt to compromise between two philosophies, the protectionism of traditional conservation and the economic development of burgeoning human societies. It is by no means clear how use can be made sustainable. Not the least problem is that economic and ecologic theory are not yet compatible. Nevertheless, few people now doubt that ocean resources are limited and that the coasts and oceans are far from pristine. The recognition is growing that many living resources, beyond just the whales and sea turtles and other large organisms that people relate to emotionally, have been overexploited, possibly to the extent that whole ecosystems have been permanently altered, especially in the most productive part of all—the coastal zone.

Science itself is also changing rapidly. "Ecosystem stress" ecology is one example of a new and important trend. Ecotoxicology is seeking new directions based on the recognition that laboratory studies are difficult or misleading to transfer to field situations.[29] Fish population dynamics of the deterministic sort are slowly but surely being replaced by models that incorporate physiology, ecosystem ecology, and population biology. In addition, technology is offering new tools. The kinds of data becoming available, the analysis and synthesis methods being developed, and the derived information needed for comprehensive management of the oceans and coastal zone are dramatically changing. Space technology now offers near-realtime data-gathering tools that can perceive the Earth in ways hardly imaginable thirty years ago. The astonishing present pace of scientific discovery about the oceans during the Marine Revolution will no doubt be noted by historians as one of the momentous events of human history. That we have achieved outer space discovery is remarkable, but I seriously doubt that the impact will be nearly as great for human society as a whole as the discovery of inner space—that is, of the hydrosphere.

If we are to match our new perspectives of the oceans, coastal zone, and humankind's impacts upon them with sustainable use, we will need, at least, to redress our terrestrial bias and meet coastal and marine management problems in their own right. We also must learn to set regional, comprehensive, sustainable-use goals that are in accord with ecosystem functioning and in full realization of our still negligible abilities to control the ecological processes of the coastal zone and oceans. Finally, we must develop ecologically sensible economic models that incorporate the full range of social values and ecological processes and address problems at the systems level.

Resource-use decisions are most often made from an economic viewpoint and by societies that have extraordinarily long time constants for change. Resource economics, acting in conjunction with ecosystem ecology, may eventually become the best equipped of all disciplines to suggest solutions to long-term environmental problems. This is a large order, for it signifies momentous alterations in human behavior, namely, the final realization of the limits of the Earth, which lie in the global ocean.

NOTES

[1] W.L. Thomas, Jr., C.O. Sauer, M. Bates, and L. Mumford, 1956, *Man's Role in Changing the Face of Earth,* University of Chicago Press, Chicago and London.

[2] M. Graham, 1956, "Harvests of the Seas," in *Man's Role in Changing the Face of Earth,* W.L. Thomas, Jr. et al (eds.), University of Chicago Press, Chicago and London, pp. 487–503.

[3] J.C. Bugher, 1956, "Effects of Fission Material on Air, Soil, and Living Species," in *Man's Role in Changing the Face of Earth,* W.L. Thomas, Jr. et al. (eds.), University of Chicago Press, Chicago and London, pp. 831–848.

[4] G.C. Ray, 1970, "Ecology, Law, and the 'Marine Revolution,' " *Biol. Cons.,* Vol. 3, No. 1, pp. 7–17.

[5] My estimate, based on the assumption of an average 10-meter thickness of terrestrial life.

[6] J.E. Lovelock, 1979, *Gaia: A New Look at Life on Earth,* Oxford University Press, Oxford.

[7] O.T. Solbrig, 1985, "Chairman's Summary: Life Systems," in *Global Change,* T.F. Malone and J.G. Roederer (eds.), Cambridge University Press, Cambridge, pp. 221–227. Here Solbrig uses "biosphere" to mean the sum of the biota, in contrast to the primary use of this term elsewhere in this book.

[8] B. Moore, III, and B. Bolin, 1986, "The Oceans, Carbon Dioxide, and Climate Change," *Oceanus,* Vol. 29, No. 4, pp. 9–15; J.J. McCarthy, P.C. Brewer, and G. Feldman, 1986, "Global Ocean Flux," *Oceanus,* Vol. 29, No. 4, pp. 16–26; M. Andreae, 1986, "The Oceans as a Source of Biogenic Gases," *Oceanus,* Vol. 29, No. 4, pp. 27–35.

[9] Andreae, 1986, *op. cit.*

[10] J.A. McGowan and P.W. Walker, 1985, "Dominance and Diversity Maintenance in an Oceanic System," *Ecol. Monogr.,* Vol. 55, No. 1, pp. 103–118.

[11] N. Myers, 1987, "Tropical Forests and Their Species: Going, Going . . . ," *Biodiversity,* E.O. Wilson (ed.), National Academy Press, Washington, DC, pp. 28–36.

[12] R.V. O'Neill, D.L. DeAngelis, J.B. Waide, and T.F.H. Allen, 1986, *A Hierarchical Concept of Ecosystems,* Princeton, Princeton University Press.

[13] G.C. Ray, 1987, "Ecological Diversity in Coastal Zones and Oceans," *Biodiversity,* E.O. Wilson (ed.), National Academy Press, Washington, DC, pp. 37–50.

[14] F. Briand and J.E. Cohen, 1984, "Community Food Webs Have Scale-Invariant Structure," *Nature,* Vol. 307, No. 5948, pp. 264–267.

[15] O'Neill et al., 1986, *op. cit.*

[16] G.C. Ray and K.S. Norris, 1972, "Managing Marine Environments," *Trans. Thirty-Seventh No. Amer. Wildlife and Nat. Res. Conf.,* The Wildlife Management Institute, Washington, DC, pp. 190–200.

[17] B.P. Hayden, G.C. Ray, and R. Dolan, 1984, "Classification of Coastal and Marine Environments," *Envir. Cons.,* Vol. 11, No. 3, pp. 199–207; and M.J. Dunbar, 1985, "The Arctic Marine Ecosystem," in *Petroleum Effects in the Marine Environment,* F.R. Engelhardt (ed.), Elsevier Applied Science Publ., London and New York, pp. 1–35.

[18] D.L. Inman and C.E. Nordstrom, 1971, "On the Tectonic and Morphological Classification of Coasts," *Jour. Geol.,* Vol. 79, pp. 1–21; B.H. Ketchum, (ed.), 1972, *The Water's Edge: Critical Problems of the Coastal Zone,* Massachusetts Institute of Technology Press, Cambridge, Massachusetts.

[19] J.J. Walsh, 1983, "Death in the Sea: Enigematic Phytoplankton Losses," *Prog. Oceanogr.,* Vol. 12, pp. 1–86.

[20] G.T. Rowe, S. Smith, P. Falkowski, T. Whitledge, R. Theroux, W. Phoel, and H.

Ducklow, 1986, "Do Continental Shelves Export Organic Matter?," *Nature,* Vol. 324, pp. 559–561.

[21] Lovelock, 1979, *op. cit.*

[22] L. Kaufman, 1977, "The Three Spot Damselfish: Effects on Benthic Biota of Caribbean Coral Reefs," in *Proceedings: Third International Coral Reef Symposium,* Vol. 1, pp. 559–564.

[23] M.A. Hixon and W.N. Brostoff, 1983, "Damselfishes as Keystone Species in Reverse: Intermediate Disturbance and Diversity of Reef Algae," *Science,* Vol. 220, pp. 511–513.

[24] S.L. Meyer, E.T. Schultz, and G.S. Helfman, 1983, "Fish Schools: an Asset to Corals," *Science,* Vol. 220. pp. 1047–1049.

[25] K.R. Johnson and C.H. Nelson, 1984, "Side-scan Sonar Assessment of Gray Whale Feeding in the Bering Sea," *Science,* Vol. 225, pp. 1150–1152; G.C. Ray, 1985, "Man and the Sea—The Ecological Challenge," *Amer. Zool.,* Vol. 25, pp. 451–468.

[26] National Academy of Sciences, 1985, *Oil in the Sea: Inputs, Fates and Effects,* Washington, DC, National Academy Press.

[27] United Nations Environment Programme, 1982, *GESAMP: The Health of the Oceans,* UNEP Regional Seas Reports and Studies No. 16.

[28] M.G. McCormick-Ray, 1987, "Hemocytes of *Mytilus edulis* Affected by Prudhoe Bay Crude Oil Emulsion," *Marine Environ. Res.,* Vol. 22, pp. 107–122.

[29] S.A. Levin and K.D. Kimball, 1984, "New Perspectives in Ecotoxicology," *Envir. Management,* Vol. 8, No. 5, pp. 375–442.

EDITORS' INTRODUCTION TO:
T.E. LOVEJOY
Deforestation and Extinction of Species

While Chapters 3, 4, and 5 emphasized the impact of life on Earth, this chapter focuses on life itself and its meaning and importance. Previously Executive Vice President of World Wildlife Fund, and currently the Assistant Secretary for External Affairs, Smithsonian Institution, the author, Dr. Thomas E. Lovejoy, is one of the world's experts on the effects of tropical deforestation. In this chapter, Dr. Lovejoy uses tropical forests as an example of one of man's most apparent effects on the global environment: extinction of species through habitat disruption. A study of remnants of such forests reveals new aspects of our effects on our global environment. Dr. Lovejoy discusses our limited knowledge about tropical deforestation and the extinction of species; the utility of remote sensing to improve this knowledge; the rates and effects of extinctions; and the importance of such extinctions to us. "In the last analysis," Dr. Lovejoy writes, "the success of man's management of the global environment can best be measured by the extent to which biological diversity is maintained."

6 DEFORESTATION AND THE EXTINCTION OF SPECIES

THOMAS E. LOVEJOY
Smithsonian Institution
Washington, DC

INTRODUCTION

Studies of the global environment will be incomplete and seriously flawed if they only concern global cycles and the interplay between biological and physical processes. For considered at the level of the biosphere, life has had an intriguing tendency over time to proliferate and diversify. There have, of course, been moments of crisis when there have been significant reductions in biological diversity, but these have been *exceptional* moments, and have been followed by renewed diversification.

Today we are at the beginning of such an exceptional moment, but one that is of our own making. What was first noted as a series of endangered species is in the process of becoming a major reduction in the variety of life on Earth in the coming decades.[1] I will not dwell here on the reasons that elevated extinction rates and net reduction in biological diversity are counter to the basic interests of human societies; these ideas have been discussed widely elsewhere. Suffice it to say that if the tendency to diversity is, as it seems, a fundamental property of life on Earth, we would do better to have life in all its variety to study this basic characteristic, rather than indulge in an uncontrolled experiment to see how long diversity might take to recover to nineteenth-century levels. Further, we do not even know today whether the Earth has 3

91

million or 30 million species—estimates vary widely. If we cannot even answer to within an order of magnitude how many species there are, it is an act of great ignorance to try to study the global environment and how we interact with it. While allowing ourselves to destroy the data with which to answer the question, we find ourselves with a paradox: life tends to diversify. There is only a superficial understanding of the significance of biological diversity. We know little about the importance of diversity for a single ecosystem, let alone for the biosphere.

Let us consider briefly evidence for elevated and rising extinction rates. Such considerations inevitably focus around that small portion of the Earth's surface occupied by tropical forests where, even in the face of science's limited exploration of biological diversity, it is apparent that half or more of all plant and animal species exist. There, driven by a variety of economic and social forces that include the aggregate effect of slash-and-burn farmers as well as global economic forces like the international debt, forests are disappearing at rapid and probably accelerating rates.

Two areas of concern illustrate our current dilemma. One area where the process of extinction has advanced the farthest is the Atlantic Forests region of Brazil, which once extended from South America's east-ernmost point to the Argentine border. The chronicle of the Atlantic Forests is best told by the history of forest cover in the state of São Paulo (Figure 6–1). Estimates of remaining Atlantic forest cover are around 2 percent of what is considered to be its original extent. It is a forest distinct from that of the much better known Amazonia, but filled with endemic species of its own, including 13 endangered forms of primates, 180 birds, and enormous numbers of plants and invertebrates. A second area of concern is the great island of Madagascar, where 90 percent of the species there occur nowhere else. The lemurs are the best known, but there are also entire families that are endemic, such as the spiny plants in the family *Dideriaceae*. Of Madagascar's forest, only about 10 percent remains.

What can these two regions tell us about the extent of the extinction problem? Science has been aware since 1835 of a relationship between the area sampled and the number of species encountered.[2] The general rule of thumb is that an increase in area by ten will double the number of species. This should work in reverse as well, although there is a time lag that is useful to conservation but that I will ignore for the moment. Applying this rough rule of thumb to Madagascar, with nine-tenths of its forests gone, that island should have already lost half its species. The Atlantic Forests of Brazil, with 2 percent remaining, are close to having

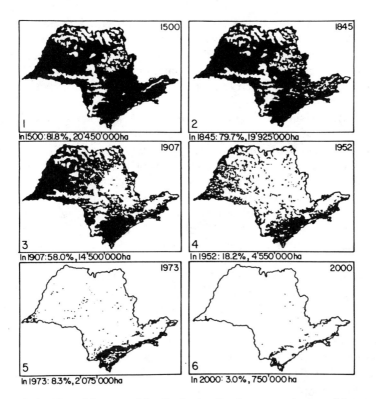

FIGURE 6–1   Map of the state of São Paulo showing the progressive loss of forest cover (black) since 1500 AD.

been reduced to one-tenth, and to yet one-tenth again, and should be reduced to three-quarters of their original species complement. Although it is not possible to give precise estimates as to what the original species complements were, it is clear that we are dealing with extinctions at the level of tens of thousands and hundreds of thousands of species.

There has been considerable discussion in the last two decades about the extent and rate of tropical deforestation, but all estimates agree that there has been a substantial net reduction. Confounding the problem is a frequent difference between reported extent and the actual extent of these forests. This is not a surprising problem because developing nations, beset with many problems, struggle to change their means of receiving and reporting such data. Another confusion, now largely clarified and understood, is between the figures produced by Norman

Myers for the United States National Academy of Science, which dealt with forest conversion from one form of vegetation to another, as opposed to figures dealing with total destruction as are measured by the United Nations Food and Agricultural Organization (FAO).[3] Conversion is not a trivial concern from the point of view of diversity in these highly complex forests, but in any case, there is a clear mandate here for the use of satellite imagery to give precise and up-to-date data on tropical deforestation on a regular and global basis (as opposed to some useful regional efforts that have taken place).

The picture worsens if one looks beyond actual habitat destruction and recognizes that it almost always includes a great deal of *fragmentation* of habitats as well. The question of the species/area relation applies anew to the remnant fragments. We know that reduction in diversity does not all occur instantly with isolation of a piece of habitat, but that it continues subsequently, and its extent is inversely related to the size of the forest remnant. The patterns of species loss from forest fragments are scarcely known nor are the processes that drive these species losses. Islands and extant patches of forest can only give us limited insights.[4] Recognition of the lack of knowledge led to the Minimum Critical Size of Ecosystems project that World Wildlife Fund and Brazil's National Institute for Amazon Research [Instituto Nacional de Pesquisas da Amazonia (INPA)] started in 1979. Taking advantage of a law requiring that 50 percent of any development project remain in forest, forest reserves of different sizes with replicates in given size classes have been laid out and studied while the forest is still intact. Then the effects of isolation have been studied as the reserves have become isolated (Figure 6–2). The smallest reserves in a sense exaggerate the effects and give rapid but simple insights, while larger reserves yield more complex and sophisticated interpretations more slowly.

The following give but glimpses of the kinds of changes that take place as a consequence of fragmentation. Birds of the understory flee the surrounding forest as it is destroyed and crowd into the remnant piece, creating overcrowding problems, apparently to the greater disadvantage of resident birds.[5] Consideration of species encounter functions shows a progressive impoverishment of the understory bird community with time—literally within a matter of months in a 10-hectare reserve.[6] Subsequent studies have shown there is also a substantial avoidance of the forest edge by understory birds, which is measurable to at least 50 meters into the forest.[7] This shows that the previously mentioned species impoverishment of the understory bird community is a mixture of edge, overcrowding, and area effect, and dictates that attention be

FIGURE 6–2   First 10-hectare reserve to be isolated in 1980 for the Minimum Critical Size of Ecosystems Project.

paid to 100-hectare reserves (as opposed to 10-hectare reserves) to gain insight into area effects.

Butterflies provide a somewhat different picture. Their numbers decline in individuals and species as the surrounding felled forest is burned to provide ash to grow pasture, but then species numbers climb to levels higher than in pristine forests. Forest glade and edge species are strongly favored in these new circumstances. Light is able to penetrate deep into the forest and these light-loving species enter the forest, increasing total butterfly species diversity, but actually adversely affecting the butterfly communities of the forest interior. These butterflies need a level of attention from conservationists that edge species do not (Figure 6–3).[8]

Primate species have differential survival patterns as well. Species with

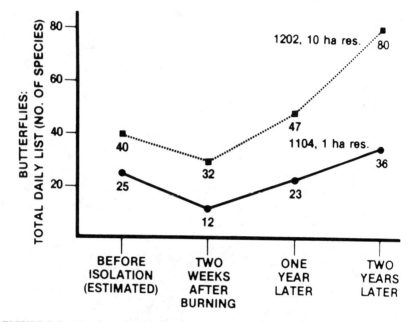

FIGURE 6–3   Number of butterfly species sampled daily in the first two reserves isolated.

large home ranges like white-bearded saki monkeys drop out of small reserves, while species with small home ranges like howlers persist. Intriguingly, bands of golden-banded tamarins, which are usually separated by some distance in natural forest, are crowded together in one isolated 100-hectare reserve.[9] So far, these have persisted together. The interesting question is whether this apparently unnatural situation can persist.

The most startling change has occurred in the woody plant community, where it was anticipated there might be changes in reproductive activity and recruitment long before there would be any change in adult trees.[10] But in a 10-hectare remnant, elevated tree mortality, particularly increased numbers of standing dead trees, were seen after only one year, and a large number of tree falls occurred on the windward side of reserves. These changes probably relate to the removal of the surrounding forest and particularly to its insulating capacity. For example, differentials have been recorded between forest margin and 100 meters into the forest of up to 4.5°C and 20 percent relative humidity.[11]

The actual physical structure of the forest is being drastically altered. To the original notion of the history of a forest fragment must be added major edge-related effects. These are effects that can be compensated for by increasing the size of protected areas, in essence providing a buffer zone for edge effects.

The minimum size project is recording complex changes induced by fragmentation. They add yet more adversity to the task of protecting the biological diversity of our planet. The one bright side, however, is that in larger areas the changes may occur over centuries, allowing for compensatory action—the easiest and most desirable of which is to add additional land to a protected area and to allow natural vegetation to return so that in the end there is a larger national park or reserve and more biological diversity can be protected with less management effort.

The predicament of biological diversity is then a matter of major concern among global environmental problems. When the global environment is changing both in its chemistry and physics, this may often come from major disruption of the biota (as in tropical deforestation and the carbon cycle), but such changes may also cause disruption of the biota (acid rain, greenhouse effect of vegetation belts). We are clearly in the initial stages of what could be a major transformation of the biology of the planet with a major loss of biological diversity. E.O. Wilson estimates current extinction rates at 400 times normal.[12] Interestingly, *all* field biologists who have addressed the problem—even those who disagree violently about other aspects of their science—agree that the problem is of great magnitude.

Given our great vested interest as living creatures in the biological sciences, the essentially heedless discard of a major segment of biological diversity can be characterized as perhaps the greatest anti-intellectual act of human history. In the last analysis, even when we have learned to manage other aspects of the global environment—even if population reaches a stable level—even if we reach a time when environmental crises have become history—even if most wastes have gone except the most long lived—even if global cycles have settled back into more normal modes—the best measure of how we have managed the global environment will be how much biological diversity has survived.

NOTES

[1] T.E. Lovejoy, 1980, in *The Global 2000 Report to the President,* G.O. Barney (Study Director), GPO, Washington, DC, pp. 327–332; N. Myers, 1979, *The Sinking Ark,* Pergamon Press, Oxford, England, 307 pp.; E.O. Wilson, 1985, *Issues in Science and Technology,* Vol. 2, pp. 20–29.

[2] H.C. Watson, 1835, *Remarks on the Distribution of British Plants,* n.p. London.

[3] N. Myers, 1980, *Conversion of Tropical Moist Forests,* National Academy of Sciences,

Washington, DC, 205 pp.; *Tropical Forests: A Call for Action,* Report of an International
Task Force convened by the World Resources Institute, The World Bank, and the United
Nations Development Programme, 1985.

[4] J. Terborgh, 1974, *Bioscience,* Vol. 24, pp. 715–722; J. Terborgh, 1975, in *Tropical
Ecological Systems: Trends in Terrestrial and Aquatic Research,* F.B. Golley and E. Medina (eds.),
Springer-Verlag, New York, pp. 369–380; R. May (ed.), 1976, *Theoretical Ecology,* Blackwell
Scientific Publishers, Oxford, England, 317 pp.; E.O. Wilson and E.O. Willis, 1975, in
Ecology and Evolution of Communities, M.L. Cody and J.M. Diamond (eds.), Belknap Press of
Harvard University, Cambridge. Massachusetts, pp. 522–534.

[5] T.E. Lovejoy, J.M. Rankin, R.O. Bierregaard, Jr., K.S. Brown, Jr., L.H. Emmons, and
M.E. Van der Voort, 1984, in *Extinctions,* M.H. Nitecki (ed.), University of Chicago Press,
Chicago, pp. 295–325; R.O. Bierregaard, Jr., and T.E. Lovejoy, "Effects of Forest Fragmen-
tation on Amazonian Understory Bird Communities," *ACTA Amazonica,* in press.

[6] T.E. Lovejoy, R.O. Bierregaard, Jr., J.M. Rankin, and H.O.R. Schubart, 1983, in
Tropical Rain Forests: Ecology and Management, S.L. Sutton, T.C. Whitmore, and A.C.
Chadwick (eds.), Blackwell Scientific Publishers, Oxford, England, pp. 377–384; R.O.
Bierregaard, Jr. "Species Composium and Trophic Organization of the Understory Bird
Community in a Central Amazonian Tere-Ferme Forest," *ACTA Amazonica,* in press.

[7] T.E. Lovejoy, R.O. Bierregaard, Jr., A.B. Rylands, J.R. Malcolm, C.E. Quintela, L.H.
Harper, K.S. Brown, Jr., A.H. and G.V.N. Powell, H.O.R. Schubart, and M.B. Hays, 1986,
in *Conservation Biology: The Science of Scarcity and Diversity,* M. Soulé (ed.), Sinauer Assoc.,
Sunderland, MA, pp. 257–285.

[8] K.S. Brown, Jr., "Comparisons Between Lepidoptera Faunas in Reserves of Different
Size and Degree of Isolation," *ACTA Amazonica,* in press.

[9] Lovejoy et al., 1986, *op. cit.;* A.B. Rylands and A. Keuroghlian, 1985, "Primate Survival
in Forest Fragments in Central Amazonia: Preliminary Results," paper presented at the
2nd Brazilian Primatology Congress, São Paulo.

[10] Lovejoy, et al., 1983, *op. cit.;* Lovejoy et al., 1984, *op. cit.*

[11] Lovejoy, et al., 1986, *op. cit.*

[12] E.O. Wilson, 1985, *op. cit.*

EDITORS' INTRODUCTION TO:
O.L. LOUCKS
Large-Scale Alteration of Biological Productivity
Due to Transported Pollutants

In the past half-century we have made rain acid and added oxidizing agents to the atmosphere; we have changed the acidity of rainfall throughout eastern North America and much of Europe, in China, and in Siberia. We are changing the biosphere. This chapter was written by Dr. Orie L. Loucks, an ecologist who has studied many aspects of forest vegetation, including the impact of acid rain, and is currently director of the Holcomb Research Institute of Butler University. He declares that the cause of acid rain appears to be emissions of sulfur and nitrogen oxides, "a process ignored entirely in 1955" at the time of the conference *Man's Role in Changing the Face of the Earth.*

This chapter documents the effects of atmospheric changes and in so doing illustrates the complex interrelationships that may occur in the biosphere—making the rain acid can affect how the biosphere responds to an increase in carbon dioxide concentration added from the burning of fossil fuels. Large-scale effects of acid rain and oxidants on the world's vegetation may be leading to a "diminished sequestering capacity" for carbon dioxide and resulting in an accumulation of carbon dioxide in the atmosphere "due to the reduction in photosynthesis." "When premature forest mortality is included, both agricultural and forest productivity are being reduced over large areas by 5 to 15 percent below historic levels."

The author concludes that "the most serious impacts appear to derive from an assumption that the atmosphere is a global 'commons' to be used as a sink, when, in fact, it is the living system at the surface of the Earth"—especially green plants—that are taking the punishment. The author also observes that the results he summarizes illustrate an important shift in the environmental outlook since the mid-1950s. Whereas the focus of the volume edited by Thomas was on "changing the face of the Earth," the focus now is on the changes induced in the Earth as a system.

7 LARGE-SCALE ALTERATION OF BIOLOGICAL PRODUCTIVITY DUE TO TRANSPORTED POLLUTANTS

ORIE L. LOUCKS
Butler University
Indianapolis, Indiana

INTRODUCTION

Although certain stresses on natural resource systems result from normal fluctuations in Earth systems (e.g., weather and natural disturbances), others are deriving now from changes in atmospheric precipitation, air chemistry, and resultant changes in soil chemistry. Although evident in 1955 for local areas,[1] the phenomenon is now global. This chapter reviews the evidence for long-term alteration of biological productivity of both crops and forests due to transported pollutants, particularly those derived from fossil fuels. Understanding of these changes is due in large part to about 30 years of progress in ecosystem research, a study area that assumed importance around 1955. The prospective alteration of ecosystem processes and of resource productivity over large areas of the world's land mass is now of urgent significance to many throughout the world and must be examined carefully. Although overreaction must be avoided, if such an alteration is taking place, the world community, scientists and public cannot afford simply to debate the risks until the experiment has run its course. Should the subtle changes we are introducing into the Earth's ecosystems actually lead to a "diminished sequestering capacity" for CO_2, the full course of the trends being evaluated is highly nonlinear and of profound consequence.

DOCUMENTED IMPACTS AT THE LOCAL SCALE

Most of the pollutants now raising concerns at continental and global scales originate as major local point sources or as local but "areawide" sources around urban centers. These sources have both local impact and cause continental-scale problems. Therefore, questions concerning effects should also be addressed at both local and continental scales.

Smelters and large coal-fired power plants are major local point sources of pollutants, particularly sulfur dioxide (SO_2). Because the concentration of SO_2 decreases with the distance from the source, these point sources allow study of changes along concentration gradients toward low background levels in most remote areas. However, comparisons between long-distance transport of acid precipitation and local effects of acid gases are complicated by the role of SO_2 in affecting vegetation versus its roles in soil acidification and the expression of effects from emissions of toxic metals and particulates.[2] In addition, major point sources emitting SO_2 represent a local source of acid fog and snow acidification.

As reviewed by Hutchinson,[3] impacts from the major smelting developments in North America have occurred since about 1870. Local effects in these early days were often visible for miles (e.g., at Sudbury, Ontario; Copper Hill, Tennessee; and Butte, Montana). Around the smelters at Copper Hill, Tennessee, approximately 6,900 hectares of vegetation were devastated; beyond this zone and radiating out from the smelter is a long gradient in which much of the plant species diversity recovers. Fifty years after closure, however, the most seriously impacted areas are still barren because of acidification and arsenic and lead in the soil.

By the 1920s, emissions has been reduced at many of the large smelters, and tall smokestacks were built to achieve long-distance dispersion of gases and metals. However, one of the largest sources is still operating: the nickel-copper smelter at Copper Cliff, near Sudbury, Ontario. Here, until the early 1980s, approximately 1.3 million tons of SO_2 per year were emitted through a smokestack 341 m tall. Emissions have been reduced recently to about 750,000 tons per year, roughly double that of large coal-fired power plants in the Ohio Valley.[4] In the late 1960s, the smelter was emitting approximately 2.5 million tons of SO_2 per year, together with approximately 907 tons of iron, 181 tons of nickel, 136 tons of copper, 18.1 tons of lead, 10.7 tons of zinc, and 4.1 tons of cobalt, all of this every 28 days.[5]

Large electric generating facilities burning high-sulfur coal emit on

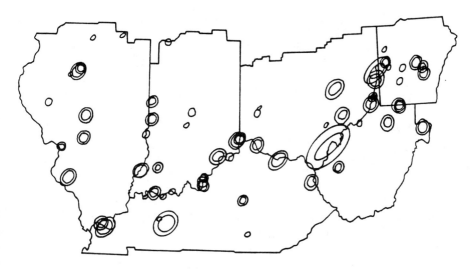

FIGURE 7–1   1985. Calculated impact areas of all power plants in the Ohio River Basin Environmental Study region with SO_2 emission rates of greater than 2000 g/sec. Outermost ellipse represents isopleth of 0.05 ppm SO_2. Other ellipses, if present, represent 0.1 ppm SO_2 isopleths.

the order of 300,000 tons of sulfur annually. The study in the Ohio Valley area[6] identified more than ten such facilities operating during the late 1970s, with total load emissions exceeding those of the Sudbury smelter during its peak operation. Figure 7–1 shows the calculated impact areas for all major SO_2 sources in the Ohio Valley area where the threshold concentrations for significant chronic effects on sensitive crop species (0.05 ppm over three hours once per month in the presence of elevated ozone concentrations) would be exceeded. This threshold must be recognized as one-tenth of the government's regulatory standard for the biota (0.5 ppm over three hours), a level that is tolerable for a few species, depending on the level of other pollutants.[7] The combined areas exceed 10 percent of the total land area. Few effects have yet been documented in the impact areas, partly because many of the sources are recent (as compared to some of the smelters), but also because virtually no studies are being carried out with the sophistication needed to determine subtle effects. Certainly, there was no basis in 1955 for establishing even the 0.5 ppm standard for protection. Detailed studies of a forest preserve near Krakow, Poland, have documented the subtle (and the not-so-subtle) alterations that take place at downwind locations from a facility of intermediate size.[8]

THE ACIDIC DEPOSITION PHENOMENON

The chemistry of precipitation throughout eastern North America and much of Europe has been changed significantly over the past half-century. The cause of the acidification appears to be the elevated emissions of sulfur and nitrogen oxides into the lower atmosphere,[9] a process entirely ignored in 1955.

Experimental studies undertaken in the late 1970s in northern Europe and during the 1980s in other temperate zone countries have shown that many kinds of acid deposition effects are possible in terrestrial and aquatic systems. In the forest, these effects range from damage to leaf surfaces to leaching of essential nutrients from soils. Acid rain impacts in sensitive watersheds may appear as reactions mobilizing soil aluminum, because the solubility of this metal increases sharply during brief periods of high acidity from atmospheric inputs. The extensive forests of Germany, Switzerland, and Czechoslovakia are subject to oxidants as well as acid deposition; effects here have been attributed to a combination of oxidants, acid leaching, low soil cations, and accumulation of aluminum to toxic levels.[10] Effects in Canada, China, the USSR, and Scandinavia are more local but are serious where they occur. Reduced annual increments in forest tree rings from poorly buffered high-elevation sites in the northern United States may be due to a combination of factors similar to those causing the effects being observed in Europe and date initially from about the same period, the late 1960s.

LOCAL PROTECTION VERSUS LONG-RANGE TRANSPORT AND TRANSFORMATIONS

Regulations promulgated under the U.S. Clean Air Act of 1970[11] established Air Quality Control Regions (AQCR) as the local geographic units within which air quality improvement strategies would be evaluated. Air quality measurements within each region are used to determine "attainment status" with respect to national ambient air quality standards for that Air Quality Control Region. These procedures focus control on local ground-level pollution problems associated in large part with local emissions. Although ozone (O_3) is known to be of secondary origin, formed during a period of residence in the atmosphere, strategies for its control also are being attempted from within Air Quality Control Regions. However, due to long-range transport of pollutants, these strategies have only limited success: as elevated concentrations in an Air

Quality Control Region are controlled (often through the use of tall stacks), reduced air quality is being observed increasingly in areas remote from local sources. Although other countries adopted similar regulations in the 1970s, enforcement remains a serious problem.

The meteorological factors that influence long-distance transport are primarily those associated with clockwise circulation of air around high-pressure systems and the counterclockwise flow of low-pressure centers. These factors tend to facilitate transport from south to north or north to south, while the prevailing air mass moves from west to east. Hence, primary emissions from the industrialized centers of the United States and Europe are precursors of the elevated pollutant concentrations observed to the north, south, and east of these regions in both continents. The concentration patterns are complicated by seasonal variations that show transport from south to northeast in the summer and a more northerly component during the winter. In the course of the long-distance transport, the concentrations of primary pollutants such as sulfur dioxide are decreased, but the concentrations of secondary pollutants can increase (especially sulfate particulates, ozone, and acidity in precipitation).

Photochemical oxidants, principally ozone, have until recently been thought of as pollutants of urban areas. However, it is now recognized that rural and forested areas some hundreds of kilometers downwind of urban centers are seriously influenced by persistent regional ozone episodes. Effects on vegetation and crops are being confirmed now over the entire eastern United States and adjacent Canada, as well as in much of Europe.

Episodes of elevated ozone concentrations are associated with warm, slow-moving high-pressure systems. As each system moves out of relatively pollutant-free areas of the northern plains in North America or from the Atlantic Ocean toward Europe, it contains between 30 and 50 parts per billion (ppb) of ozone. However, ozone is formed from oxidation reactions involving nitrogen oxides (NO_x) and hydrocarbons and, despite some scavenging by vegetation, ozone accumulates within the air mass.[12] With the slow west-to-east movement of these weather patterns in the summer, air-mass stagnation occurs and ozone concentrations three to eight times greater than natural background are observed.[13] The long-distance transport processes during these episodes bring elevated ozone concentrations into remote regions of both North America and Europe, and probably to many areas in Asia and South America where major urban centers occur.

EFFECTS OF OZONE ON AGRICULTURAL PRODUCTIVITY

Photochemical oxidants are recognized as the air pollutant causing the most significant losses to agriculture in the United States. A two-step approach was adopted by Loucks and Armentano[14] to estimate regional impacts: 1) a synthesis of experimentally determined crop loss data in relation to measured pollutant exposures and 2) regionwide determination of the comparable dose being experienced in the field. Where moderate pollutant exposures extend over the growing season, as in the case of ozone, growth rates and yield reductions tend to correlate with the total exposure accumulated over the growing season. Summation of the exposure as a cumulative dose (pollutant concentration times duration of exposure hours) or a mean concentration based on this sum has been adopted as a useful measure of prospective biological effects, primarily grain yield.

Upper and lower bounds in the indicated yield reduction due to both sulfur dioxide and ozone exposures in the Ohio River Basin, were developed[15] from existing data bases and were expressed as loss coefficients (percent reductions). These could be extended to the field directly through existing monitoring of ozone and the known crop acreage to compute total losses (or benefits) from abatement.

More recently, the Office of Technology Assessment of the U.S. Congress utilized data developed by the National Crop Loss Assessment Network (NCLAN),[16] in combination with current crop and ozone data, to determine more accurately the impact of pollutants on the productivity of corn, wheat, soybean, and peanut crops throughout the United States. Effects were estimated using new NCLAN dose-response relationships, 1978 ozone monitoring data, and 1978 agricultural statistics. The selected crops represented a range of ozone sensitivity from sensitive (peanut), to sensitive-intermediate (soybean), to intermediate (wheat), to tolerant (corn).[17]

The dose-response equations predict percent yield reduction for given ozone exposures represented as the maximum 7 hour per day (h/d) seasonal mean in parts per billion (ppb) with zero reduction at the background concentration of 25 ppb. Hourly ozone measurements from approximately 300 selected monitoring stations were used to estimate nonurban concentrations. Maximum 7-h/d ozone data were used to estimate monthly mean concentrations, which then were used to calculate growing season means. The 1978 crop data represent yields that have been reduced due to current levels of atmospheric pollutants, including ozone. Therefore, the dose-response equations were used to

TABLE 7–1   POTENTIAL GAINS IN MILLIONS OF DOLLARS PROJECTED IF
OZONE LEVELS WERE REDUCED FOR 1978 U.S. CROP PRODUCTION, PRICES,
AND OZONE DATA (OTA 1982).

Major States	Wheat	Corn	Soybeans	All Crops
Arkansas	2	0	161	163
California	5	2	0	7
Colorado	15	6	0	21
Georgia	1	5	32	118
Illinois	7	87	339	433
Indiana	5	53	161	218
Iowa	0	106	269	374
Kansas	63	13	36	112
Kentucky	1	9	41	51
Louisiana	0	0	81	81
Michigan	2	10	14	25
Minnesota	20	36	103	159
Mississippi	1	0	107	107
Missouri	6	17	191	214
Montana	32	0	0	32
Nebraska	17	56	38	111
North Carolina	1	14	61	121
North Dakota	62	1	3	66
Ohio	6	27	124	157
Oklahoma	32	0	7	50
South Carolina	1	3	46	52
South Dakota	15	10	10	35
Tennessee	1	4	62	67
Texas	15	10	18	61
Virginia	1	5	18	49
Washington	28	0	0	0
Wisconsin	0	19	0	26

estimate potential gains in productivity that could be achieved if ozone
levels were reduced to 25 ppb.

Crop gains were then calculated as the difference between current
yields and the estimated yield, assuming a reduction in ozone levels to
background concentrations (25 ppb). Table 7–1 summarizes the dollar
values of the crop gains for three of the crops considered. The
assessment estimates that approximately three billion dollars of productivity could be gained if current maximum 7-h/d ozone levels were
reduced below 25 ppb. Dollar values are based on 1978 crop prices,
without accounting for price effects, to provide an overall estimate of the
impact. Of the estimated dollar impact, soybeans represents 64 percent,

corn 17 percent, wheat 12 percent, and peanuts 7 percent. The corn-belt states of Illinois, Iowa, and Indiana experienced the greatest impacts.

The Office of Technology Assessment report[18] notes that assumptions and caveats must be considered when interpreting the level of uncertainty associated with this or similar assessments; however, important findings can be drawn from the assessment. First, maps of ozone levels show that during the growing season a significant portion of the agricultural land area of the United States is experiencing levels of oxidant capable of reducing crop yields. Since similar (or higher) oxidant levels are found in Europe, the same conclusion probably holds. Second, potential impacts, as determined by the Office of Technology Assessment method, further support the earlier estimates by Heagle and Heck[19] indicating annual losses in the United States alone of approximately two billion dollars.

ESTIMATING FOREST GROWTH LOSSES DUE TO OZONE AND ACIDS

A complete analysis of the prospective effects of acidic deposition on forest growth must consider the four pathways by which effects can be expressed: 1) direct effects on growth from exposure to oxidants; 2) the leaching of chemical nutrients from leaves by acid rain; 3) indirect effects on tree growth that may be attributable to a reduction in the stock of nutrient cations in the soil or associated alterations of the cycling of nutrients such as N, P, K, and S from acid rain; and 4) the extent to which episodic mobilization of toxic forms of aluminum in the soil can affect tree roots and, therefore, affect above-ground growth.[20] All four factors can operate simultaneously and a potential exists for synergistic effects.

Questions regarding the effects of multiple pollutants on forests and other nonagricultural species require an approach much different from that applied to crops. The eastern United States has on the order of 50 native forest tree species, with many more introduced as horticultural plantings, while European forests support about 15 species. These long-lived trees are exposed year after year to atmospheric pollutants, and effects on tree vigor are expressed cumulatively over years rather than over a single growing season. The "product" value of these trees is the accumulated woody growth rather than an annual harvest that can be removed from exposure each year.

Because of differences in duration of exposure, number of species, and the limited experimental information, it is more difficult to differentiate effects on forests from gaseous acid emissions (SO_2 and NO_x), as opposed to the effects of secondary pollutants such as oxidants (ozone) and acidic deposition (SO_4). The differences can be distinguished where one or another pollutant dominates, but over a large area the effects expressed tend to be those of multiple-factor impacts. Adding to the difficulty are the simultaneous effects on trees from periodic droughts, infertile soils, or elevated insect populations that take advantage of weakened host plants.[21] Air pollutants also have direct effects on natural insect control agents such as birds and hymenopteran parasites.

Research on the effects of air pollutants on forest growth in the San Bernardino Mountains of southern California began in the early 1970s and show a region of long-term exposure to ozone at moderate-to-high concentrations. Here, ponderosa pine growth losses approach 37 percent reduction in radial growth in trees under age 30, with annual mortality rates averaging nearly 3 percent per year in ponderosa and Jeffrey pine. Studies in the most affected forests indicate an 83 percent loss of merchantable volume growth from 30-year-old trees.[22] Forest growth losses in the eastern United States have been less well documented. Several studies have shown, however, that air pollutants markedly reduce annual growth ring widths. For example, Skelly et al.[23] found that photochemical oxidants (including ozone) have increased mortality rates of eastern white pine in the Shenandoah National Park, the Blue Ridge Parkway, and Great Smoky Mountains National Park by 11 percent per year, a greater loss than has been observed in southern Californian yellow pine forests.[24]

By the 1980s we had become aware and had evidence of the synergistic effects of several pollutants acting together. For example, white pine exposed to ozone and acid-gas in 13 stands sampled from central Wisconsin to southern Indiana show foliar chlorosis, necrosis, and tip-burn associated with these gaseous pollutants.[25] The SO_2 levels were high enough to interact synergistically with ozone in urban areas and near industrial sources; but the occurrence of typical ozone foliar injury symptoms at rural Wisconsin and Indiana study sites was attributed primarily to the regional ozone exposures. The spectrum of disease symptoms known to be usually associated with air pollutants was used to formulate a "disease index" for the various locations in Indiana and Wisconsin. The differences in disease indices at the various locations correspond to observable differences in canopy density; when trees lose foliage prematurely, canopy density deteriorates and the trees experience a decline in stem vigor. Mortality after the 1980 foliage survey has been severe in the stands with the highest disease index.

Given the widespread evidence accumulated in the last two decades of the sensitivity of tree species to air pollutants, researchers have sought to document the potential for a similar response to deposition of acidic substances in rainfall. Studies of the effects on forest growth have been underway in Sweden and Norway, and the early results have been summarized by Abrahamsen.[26] Early work in Sweden using tree-ring analysis methods indicated possible effects, but was couched in considerable uncertainty. More recent Swedish research has found negative effects of tree growth under certain sensitive soil conditions.[27]

Studies in Germany[28] have led to the conclusion that multiple factors associated with acidic deposition have operated together to produce loss of growth and extensive mortality of spruce and pine throughout much of central Europe and to produce more limited effects on beech and oak. In the United States, similar effects on diameter growth of pitch pine in New Jersey have been shown in association with fluctuations in rainfall acidity.[29] Other studies show greatly altered growth rates and considerable mortality for red spruce in New York and Vermont.[30] Although no definitive explanation has been presented, all of the factors found to be significant in Europe seem likely to be involved in North America as well, probably in different combinations due to the differences in species.[31]

EFFECTS ON PRODUCTIVITY OF THE BIOSPHERE

The idea of stable biological productivity of the continents has figured in several previous assessments of regional and global resource status. Viewpoints range from those making an implicit assumption of long-term essentially unchanging biological productivity, even in the face of radical changes in landscape condition, to those viewing the biological production process as extremely sensitive to subtle changes in the environment.

Questions regarding the possible negative impact of industrial emissions on biological productivity require a different assessment strategy from that of enhancement through increasing CO_2 concentrations. Although extensive research, largely on agricultural crops, is available on the eastern United States, intensive studies and similar results also are becoming available from Europe. As already indicated, several studies have tried to estimate the general impact of air pollutants on forests.

In 1980, a preliminary assessment was completed of prospective pollutant impacts from expanded power-generating capacity in the Ohio River Basin.[32] Softwood species (including air-pollution-sensitive white,

Virginia, and short-leaf pines) comprise 4.2 percent of the Ohio River Basin forests, with hardwoods making up the remainder. Estimates in 1980 indicate that the sensitive softwood species experienced a reduction in annual volume increment of 2.6 to 11.7 percent of the potential annual growth or 0.078 to 0.207 million m^3/yr. This range encompasses Linzon's[33] estimate of 5.6 percent for areas considerably downwind from Sudbury, but it does not consider the effects of softwood mortality that can be induced by long-term reduction in growth.

Utilizing early work,[34] the annual losses from the three sensitive hardwood species (black walnut, black cherry, and black locust) were estimated to be similar to those of softwoods (i.e., 2.6 to 11.7 percent per year). The ten species of intermediate sensitivity were judged to sustain losses of 1.1 to 5.9 percent per year or 0.137 to 0.728 million m^3/yr. The resistant hardwoods were assumed to suffer no growth losses. The total estimated reduction was between 0.31 to 1.44 million m^3/yr, but this estimate does not include growth losses due to the mortality of hardwoods induced, in part, by the pollutant and acid deposition exposure. In addition, the assumption that "resistant" hardwoods suffered no growth losses seems now to have been unduly conservative.

As a result of these large-scale effects, a new term emerging in studies of the global carbon balance is "diminished sequestering capacity."[35] This term describes an increase in carbon dioxide accumulation in the atmosphere due to the reduction in photosynthesis attributable to air pollutant exposures. It is best discussed in relation to the effects on carbon exchange due to the increases in ozone concentrations and the impacts of acidic substances. For example, by 1978–1979, the recurring "episode peak" value for ozone in the Ohio Valley was 0.15 ppm, an increase of 0.02 ppm per decade over the period since the 1920s when anthropogenic increases in ozone appear to have been negligible.[36] This rise in ozone probably explains the comparative leveling off of a historic upward trend in agricultural yields—low and relatively unchanged until the 1930s when a sharp increase began and continued until 1968. (The first indications of ozone damage to a then very sensitive crop, tobacco, was in the middle 1950s; only ozone-tolerant varieties are grown today.) The "plateau effect,"[37] in which yields no longer increased as much in response to continuing technological inputs and improvements in agricultural practice, became established in the late 1960s (see Figure 7–2). This was a time when several gaseous pollutants are now known to have reached biologically significant levels in the large agricultural regions of the U. S. Midwest, including Michigan. The difference between the average observed yield in 1980 and the yield that might have been

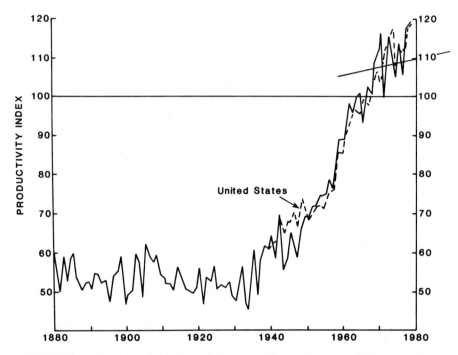

FIGURE 7–2   Historic trends in agricultural crop yields per hectare in Michigan between 1880 and 1978. Dotted line represents total U.S. agricultural productivity (1967 = 100). (Adapted from Wittwer, 1979.)

expected is about 15 percent, larger by only a factor of 1.5 than the average yield reductions reported for the late 1970s in the National Crop Loss Assessment.[38]

To determine the long-term implication of oxidants and acidic deposition for global productivity and carbon storage, one needs to evaluate the large-scale effects of these substances on the major ecosystems and carbon pools, forests as well as agriculture. Evidence of ozone-induced foliage pathology, reduced growth, and mortality of selected forest species has been noted by many authors, not only for the eastern United States, but also for areas in California, Europe, and the Far East. Where the original forests already have been greatly depleted, as in California, a species conversion has been induced—in this case, toward chaparral species of lower productivity.

The data cited earlier suggest that, when premature forest mortality is included, both agricultural and forest productivity are being reduced over large areas by 5 percent to 15 percent below historic levels.

Assuming that this impairment of forest productivity applies to the eastern United States and to most of central Europe, the "diminished sequestering capacity" in these productive temperate zone forests would be on the order of 0.5 to 1.0 Gt (gigatons/yr) or from 10 to 20 percent of the total CO_2 release from the present global fossil fuel combustion (ca. 5.0 Gt). The possibility that this new source of CO_2 from the apparent impairment of biological productivity and carbon sequestering may constitute a potential increase in the rate at which CO_2 accumulates in the atmosphere, will have to be investigated carefully.

CONCLUSIONS

Many laboratory experiments have shown that the growth rate of plants can increase logarithmically in the presence of increased CO_2 concentrations. This response can be viewed as a *potential growth rate,* in the absence of limiting nutrients or gaseous pollutants, that may counterbalance the impairment of productivity in polluted areas. The upper limit for enhanced CO_2 uptake of individual leaves is about a 0.5 percent increase for each 1 percent increase in CO_2 concentration under closed-chamber, experimental conditions.[39] Since atmospheric CO_2 has increased 26 percent since the mid- to late-nineteenth century, [40] some CO_2 uptake rates may be found to have increased as much as 13 percent.

Field evidence for increasing biospheric exchange due to increasing CO_2 is limited at this time. Suggestive results have emerged from two lines of inquiry. First, two studies of annual wood increments of trees have concluded that increased growth in remote regions is consistent with rising CO_2 concentrations. LaMarche et al.[41] observed increased growth since the mid-nineteenth century in subalpine trees at several locations in the Rocky and Sierra mountains. The increase in growth exceeds rates expected as a response to a climatic amelioration observed since the mid-nineteenth century.

The second source of evidence arises from the annual cycle in atmospheric CO_2 concentration at Mauna Loa, which many observers believe to result from an increased warm-season uptake of CO_2 during terrestrial photosynthesis, followed by net respiratory release later in the year. Recent investigations have found an amplitude increase of 20 percent over the period 1958 to 1982, particularly since 1975.[42] Because the atmospheric CO_2 concentration has increased by only 9 percent over the period 1959–1982, compared to a 20 percent increase in the annual amplitude in the Mauna Loa research, enhanced photosynthesis probably cannot be the entire cause. The strong temperature dependence of respiration and the observed warming of the northern hemisphere over recent decades suggest that increased respiration could be important in the amplitude increase. Some investigators (e.g., Houghton, personal communication) believe the increasing amplitude requires *both* an increase in system respiration and the increase in carbon uptake, with no net change in net ecosystem production on an annual basis. New, well-focused research will be required to remove the uncertainty around these relationships and to develop numerical estimates of the possible interaction between CO_2 uptake, respiration, and air pollution effects over large areas.

While uncertainty remains regarding many details of productivity loss from air pollutants or of CO_2 enhancement, one conclusion is clear: Regional and global productivity is not presently in a state even approaching an equilibrium, and the destabilizing factors are entirely attributable to human activities. The most serious impacts appear to derive from an assumption that the atmosphere is a global "commons" to be used as a sink, when, in fact, it is the living system at the surface of the Earth that is presently functioning as a sink. These source-sink relationships, now well understood at the local level, need to be quantified more fully at regional and continental scales if human self-interest at the global level is to find expression.

The results summarized here illustrate an important shift in the environmental outlook since the mid 1950s. Whereas the focus of the volume edited by Thomas[43] was on "changing the face of the earth," our focus now is on the changes induced in the Earth as a system. This shift is only partly due to our more complete understanding of environmental systems; more than anything, the results presented here and elsewhere show that the nature of the shift in environmental outlook and our potential to alter earth processes has profound implications for further, and almost unavoidable, change in the future. Most important, the institutional responses by scientific and governmental leadership must begin to function not only within the framework of correcting past mistakes, but also by anticipating future degradation and avoiding it.

NOTES

[1] W.L. Thomas (ed.), 1956, *Man's Role in Changing the Face of the Earth,* University of Chicago Press, Chicago.

[2] T.C. Hutchinson, 1982, "The Ecological Consequences of Acid Discharges from Industrial Smelters," in *Acid Precipitation: Effects on Ecological Systems,* F.M. D'Itri (ed.), Ann Arbor Science, Ann Arbor, Michigan, pp. 105–122.

[3] *Ibid.*

[4] O.L. Loucks (ed.), 1980, *Crop and Forest Losses Due to Current and Projected Emissions from Coal-fired Power Plants in the Ohio River Basin*, Report to the Ohio River Basin Energy Study, The Institute of Ecology, Indianapolis.

[5] Hutchinson, 1982, *op. cit.*

[6] Loucks (ed.), 1980, *op. cit.*

[7] G.T. Wolff, P.J. Lioy, and G.D. Wright, 1980, "Transport of Ozone Associated with an Air Mass," *Journal Env. Sci. and Health*, Vol. A15:2, pp. 183–189.

[8] W. Grodzinski, J. Weiner, and P.F. Maycock (eds.), 1984, *Forest Ecosystems in Industrial Regions*, Springer Verlag, Berlin, Heidelberg.

[9] G.E. Likens, 1976, "Acid Precipitation," *Chem. and Eng. News*, Vol. 54, pp. 29–44; National Research Council, 1983, *Acid Deposition: Atmospheric Processes in Eastern North America. A Review of Current Scientific Standing*, National Academy Press, Washington, DC, 377 pp; Office of Technology Assessment, 1984, *Acid Rain and Transported Air Pollutants: Implications for Public Policy*, Congress of the United States, Washington, DC, 232 pp.

[10] G.H.M. Kraus, U. Arndt, C.J. Brandt, J. Bucher, G. Kenk, and E. Matzner, 1985, "Symptoms of Forest Decline," Abstracts, Muskoka Conference, International Symposium on Acidic Precipitation, 282 pp.

[11] Environmental Quality: The Third Annual Report of the Council on Environmental Quality, U.S. Government Printing Office, Washington, DC, 476 pp.

[12] Wolff et al., 1980, *op. cit.*

[13] *Ibid.*

[14] O.L. Loucks and T.V. Armentano, 1982, "Estimating Crop Yield Effects from Ambient Air Pollutants in the Ohio Valley," *Journal Air Pollution Control Assn.*, Vol. 32, pp. 146–150.

[15] Loucks (ed.), 1980, *op. cit.*

[16] Office of Technology Assessment, 1984, *op. cit.*

[17] The dose-response functions for biological productivity are linear equations utilizing results from open-top field chamber studies.

[18] Office of Technology Assessment, 1984, *op. cit.*

[19] A.S. Heagle and W.W. Heck, "Field Methods to Assess Crop Losses Due to Oxidant Air Pollutants," in *Assessment of Losses Which Constrain Production and Crop Improvement in Agriculture and Forests*, P.S. Teng and S.V. Krupa (eds.), Proceedings, E.C. Stackman Commemorative Symposium, Misc. Publ. #7, Agric. Exp. Station, University of Minnesota.

[20] U.S. EPA, 1984, *The Acidic Deposition Phenomenon and Its Effects. Critical Assessment Review Papers*, Vol. II *Effects Sciences*, Ecological Research Series EPA-600/8-83-016BF, Office of Research and Development, Washington, DC.

[21] Ibid.

[22] U.S. EPA, 1977, *Photochemical Air Pollution Effects on a Mixed Conifer Forest Ecosystem—A Progress Report*, Ecological Research Series EPA-600/3-77-104, Environmental Research Laboratory, Corvallis, Oregon.

[23] J.M. Skelly, S. Duchell, and L.W. Kress, "Impact of Photochemical Oxidant to White Pine in the Shenandoah, Blue Ridge Parkway, and Great Smoky Mountains National Park," Second Conference on Scientific Research in the National Parks, American Institute of Biological Sciences, and National Park Service.

[24] U.S. EPA, 1977, *op. cit.*

[25] R.W. Usher and W.T. Williams, 1982, "Air Toxicity to Eastern White Pines in Indiana," *Plant Disease*, Vol. 66, pp. 199–204.

[26] G. Abrahamsen, 1980, "Acid Precipitation, Plant Nutrients and Forest Growth," in

Ecological Impact of Acid Precipitation, Proceedings of an International Conference, D. Drabls and A. Tollan (eds.), SNSF Project, Oslo, Norway, pp. 58–63.

[27] F. Andersson, 1985, "Air Pollution and Effects in Nordic Europe—the Forests of Fennoscandia," Abstracts, Muskoka Conference, International Symposium on Acidic Precipitation, 344 pp.

[28] B. Ulrich, R. Mayer, and P.H. Khanna, 1980, "Chemical Changes Due to Acid Precipitation in a Loess—Derived Soil in Central Europe," *Soil Sci.*, Vol. 130, No. 4, pp. 193–199; D.A. O'Sullivan, 1985, "European Concern About Acid Rain Is Growing," *Chem. and Eng. News*, Vol. 63, No. 4, pp. 13–18; Kraus et al., 1985, *op. cit.*

[29] A.H. Johnson, T.G. Siccama, D. Wang, R.S. Turner, and T.H. Barringer, 1981, "Recent Changes in Patterns of Tree Growth Rate in the New Jersey Pinelands: A Possible Effect of Acid Rain," *Journal Env. Quality*, Vol. 10, pp. 427–430.

[30] F.H. Bormann, 1985, "Air Pollution and Forest: An Ecosystem Perspective," *Bio-Science*, Vol. 35, pp. 434–441.

[31] O.L. Loucks, 1988, "Impacts on Agriculture and Forestry from Fossil Fuel Combustion Residuals," in *Learning for Tomorrow's World*, Wilkinson and Wyman (eds.), The Althouse Press, The University of Ontario, London, Ontario, pp. 149–160.

[32] Loucks (ed.), 1980, *op. cit.*

[33] S.N. Linzon, 1971, "Economic Effects of Sulfur Dioxide on Forest Growth," *Journal Air Pollution Control Assn.*, Vol. 21, pp. 81–86.

[34] D.C. West, S.B. McLaughlin, and H.H. Shugart, 1980, "Simulated Forest Response to Chronic Air Pollution Stress," *Journal Env. Quality*, Vol. 9, pp. 43–49.

[35] O.L. Loucks, 1981, "Recent Results from Studies of Carbon Cycling in the Biosphere," in *Carbon Dioxide Effects Research and Assessment Program*, Proceedings of the Carbon Dioxide and Climate Research Program Conference, Washington, DC, US Dept. of Energy, CONF-8004110, pp. 3–142.

[36] Loucks (ed.), 1980, *op. cit.*

[37] S.H. Wittwer, 1979, "Future Trends in Agriculture–Technology and Management," in *Long-Range Environmental Outlook*, National Research Council, Commission on Natural Resources, Washington, DC, pp. 64–107.

[38] W.W. Heck, O.C. Taylor, R. Adams, G. Bingham, J. Miller, E. Preston, and L. Weinstein, 1982, "Assessment of Crop Loss from Ozone," *Journal Air Pollution Control Assn.*, Vol. 32, pp. 353–361; Office of Technology Assessment, 1984, *op. cit.*

[39] L.H. Allen, Jr., 1979, "Potentials for Carbon Dioxide Enrichment," in *Modification of the Aerial Environment of Crops*, Barfield and Gerber (eds.), ASAE.

[40] V.C. LaMarche, Jr., D.A. Graybill, H.C. Fritts, and M.R. Rose, 1984, "Increasing Atmospheric Carbon Dioxide: Tree Ring Evidence for Growth Enhancement in Natural Vegetation," *Science*, Vol. 225, pp. 1019–1021.

[41] *Ibid.*

[42] R.B. Bacastow and C.D. Keeling, 1981, "Atmospheric Carbon Dioxide Concentration and the Observed Airborne Fraction," in *Carbon Cycle Modeling, Scope 16*, B. Bolins (ed.), John Wiley and Sons, New York, pp. 103–112; C.D. Keeling, A.F. Carter, and W.G. Mook, 1984, "Seasonal Latitudinal and Secular Variations in the Abundance and Isotopic Ratios of Atmospheric Carbon Dioxide," *Journal Geophs. Res.*, Vol. 89, pp. 4615–4628.

[43] Thomas (ed.), 1956, *op. cit.*

EDITORS' INTRODUCTION TO:
S. H. WITTWER
Food Problems in the Next Decades

Food production is a global concern and changes in the global environment affect food production. This chapter, by Dr. Sylvan H. Wittwer, one of the leading agricultural scientists of the United States, places the food problem in a global context. Dr. Wittwer is director emeritus of the Agricultural Experiment Station, Michigan State University, and the author of the recently published book *Feeding a Billion: Frontiers of Chinese Agriculture*. Dr. Wittwer writes that the world food problems for the next decade are not caused by inadequate production but rather by faulty distribution and utilization systems. However, beyond the next decade, a number of factors require an increase in the capacity to produce more food. These include rising human populations, declines in production per acre as a result of large-scale air pollution such as acid rain and ozone (which were discussed in the previous chapter), increases in soil erosion and losses in soil fertility, and fresh water supply limitations. Social factors such as increasing purchasing powers and demands for improved diets will also influence whether future food supplies are adequate. On the positive side, new technologies offer great promise for an agriculture that has less detrimental effects on the environment, increases yield per unit area, and allows crop production in previously unused environments. By bringing all of these issues together and placing them within a global context, Dr. Wittwer provides us with an important perspective on food production.

8 FOOD PROBLEMS IN THE NEXT DECADES

SYLVAN H. WITTWER
Michigan State University
East Lansing, Michigan

INTRODUCTION

The world in 1985 was beset with contrasting dilemmas at opposite ends of its food-producing systems. At one end was widespread hunger, disease, and brutal deaths in a number of Sub-Saharan countries, as Africa's worst drought in this century continued. At the other end was overproduction, particularly of grain, in the United States, Canada, Brazil, Argentina, Australia, and more recently in China, India, England, France, and Thailand.[1] Surplus rice now plagues the markets in Taiwan, Thailand, Japan, and Indonesia, and mountains of surplus dairy products have accumulated in the United States and the European community. These countries are in an overproduction trap. China and Indonesia are witnessing some of the most impressive gains in food production in history. For the first time in several decades, China is now concerned with marketing, handling, and storing food surpluses. As of 1985 and through 1987, China has become the world's leading producer of rice, wheat, and cotton. Within four years, Indonesia has moved from the world's largest importer of rice to a country not only self-sufficient but also with the world's largest rice reserves. There is surplus cassava, corn, and rice in Thailand. Many countries in the European community are moving in as strong exporters of wheat, corn, and dairy products. The only areas now on Earth with endemic food insufficiencies are the

Soviet bloc, some elements of the Middle East, and a few African countries. Even in Africa, as of 1986, only five nations suffered food shortages, compared to 25 in 1985, and for the first time in recent history food production in Africa grew by more than 15 percent and outstripped population increases. Some African countries are now struggling with food surpluses.

Global food production problems in the 1980s shifted from Asia to Africa, where the need for food is now most critical.[2] Concurrent with the major improvements in food production is the continued existence of extensive malnutrition, poverty, and starvation, especially in Africa.

The spectacle of world impotence toward too much food in some places and too little in others is especially shocking since it is caused by human beings. It has been said that while half the people in the world are starving the other half are dieting. Hundreds of millions of people suffer for want of food and from malnutrition in a world that has, in total, more than enough to feed everyone and more food per capita than at any time in recent history.

Meanwhile, there was never a greater avalanche of reports from conferences, workshops, commissions, symposia, and individually written books and articles on issues of the adequacy, security, sustainability, safety, health aspects, strategic values, and dependability of our food supplies and agricultural productivity.[3] A new series of buzzwords and phrases relating to resource conservation and management—*sustainable, regenerative, alternative, agro-ecological, biological farming, ecologically healthy systems, holistic, closed system agriculture,* and *stewardship* has found a place in our vocabularies. All this may seem ironic when agriculture is currently beset with global surpluses and record crop yields. The governments of at least a third of the world's nations subsidized the production of surplus commodities to a total of $150 billion in 1986.

The determinants for future food production are new technologies, economic incentives, and resources for inputs.[4] Many uncertainties reign as to our capacity to provide and maintain the agricultural output necessary to meet the food needs of future generations. These uncertainties relate, in large part, to the current inadequacies for preservation, management, protection, diversion, and utilization of our natural resources of air, soil, land, water, energy, and the Earth's plant and animal genetic resources. One of our future challenges will be to provide food for people while at the same time protecting our natural resource base that supports food production. Alternative production systems must be vigorously explored.[5]

The current world food situation may be summarized as follows. During the next decade, the problems will be distribution and utilization of existing production and increases in resource conservation, especially of soil and water. Beyond the next decade, there will be a need for increased output, since food production per unit will otherwise decline from global pollution such as ozone, nitrogen oxides, and acid rain; soil losses from erosion; overdraft of nonrenewable groundwater; possible climatic change from rising levels of carbon dioxide and other atmospheric gases; (today, areas with good agricultural soils tend to also have good food-producing climates; human-induced climate change could alter that co-occurrence); saltation of soils; and increases in the total human population. The keys to sustained or increased food production include preservation and wise management of the resource base, careful use of genetic engineering; exploration for new crops among the world's existing vegetation; and control of land, water, and air pollution. This chapter discusses the current and future dilemmas, and possible solutions, especially as they are viewed within the social-political framework of our current knowledge and society.

NATURAL RESOURCE MANAGEMENT

The importance of an overall view of the life-supporting resources of the Earth and an inventory of their changes has been manifested by the mission of the National Aeronautics and Space Administration designated as "global habitability" or "mission to planet Earth" and by the International Council of Scientific Unions programs on "Global Change."[6] There is the recognition that in some ways we know more about our neighboring planets than we do about Earth. Future Earth habitability, and how we can best monitor and assess changes in the global environment are essentially the theme of this book. These global environmental concerns include changes in the global hydrologic cycle; biogeochemical cycles in the water, soil, and air; biological productivity of crops and livestock; and changes in the land surface characteristics.[7] Human activities, fueled by increasing numbers of people, their demands for goods and services, and their rising consuming powers, are affecting the biosphere. There are changes in land cover, biological productivity, distribution and preservation of genetic resources, soil moisture and groundwater reserves, biogeochemical cycles, atmospheric CO_2 and other so called "greenhouse gases," trace compounds, including pollutants, and toxic substances in the environment. Some of these world environmental trends have been of great concern during the past decade.[8] Many of these changes can now be recorded by remote sensing technologies.[9]

Some of these global changes are affecting agricultural production.

Global problems relating to the sustainability and productivity of agriculture, along with the adequacy and dependability of our food supplies, are associated with changes in the gaseous and particulate composition of the atmosphere; the nature of land surfaces; the abundance, availability, distribution, and quality of our water resources; our fossil fuel reserves; and the increasing rate of extinction of plant and animal genetic resources.

AIR OR ATMOSPHERE

Ozone (O_3) alone or in combination with sulfur dioxide (SO_2) and nitrogen dioxide (NO_2) accounts for 90 percent of the crop losses in the United States caused by air pollution. Annual crop losses range between 2 and 4 percent.[10] Legumes are particularly susceptible to air pollutants. No one can yet accurately predict the effects of acid deposition (rainfall) on agricultural productivity.[11] (See the discussion of these effects in Chapter 7.) The effects of a buildup of CO_2 along with other "greenhouse gases" such as methane,[12] carbon monoxide,[13] and chlorofluorocarbon trace gases may not be so much through the widely publicized warming per se but through a large-scale disruption of the global weather machine.[14] This may have profound effects on agricultural productivity. With the man-made rising levels of atmospheric CO_2 and other gases, we are inadvertently conducting a great biological and physical experiment of global proportions, the outcome of which is not known but has been predicted to affect markedly (both directly and indirectly) the total biological and agricultural productivity of the Earth.[15]

Four points are important with respect to agricultural productivity and air pollutants. The effects are 1) regional, 2) subtle in magnitude, 3) multiple, and 4) interact with man-made and environmental constraints and stresses. No major food-producing area on Earth is immune from the effects of air pollution. The problem could become critical in most of the United States, the Yangzi River Basin (the rice bowl) of China, in western Europe, Japan, Korea, Taiwan, much of India and Thailand.

LAND

Several global problems are associated with ongoing changes in the global land surface. These include deforestation (estimated to range from 11 to 20 million hectares per year), desertification, soil erosion, excess tillage, salinization, a lack of drainage, aluminum toxicity, and accumulation of heavy metals and other toxic materials. Included are changes in tillage practices, marginal soils, conversion of noncropland, and management of range land resources.[16] Neither the magnitude nor the rate of these changes are known, yet volumes have been published on the future hazards of some of these practices and phenomena on supporting systems for global habitability. Estimates of annual losses from soil erosion range from 4 billion tons annually in the United States (mostly from row cropped land) to 25 billion tons globally.[17] During the past two decades, there has been a major shift in the grain belt of the United States from mixed crop-livestock systems to cash or row cropping, prompted by future hopes of an expanding export market. The rise of corn (maize) production during the past two decades in western Europe is even more striking.

Meanwhile, no soil management technology has moved more rapidly for conservation of the nation's soil, water, energy, and organic matter than reduced or conservation tillage, which now stands at approximately 50 million hectares or one-third of the total cultivated crop area of the United States.[18] Combined with allelopathic properties of plant residues,[19] conservation tillage or no-till[20] or alley cropping[21] may provide the long sought after means for continuous cropping of many of the shallow, fragile, and easily erodible soils in the lowland humid tropics.[22] Globally, it is estimated that we are annually losing 8 million hectares of land from nonagricultural conversions, 3 million to soil erosion, and 2 million hectares each to desertification and toxification.[23]

Seldom has such a totally new set of competitive forces been unleashed on our resources as is now occurring. Yet, this is coupled with a marvelous opportunity to take corrective measures to conserve our soil. With a surplus of most agricultural products and declining public support for farm subsidies, there is increasing support for resource conservation. It is time we began to devote highly erosive, fragile, and shallow cropland to other uses such as grazing, wood fuel production, and reserves for wildlife and recreation. Such measures would simultaneously conserve our soil, land, and water resources, and bring agricultural production down to a level where prices will be profitable to the farmer and acceptable to the consumer.[24] Can our leadership merge the two problems and does leadership to do so now exist?[25]

WATER

The management of water resources as to availability and quality will be particularly critical for the sustainability, productivity, and dependability for food production in the decades ahead.[26] It is estimated that agriculture consumes, mostly through irrigation, 80 to 85 percent of fresh water resources in the United States. This is far greater than the percent of the nation's energy used for agriculture. The current overdraft of groundwater resources, again, mostly for crop irrigation in the United States is estimated at 20 to 25 million acre feet per year. Land subsidence from these overdrafts is becoming serious in some of the valleys of the western United States and in many other areas throughout the world. One-third of the world's food supplies are now grown on the 18 percent of the cropland that is irrigated.[27]

The amount of irrigated cropland both in the United States and globally is increasing. Since 1950, irrigated land has almost tripled and now totals nearly 270 million hectares.[28] An estimated three-fourths of the potentially arable land in the tropics has limited production capacity because of insufficient moisture. Irrigation is the one option that nations have for increasing agricultural output and assuring dependability of supplies. Water for irrigation requires energy for lifting, transporting, and pressurizing. Most of the orchard and vegetable land on Earth is now under irrigation, and the irrigated areas are increasing yearly. There are now approximately 1 million acres of drip irrigation in the United States, Israel, and Australia. Of the major food crops, rice is the most intensive user of water. Because of the high water requirement of rice, one may expect a shift in the decades ahead to a greater proportion of wheat, sorghum, millet, corn, and other cereal grains that have only half or less than half the water requirement of rice. Even further shifts may occur in going from corn to sorghum and millet.

Water is currently the limiting resource input for agricultural production in each of the five most populous countries on Earth—China, India, the USSR, the United States, and Indonesia. It is recognized as the most critical natural resource for future agricultural development in the Middle East, southern Europe, Egypt, the Sudan, Sub-Saharan Africa, Taiwan, Pakistan, Australia, Argentina, Brazil, most of Canada, and Central and South America. Much of the relative stability as well as the magnitude of agricultural production in China, the most populous country on Earth, relates to the fact that almost 50 percent of the cultivated cropland is irrigated.

GENETIC RESOURCES

There are an estimated 5 to 10 million plant and animal species now on the Earth. The extinction rate is high (perhaps 1,000 per year) and rising.[29] By the end of the century, we could lost one million of the five to ten million species. Five thousand plant species historically have fed the human race. Today, about 150 plant species, with a quarter of a million local races, meet most of the calorie needs of people. Plant products derived from fewer than 30 species provide more than 90 percent of the human diet.[30]

Only 5 to 10 percent of the 250,000 to 750,000 existing species of higher plants have been surveyed for biologically active compounds. The plant kingdom has received little attention as a resource of potentially bioactive materials.[31] Gene banks in developing countries may soon outnumber those in industrialized nations since 70 percent of the Earth's species are in the developing countries of the tropics. Our Earth has 3.8 million square miles of soils too salty to grow conventional crops. Yet there are many crops that tolerate salt, including relatives of commercial barley, wheat, sorghum, rice, millet, sugar beets, tomato, date palm, and pistachio. Also, there are many equally salt-tolerant forage plants for livestock feeding such as alfalfa, Ladino clover, creeping bent grass, Bermuda grass, and various reeds and rushes. Most of our so-called world collections of crops (genetic resources) are sadly deficient in wild races.[32]

We have an International Board for Plant Genetic Resources and hundreds of thousands of collected samples are now found in various parts of the world, particularly in the United States, the USSR, and China. There is, however, no organized program, either in the United States or internationally, to sample, evaluate, preserve, and utilize exotic and endangered species (or sources) of chicken, turkey, swine, sheep, goat, and cattle germ plasm. Existing research and service organizations do not have the financial resources to undertake this expensive long-term program, which is of vital importance to the future of the world.[33]

THE CAPACITY TO PRODUCE FOOD

In the face of uncertainty and instability, an increased production capacity, whether or not that capacity is actually used, must be called for. All projections indicate that because of international competition in commodity markets, coupled with increasing constraints for limited energy, water, and land resources, rising costs of labor, demands for improved diets, and population growth, more science and technology must be put into agriculture to double food production in the next half-century. This must be accompanied by improved institutions and policies, an expanded capital base, and greater entrepreneurial and managerial skills. The capacity to produce food through an adequate resource base and the use of computers and sensors may or may not expand production in periods of food shortages; their use should not be suppressed in periods of food surpluses. A well-managed resource base of soil, water, energy, and air, with preservation of plant and animal genetic resources and the development of human capital, must accompany either an expansion or suppression of agricultural production for the decades ahead.[34] Continuing support of agriculture research for the development of new technologies, advances in crop and livestock productivity, and improvements in food utilization technologies, both in times of plenty and shortages, is essential. Yet, this is seldom understood by decision makers.

FOOD CROPS

Some 20 crops essentially stand between people and starvation.[35] As past trends and future prospects are reviewed, increasing crop yields is the key to greater agricultural and food crop production. There is little evidence of yields plateauing except when constrained by socioeconomic and political factors. Genetic breakthroughs have been significant, and progress in genetic improvements will continue. These breakthroughs and improvements will include: a greater carbon dioxide fixation in photosynthesis; enhanced biological nitrogen fixation; using a new generation of chemical substances both to control parasitic fungi and regulate plant growth for intensive crop management; a renewed emphasis on the root zone including the potential benefits of root colonizing and symbiotic bacteria and fungi; improved efficiency in producing new crops through anther, cell, and tissue cultures; and the development of herbicide-resistant crops and plants that have built-in pesticides.[36] The current constraints on food production are biological, environmental, cultural, and political.

One of the future scientific challenges, important for practice as well as for theory, will be to assemble and demonstrate through maximum yield trials the appropriate technologies and resource inputs necessary to optimize stable and enhanced crop productivity. Record crop yields should not be regarded as abnormal occurrences. They occur when environmental and other biological stresses are minimal and resource inputs and technologies are optimal.[37]

Both biological and environmental constraints are currently limiting global food crop production. Important strategies must include integrated pest management coupled with pesticide resistance management and the use of allelopathic responses, chemicals that are the herbicides naturally produced by plants. These methods not only aid crop protection, but also reduce production costs, assure food safety, provide resource-sparing technologies, minimize impacts on the environment, and have no adverse effects on human health.

FOOD ANIMALS

Animals are natural protein factories that harvest vast food resources, otherwise of little value, and convert these to milk, meat, eggs, and other useful products. They are living storehouses of mobile food. They constitute a global food reserve that approaches the grain reserves and that is distributed more evenly and transported more easily. They also serve as buffers for grain prices and supplies. Per capita meat consumption during the past two decades has increased in all nations, with severalfold increases in Japan and Israel. Immediate research objectives for increasing the production capacity and improving the nutritional values of animal products include resource conservation, environmental adaptation, improved reproduction efficiency, more lean and less fat meat, and genetically engineered vaccines for disease control.[38]

The current widespread use of feed additives, growth hormones, steroids, antibiotics, chemotherapy, and pesticides for food animals will reach a crisis within the next two decades from an objecting public concerned with environmental issues, food safety, human health, and animal welfare. Alternatives to the use of the many biological and pharmaceutical products for pigs, chickens, and cattle will include the emphases on built-in permanent genetic resistance to disease coupled with improved management and housing.[39]

FOOD UTILIZATION

Vastly improved food utilization programs are the great challenge ahead.[40] The capacity to produce an abundance of food has been demonstrated, not only in the United States, but also in India, Pakistan, Taiwan, Japan, France, England, Scandinavia, Brazil, Canada, Australia, Argentina, New Zealand, Mexico, Kenya, Nigeria, China, Indonesia, and Thailand. The problem is utilization. Densely populated countries, such as India, Indonesia, Thailand, and China, encounter enormous food wastes and losses in nutritional value in what is produced because of insufficient and inadequate storage, packaging, and distribution systems, and because the surpluses are not stored or processed. Food problems on a global scale cannot be managed without the utilization component. This may require significant input from the private or industrial sectors, which will be difficult in those nations with centrally controlled economies.

Successful production, especially of perishable food products for fresh markets, must be accompanied by a processing or utilization component to absorb surpluses, avoid wastes, and stabilize supplies. The lack of port-harvest handling, storage, processing, and utilization facilities and technologies is a severe constraint in the introduction of new crops. This is especially true for tropical countries in the Caribbean, Central and South America, Southeast Asia, and in China.

THE POLITICS OF FOOD PRODUCTION

A decade ago David Hopper, now of the World Bank, make this perceptive observation. "There was never a greater opportunity for food abundance, but the exploitation of that opportunity was never more vulnerable to the uncertain responses of human political institutions." The conclusion of the National Academy of Science's World Food and Nutrition Study[41] of 1977, in which 1,500 scientists participated, was "If there is the political will in the USA and abroad, it should be possible to overcome the worst aspects of widespread hunger and malnutrition within one generation."

THE AFRICAN SITUATION

Food problems today in south Saharan Africa, the Middle East, and the Soviet bloc nations are indicative of those to be encountered in the decades ahead—climate, with its shortages of fresh water, and government policies predominate as causal factors, but so also does a shortage of yield-increasing technologies. Food problems are not now universal in all of Africa and they are not new in the south Saharan region where they are now concentrated and have recently been given such wide visibility. Major contributing factors to food problems are short-sighted government policies with little or no priority or economic incentives for food production; minimal or no support for agricultural research and education programs; a continuing goal of producing high value non-food cash crops for export; little, if any, concern for food security, abolishing hunger, and a failure to recognize nutritional problems and take measures to correct them. Coupled with these constraints are civil strife, low labor productivity, inadequate or nonexistent transportation and storage facilities, and soils that are low in fertility and often shallow, fragile, and easily erodible.[42]

Significant high-yielding food-producing technologies for Africa are now on the horizons, many having their origin at the International Institute for Tropical Agriculture (IITA). They may portend a green revolution for Africa.[43] They include pest-resistant cowpeas that will mature in 60 days, thus requiring a minimum of moisture and pesticide spraying. A good second crop will be produced in wetter and moisture-retaining areas even during the dry season; and a drought-resistant sorghum with twice the productivity of earlier varieties. Soybeans have been developed with natural nodulation in the tropics and with a seed longevity that has been extended from 3 to 8.5 months, so that seed that is harvested for one crop can be carried forward for planting the next. There is now a mosaic-resistant cassava that has inadvertently spread beyond the boundaries of IITA to over a million hectares in neighboring Nigeria with yields threefold above native strains. Sweet potatoes have been developed that will mature in $3\frac{1}{2}$ months and yield up to 15 tons per hectares without fertilizer or 30 tons if fertilizer is added. The successful propagation of yams by new minisett and microsett techniques has reduced seed costs to a fraction of previous outlays. Tissue culture programs are now operational for rapid propagation and distribution throughout Africa of superior selections of sweet potato, yams, coco yams, cassava, and plantains. Significant progress has been made in high-yielding rice varieties that are blast-resistant and adapted to both paddy and upland conditions. There are now hybrid tropical maize varieties resistant to the streak virus and with yield performance records of 4.5 to 9.5 tons per hectare. Finally, tropical soil management schemes of conservation tillage, no-till systems, and alley cropping coupled with the use of the "rolling injection planter" and occasional plastic soil mulches will allow continuous cropping without soil erosion, in contrast to traditional shifting cultivation or the slash-and-burn techniques, which allows cropping only once or twice in 10 to 12 years.

The current problem now facing Africa for a solution of its food problems is an effective agricultural extension service to carry the results of research to the farmers.[44] Without those incentives, farmers will produce only enough for themselves and will not seek out yield-improving technologies. They have little, if any, interest in the management and conservation resources and have established no food reserves. When the rains fail, people and nations are in great jeopardy.

The current world food problem is not a production problem—that has been amply demonstrated in the real world of today. It's getting food or producing food where people are. It's providing economic incentives for farmers to produce food. It's delivery and it's purchasing power. Only poor people go hungry. What exists in the world today in terms of hunger, poverty, starvation, famine, malnutrition, and food shortages—whether in India, Latin America, or south Saharan Africa—is in large part a decision of governments and leaders of nations. It's a lack of priority with respect to food production, and the lack of priority with respect to agriculture research. It's a failure to provide economic incentives for farmers to produce.

SUMMARY

More food must be produced. It will be a response to rising populations, demands for improved diets, increased purchasing power, desires for food security, and improvements in international trade. This will be achieved primarily by yield increases with slight benefits from more crops per year and more land under cultivation.

Greater stability and dependability in food production will be sought. Stability of production will parallel, in priority, production increases derived from expanded irrigation of cultivated crops, improved pest control, advances in protected environments for both crops and live-stock, alleviation of climatic and environmental stresses through genetic improvements, management and chemical treatments, and emphasis on food reserves and home storage. Plants must be developed that can be productive on nutrient-poor soils and under excess of toxic elements and salt.

Resource conservation will become a byword for land, water, and energy. There will be renewed emphasis on the production and utilization of renewable resources, of which food is number one. Food production must move toward a series of more science-based rather than resource-based (land, water, energy) technologies. Land and water will assume increasing importance for food production.

New high pay-off food-producing technologies for the future will be those that result in more dependable production coupled with higher yields, are more labor than capital intensive, are crop intensive, and are sparing of natural resources. They should be nonpolluting. They will be the ones that offer solutions to the global problems of hunger, poverty, malnutrition, inflation, deforestation, and soil erosion; increase the demand for underutilized labor resources and put more people to work more days per year and more productively.

Environmental issues and costs of inputs, as they impinge on the use of science and new technology for food production will continue to mount. Concerns about food safety and toxic substances in the environment and biologicals for feeding livestock and chickens will multiply.

Constraints will increase as to the use of chemicals for pest control, disease prevention, and growth regulation. The options for chemical usage will continue to decrease as the needs for food increases. Alternatives to the use of hormones and antibiotics for pigs, chickens, and cattle will be built-in permanent genetic resistance to disease, coupled with improved management and housing.

How we can commit to aggressively expanded food production programs, while at the same time achieve nondegradation of soils, reduce overdraft of groundwaters, and avoid modifications and pollutions of the environment through increased use of irrigation, fertilizers, and pesticides, stands as the supreme challenge for future agricultural science and technology.

The development of increased capacities to produce food is essential to meet current needs and establish future food security. This must be accompanied by improved institutions and policies, an expanded capital base, and greater entrepreneurial and managerial skills. The capacity to produce, and the use of such capacity, are two different things, but the capacity must be at hand. New technologies may or may not expand production in periods of shortages, and their development and use should not be suppressed in periods of surpluses.

Vastly improved food utilization programs constitute a great future challenge. The capacity to produce enough food and some to spare has been demonstrated. The problem is utilization, especially surplus production. Densely populated countries now encounter enormous wastes and losses in nutritional values of what is produced because of inadequate storage, packing, and distribution systems, and because the surpluses are not stored or processed. Food problems on a global scale cannot be managed without the utilization component.

Current food and nutrition problems in south Saharan Africa are largely political. They relate to governments with little priority for food production and low economic incentives for farmers to produce or for agricultural research. New biological advances recently achieved in food production have the ingredients, if adopted by farmers, to initiate a green revolution for south Saharan Africa and in other nations now witnessing food production problems.

NOTES

[1] B. Insel, 1985, "A World Awash in Grain," *Foreign Affairs*, Vol. 63, pp. 892–911.

[2] J.W. Mellor, 1985, *The Changing World Food Situation*, International Food Policy Research Institute, Washington, DC.

[3] *BioScience*, Vol. 35, July/August 1985. The entire issue is devoted to soil (land), air and

plant resources; L.R. Brown, W.U. Chandler, C. Flavin, C. Pollock, S. Postel, L. Starke, and E.C. Wolf, 1985 and 1986, *State of the World*, W.W. Norton and Co., New York; L. Busch and W.B. Lacy, 1984, *Food Security in the United States*, Westview Press, Boulder, Colorado; G.A. Douglas (ed.), 1984, *Agricultural Sustainability in a Changing World Order*, Westview Press, Boulder, Colorado; R. Henkes, 1985, "The Mainstreaming of Alternative Agriculture," *The Furrow*, Vol. 96, pp. 10–13, John Deere, Moline, Illinois; G.L. Johnson and S.H. Wittwer, 1984, "Agricultural Technology Until 2030: Prospects, Priorities and Policies," Michigan State University Agricultural Experiment Station Special Report 12; J.P. Madden, 1984, "Regenerative Agriculture: Beyond Organic and Sustainable Food Production. The Farm and Food System in Transition. Emerging Policy Issues," Michigan Cooperative Extension Service, Michigan State University, East Lansing, Michigan; D. Pimentel and C.W. Hall (eds.), 1984, *Food and Energy Resources*, Academic Press, Inc., New York; J.G. Speth, L. Fernandez, and N.C. Yost, 1985, "Protecting Our Environment: Toward a New Agenda," *Alternative for the 1980s, No. 18*, Center for National Policy, Washington, DC; and S.H. Wittwer, 1985, "Carbon Dioxide Levels in the Biosphere: Effects on Plant Productivity," CRC *Critical Reviews*, Vol. 2, pp. 171–198.

[4] S.H. Wittwer, 1980, "Food Production Prospects: Technology and Resource Options," in *The Politics of Food*, D. Gale Johnson (ed.), The Chicago Council of Foreign Relations, Chicago, Illinois, pp. 60–99.

[5] Douglas, 1984, *op. cit.*; Henkes, 1985, *op. cit.*; Insel, 1985, *op. cit.*; Wittwer, 1980, *op. cit.*; and S.H. Wittwer, 1985, "Food Production Interrelationships to Nutrition," in *Proceedings of a Symposium—Workshop on Diet and Health*, University of California, Davis, September 19.

[6] B.I. Edelson, 1985, "Mission to Planet Earth" (editorial), *Science*, Vol. 227, p. 367.

[7] National Aeronautics and Space Administration, 1983, *Land-Related Global Habitability Science Issues*, NASA Technical Memorandum 85841, Washington, DC; and C.J. Tucker, J.R.G. Townshend, and T.E. Goff, 1985, "African Land-Cover Classification Using Satellite Data," *Science*, Vol. 227, pp. 368–375.

[8] M.W. Holgate, M. Kassas, and G.F. White, 1982, "World Environmental Trends between 1977 and 1982," *Environmental Conservation*, Vol. 9, pp. 11–29.

[9] A.F.H. Goetz, G. Vane, J.E. Solomon, and B.N. Rock, 1985, "Imaging Spectrometry for Earth Remote Sensing," *Science*, Vol. 228, pp. 1147–1153; Tucker et al., 1985, *op. cit.*

[10] W.W. Heck, O.C. Taylor, R. Adams, G. Bingham, J. Miller, E. Preston, and L. Weinstein, 1982, "Assessment of Crop Loss from Ozone," *Journal of the Air Pollution Control Assoc.*, Vol. 32, pp. 353–361.

[11] C.H. Epstein and R.E. Yuhuke, 1985, "Acid Deposition, Smelter-emissions, and the Linearity Issue in the Western United States," *Science*, Vol. 229, pp. 859–862; Interagency Task Force on Acid Precipitation, 1983, National Precipitation Assessment Program Annual Report, Washington, DC; M. Shepard, 1985, "Forest Stress and Acid Rain," *EPRI Journal*, Vol. 10, pp. 16–25; M. Sun, 1985, "Possible Acid Rain Woes in the West," *Science*, Vol. 228, pp. 34–35; World Resources Institute, 1985, *The American West's Acid Rain Test*, World Resources Institute, Washington, DC.

[12] R.A. Keer, 1984, "Doubling of Atmospheric Methane Supported," *Science*, Vol. 226, pp. 954–955.

[13] M.A. Khalil and R.A. Rasmussen, 1984, "Carbon Monoxide in the Earth's Atmosphere: Increasing Trend," *Science*, Vol. 224, pp. 54–56.

[14] Hansen, J.G. Russel, A. Lacis, I. Fung, and D. Rind, 1985, "Climate Response Times: Dependence on Climate Sensitivity and Ocean Mixing," *Science*, Vol. 229, pp. 857–859; Speth et al., 1985, *op. cit.*

[15] E. Lemon (ed.), 1983, CO_2 and Plants. *The Response of Plants to Rising Levels of*

Atmospheric Carbon Dioxide, Westview Press, Boulder, Colorado; National Academy of Sciences, 1983, "Changing Climate, Report of the Carbon Dioxide Assessment Committee," National Research Council, National Academy Press, Washington, DC; Wittwer, 1980, *op. cit.;* S.H. Wittwer, 1982, "U.S. Agriculture in the Context of the World Food Situation," in *Science, Technology, and the Issues of the Eighties: Policy Outlook,* A. Teich and R. Thornton (eds.), Westview Press, Boulder, Colorado, pp. 191–214.

[16] W.L. Hargrove, F.C. Boswell, and G.W. Langdale (eds.), 1985, "The Rising Hope of Our Land," Proceedings of the 1985 Southern Region No-Till Conference, July 16–17, Griffen, Georgia; W.E. Larson, F.J. Pierce, and R.H. Dowdy, 1984, "The Threat of Soil Erosion in Long-term Crop Production," *Science,* Vol. 219, pp. 458–465.

[17] Speth et al., 1985, *op. cit.*

[18] L.R. Brown and E.C. Wolf, 1984, "Soil Erosion: Quiet Crisis in the World's Economy," *Worldwatch Paper* 60, Worldwatch Institute, Washington, DC.

[19] A.R. Putnam, 1983, "Allelopathic Chemicals, Nature's Herbicides in Action," *Chemical Engineering News,* Vol. 61, pp. 34–45.

[20] Holgate et al., 1982, *op. cit.*

[21] B.T. Kang, G.F. Wilson, and T.L. Lawson, 1984, "Alley Cropping. A Stable Alternative to Shifting Cultivation," International Institute for Tropical Agriculture, Ibadan, Nigeria.

[22] R. Lal, 1987, "Managing the Soil of Sub-Saharan Africa," *Science,* Vol. 236, pp. 1069–1076; and B.N. Okigbo, 1984, "Improved Permanent Production Systems as an Alternative to Shifting Intermittent Cultivation," FAO Soils Bulletin 53, Food and Agricultural Organization of United Nations, Rome.

[23] Speth et al., 1985, *op. cit.*

[24] Brown and Wolf, 1984, *op. cit.*

[25] Technology, Public Policy and the Changing Structure of American Agriculture: A Special Report for the 1985 Farm Bill," U.S. Congress, Office of Technology Assessment, OTA. F–272, March, Washington, DC.

[26] Council for Agricultural Science and Technology, 1985, *Agriculture and Groundwater Quality,* Ames, Iowa; M.E. Jensen, 1982, "Water Resource Technology and Management," paper presented at the Resource Conservation Symposium, December 5–9, Washington, DC; Office of Technology Assessment, 1983, *Water-related Technologies for Sustainable Agriculture in U.S. Arid/Semi-Arid Lands,* Office of Technology Assessment, Washington, DC; S. Postel, 1984, *Water: Rethinking Management in an Age of Scarcity,* World Paper 62, Worldwatch Institute, Washington, DC; G. Slogget, 1981, "Prospects for Groundwater Irrigation," Agricultural Economics Report 478, Economic Research Service USDA, Washington, DC; U.S. Water Resources Commission, 1978, "The Nation's Water Resources 1975–2000," Vol. 1 Summary, Washington, DC.

[27] S. Postel, 1985, *Conserving Water: The Untapped Alternative,* Worldwatch Paper 67, Worldwatch Institute, Washington, DC.

[28] W.R. Rangeley, 1985, "Irrigation and Drainage in the World," paper presented at the International Conference of Food and Water, College Station, Texas, May 26–30.

[29] N. Meyers, 1983, *A Wealth of Wild Species,* Westview Press, Boulder, Colorado; Speth et al., 1985, *op. cit.*

[30] Council for Agricultural Science and Technology, 1984, *Development of New Crops: Needs, Procedures, Strategies and Options,* Ames, Iowa; D.L. Plucknett, N.J.H. Smith, J.T. Williams, and N. Murthi Anishetty, 1983, "Crop Germplasm Conservation and Developing Countries," *Science,* Vol. 220, pp. 163–169; G. Wilkes, 1983, "Current Status of Crop Plant Germplasm," *CRC Critical Reviews,* Vol. 1, pp. 133–181.

[31] M.F. Balandrin, J.A. Klocke, E.S. Wurtele, and W.H. Ballinger, 1985, "Natural Plant Chemicals: Sources of Industrial and Medicinal Materials," *Science*, Vol. 228, pp. 1154–1160.

[32] Meyers, 1983, *op. cit.*

[33] Council for Agricultural Science and Technology, 1984, *Animal Germplasm Preservation and Utilization in Agriculture*, Ames, Iowa.

[34] Johnson and Wittwer, 1984, *op. cit.;* Wittwer, 1980, *op. cit.*

[35] Wittwer, 1985, *op. cit.*

[36] Board on Agriculture, 1985, *New Directions for BioScience Research in Agriculture*, Committee on BioScience Research in Agriculture, National Research Council, National Academy Press, Washington, DC; Teich and Thornton, 1982, *op. cit.*

[37] Wittwer, 1985, *op. cit.*

[38] W.G. Pond, R.A. Merkel, L.D. McGilliard, and V.J. Phodes (eds.), 1980, *Animal Agriculture. Research to Meet Human Needs in the 21st Century*, Westview Press, Boulder, Colorado.

[39] Wittwer, 1985, *op. cit.*

[40] D. Knorr and A.J. Sinskey, 1986, "Biotechnology in Food Production and Processing," *Science*, Vol. 229, pp. 1224–1229.

[41] National Academy of Sciences, 1977, *World Food and Nutrition Study*, National Academy Press, Washington, DC.

[42] N.C. Brady, 1985, "Toward a Green Revolution for Africa," *Science*, Vol. 227, p. 1159; C.K. Eicher, 1985, "Agricultural Research for African Development: Problems and Priorities," paper prepared for a World Bank conference on Research Priorities for Sub-Saharan Africa, Bellagio, Italy, February 25–March 1; Knorr et al., 1986, *op. cit.;* J.R. Morris, 1983, "Reforming Agricultural Extension and Research in Africa," Discussion Paper 11, Agricultural Administration Unit Overseas Development, London.

[43] Brady, 1985, *op. cit.;* O.M. Solandt, S.H. Wittwer, R. Cowan, and N.E. Mumba, 1985, "Report of the In-house Review Panel," International Institute of Tropical Agriculture, January 21–25.

[44] Morris, 1983, *op. cit.*

EDITORS' INTRODUCTION TO:
R. LAL
Soil Degradation and Conversion of Tropical Rainforests

One of the major effects our civilization has had on the land is an increase in erosion accompanying forest clearing and the introduction of farming. The effects of these activities have long been known for temperate zone lands; we can find references to the destructive effects of deforestation in the writings of the ancient Greeks. George Perkins Marsh discusses these problems at length in his classic nineteenth-century book, *Man and Nature*. However, during the past several decades the scientific study of erosion has expanded greatly so that it is now possible to provide some facts about the rates of erosion under different land use practices. These facts are especially important today for the tropics, which, as we learned in Chapter 6 by Dr. T.E. Lovejoy, have been subject to rapid deforestation in the last decades. In this chapter, Dr. R. Lal, an expert on tropical soils and deforestation, discusses the status of our current knowledge about these issues, pointing out what we know and what we do not know, and placing this information in a global context.

Dr. Lal tells us that the rapid decline in soil quality and its capacity for agricultural production in the tropics is partly due to inherently low soil fertility and to a harsh climate. However, land use practices and soil and crop management also play an important role. Removal of protective vegetation cover drastically alters water and energy balances and disrupts chemical cycles. Mechanical methods of deforestation, especially the use of the bulldozer, have adverse effects on soils. On the positive side, Dr. Lal discusses the technology that exists that allows us to manage tropical soils and water resources judiciously so that we can obtain a high but sustainable productivity that will not cause severe damage to soil and to the environment.

Formerly associated with the International Institute of Tropical Agriculture and now on the faculty of Ohio State University in the Department of Agronomy, Dr. Lal is an expert on soil erosion in tropical areas.

9 SOIL DEGRADATION AND CONVERSION OF TROPICAL RAINFORESTS

R. LAL
The Ohio State University
Columbus, Ohio

INTRODUCTION

The tropics cover about 40 percent of the Earth's surface and, with few exceptions, are distinguished by agriculturally low yields, a high rate of soil degradation, a rapid decline in productivity following initial cropping, widespread poverty, malnutrition, and low standards of living. However, given the land and water resources and the productive potential of different ecological regions of the wet and dry tropics[1], there should be no reason for an overwhelming food shortage common in these regions, particularly in Sub-Saharan Africa. For example, it is estimated that out of three billion potentially cultivable hectares on the Earth, at least 64 percent lie in the tropics and only 37 percent of those have ever been cultivated. The estimates of potentially cultivable land area alone range from 1,445 to 1,648 million hectares. Considering the possibilities of multiple cropping and growing more than one crop every year, the gross arable land area could be two or three times higher.

Yet per capita food production has not kept pace with the demand in most tropical regions. In comparison to the 1970 level, world food demand was expected to rise 44 percent by 1985 and 112 percent by 2000.[2] Sub-Saharan Africa has had a drastic decline in per capita food production over the last two decades. For example, the per capita food

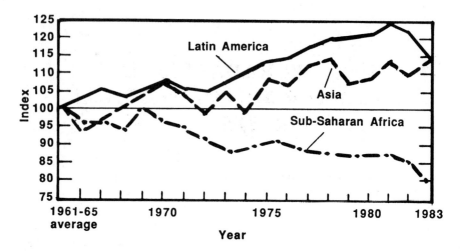

FIGURE 9–1   Index of per capita food production in Sub-Saharan Africa, 1961–65 to 1983 (1961–65 Average = 100). [Note: World Bank, 1984, "Toward Sustained Development in Sub-Saharan Africa," Washington, DC.]

production in tropical Africa declined by as much as 9.6 percent in 1981 compared with 1970 (Figure 9–1).

One of the reasons for the discrepancy between the potential and actual productivity in tropical Africa is that the natural resources have been overestimated. Even under the best known agricultural technologies, the food production of most forested tropical lands is inferior to lands in temperate regions as well as tropical lands already being farmed. Most existing forested lands are of marginal quality, being either too steep, too shallow, or having rocky soils of low inherent fertility. Bringing marginal lands under cultivation creates problems of accelerated erosion and general degradation of the natural resource. It is not just the high population density but the ratio of developed and potential resources to the population and the population carrying capacity that are crucial to maintaining the favorable balance among land, food, and people.

DEFORESTATION AND BRINGING MORE LAND UNDER CULTIVATION

The existing reserves of tropical rainforests are shrinking rapidly because bringing new land under cultivation is considered an easy method of increasing food production. The annual rate of deforestation in the tropics is estimated to be 11 million hectares and an additional 20 million hectares of forest is estimated to decline annually to more degraded secondary forest. About 100,000 km^2 of forest in South and Central America are cleared annually. In Thailand, Malaysia, Indonesia, Vietnam, Laos, and Cambodia, more than 150,000 km^2 of forest are cleared annually. More than one million km^2 of forest has already been removed in equatorial Africa.[3] Overall, about 3.6 million hectares of African open woodland and forest have been denuded over the last decade. Since 1950, the total loss of forest in Africa is estimated to be over 200 million hectares, reducing the size of African forest from 901 million hectares in 1950 to 690 million hectares in 1985.[4] Most of the 500,000 km^2 of the Atlantic coastal forest of Brazil has disappeared except for a few patches of the forest reserves. Also lost to expanding agricultural activity are the 3.5 to 4 million hectares of the deltaic tropical rainforest of lower Burma.[5]

These well-intended efforts to increase food production are often unsuccessful[6] and usually result in great environmental damage. The consequences of deforestation in terms of soil erosion, land degradation, nutrient loss, and the disruption of the delicate equilibrium among soil, plant, and atmosphere can be seen in the vast tracts of barren and unproductive land where lush green forest once grew.

Another important reason for forest conversion is to meet the fuel wood demand. The rural population in the tropics depends almost exclusively on fuel wood as their primary source of energy. It is estimated that the percent energy requirement by fuel wood in rural areas is 90 percent in Kenya, 88 percent in Zambia, and 95 percent each in Nepal, Sri Lanka, and Thailand.[7] The estimated fuel wood consumption ranges from 0.8 to 2.0 m^3/capita/year, with an average consumption of 1.5 m^3/capita/year. Deforestation will continue unabated unless alternative, equally reliable, and economic sources of fuel are made available to the rural population.

METHODS OF FOREST CONVERSION

The diverse methods used for forest conversion are classified into manual, mechanical, and chemical (Table 9–1). In traditional farming systems, land is cleared manually with native tools and the felled biomass is burnt in place. Fire is an important tool in traditional farming and is widely used throughout the tropics. Traditional clearing is generally incomplete and stumping is often not done. Where incomplete clearing is acceptable, even large trees are killed by ring barking followed by burning of the tree base.

Establishment of pasture in subhumid and semiarid regions can be

TABLE 9–1 METHODS OF FOREST CONVERSION

	Methods	
Manual	**Chemical**	**Mechanical**
Native tools (axe, cutlass)	Defoliants	No stumping Shear blade Tree crusher
Improved tools (chain saws)	Systemic	Stumping Dozer blade Tree pusher Tree extractor Root rake Chaining

done by using chemicals for killing the existing vegetation cover. Some commonly used chemicals for tree poisoning are 2, 4, 5-T, sodium arsenite, Tordon 105, Silvisor 510, sodium nitrite, and sodium chlorite. Chemicals are applied as foliar spray, by injection into the trunk, or sprayed on the bark of the tree trunk. If properly planned, chemical clearing can be used in the savanna region with less tree density. Some of the severe problems of using chemicals are the environmental pollution and health risks. Little is known about the retention, movement, degradation, and pathways of these chemicals in tropical environments.

Motorized clearing using heavy earth-moving equipment is widely used for large-scale deforestation throughout the tropics, especially in regions with acute labor shortage. Commonly used attachments for motorized clearing are of two types, depending on whether roots and stumps are removed. Shear blades and tree crushers leave roots and stumps in the ground, whereas chaining, tree extractors, tree pushers, root rakes, and root plows do not. The front-mounted dozer blade is perhaps one of the most destructive methods of land clearing because it causes considerable soil disturbance, removal of fertile top soil to the windrows and boundaries, and compaction of the exposed subsoil, which restricts water acceptance and root penetration.

ENVIRONMENTAL IMPACT OF DEFORESTATION

The possible ecological implications (Table 9–2) of large-scale deforestation now taking place in humid and subhumid tropics have caused much concern among scientists, environmentalists, and planners around the world. However, fears of environmental changes, local and

TABLE 9–2   ALTERATIONS IN SOIL AND MICRO-CLIMATIC
ENVIRONMENTS BY DEFORESTATION AND INTENSIVE CULTIVATION OF
TROPICAL SOILS

Hydrologic cycle

Decrease in interception by vegetation
Decrease in the water transmission and retention characteristics of the soil
Decrease in water uptake from subsoil below 50-cm depth
Increase in evaporation
Increase in surface runoff
Increase in the interflow component

Microclimate

Increase in temperature amplitude
Decrease in the mean relative humidity
Increase in the incoming radiation reaching soil surface

Energy balance

Increase in the fluctuations in soil temperature
Change in the heat capacity of the soil
Change in the phase angle, periodicity, and damping depth

Nutrient status

Decrease in organic matter
Decrease in base status
Decrease in nutrient recycling

Soil flora and fauna

Decrease in biological activity of macro- and microorganisms, notably earthworms
Shift in the vegetation type from broad leaves to grasses and from perennials to
    annuals
Shift in climatic climax

R. Lal, 1981a, "Deforestation and Hydrological Problems," in *Tropical Agricultural Hydrology*, R. Lal and E. W. Russel (eds.), J. Wiley & Sons, Chichester, UK.

global impact on climate, and degradation of fragile soil resources are easily exaggerated in the absence of solid data from well-designed and adequately equipped long-term studies—planned to quantify the effects of deforestation. Some effects of forest conversion are described here.

SOIL DEGRADATION

Soil degradation is a universal problem.[8] Worldwide degradation of agricultural land causing irreversible loss of productivity is estimated to be 6 million hectares per year.[9] It is estimated that for the 1,500 million hectares of land now being used for crop production, an additional 2,000 million hectares of once fertile land has gone out of production due to degradation in soil quality. The current rate of loss of arable land due to different degradative processes is estimated to be 5 to 7 million

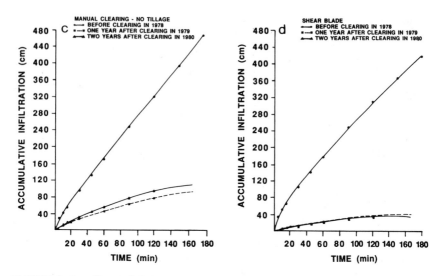

FIGURE 9–2   Effects of deforestation by tree pusher-root rake attachments and subsequent cultivation involving plowing and harrowing on water infiltration in an Alfisol.

hectares per year. Consequently, the per capita arable land area of
0.31 hectare in 1975 is expected to decrease to 0.15 hectare by the end of
the twentieth century.

Mismanagement-caused productivity declines on existing arable lands
is one of the factors responsible for the high rate of forest conversion.
Soil properties under primary or a high rainforest cover are generally
favorable, especially the soil physical properties. Mechanized deforestation and use of motorized equipment for soil and crop management lead to soil compaction, decrease in total and macroporosity, and
decline in infiltration capacity. Experiments conducted on an Alfisol soil at
the International Institute of Tropical Agriculture (IITA) in Ibadan,
Nigeria, showed drastic increase in soil bulk density and penetrometer
resistance and decrease in infiltration capacity.

There are also measurable differences in these properties among
different methods of land clearing.[10] Alterations in soil properties are
more drastic for mechanized than manual clearing and traditional
farming systems. The available results show significant reductions in
infiltration rates when the forest vegetation is removed by heavy
equipment. For example, the data in Figure 9–2 show that the cumulative water intake at 180 minutes decreased from 600 cm under forest to
40 cm one year after clearing and to 35 cm two years after vegetation

TABLE 9–3   EFFECTS OF LAND CLEARING METHODS ON WATER
INFILTRATION RATE OF AN ALFISOL SOIL IN SOUTHERN NIGERIA
(MEASURED BY THE DOUBLE-RING INFILTROMETER)

Land clearing method	Accumulative water intake cm/3 hr.		Equilibrium infiltration rate cm/ha	
	Before clearing	One year after	Before clearing	One year after
Manual clearing	465	96	146	18
Shear blade	416	42	102	3
Tree pusher-root rake	193	55	54	11

removal. The data in Table 9–3 also show that in comparison with traditional farming, the decline in accumulative infiltration was greatest in tree pusher clearing and next greatest in conventional plowing. Considerable disturbance of the soil surface by the tree pusher is a major factor responsible for deterioration in soil structure.[11] Two of the many interacting factors responsible for rapid deterioration in soil physical properties are a decline in soil organic matter content and a decrease in activity of soil fauna (e.g., earthworms). Indiscriminate use of some agrochemicals have drastic adverse effects on soil fauna. In addition to decline in soil physical properties, deforestation and burning also cause depletion of the soil fertility and lead to nutrient imbalance.[12]

SOIL EROSION

According to some estimates, the present rate of soil loss through erosion alone is about a half-ton of soil for every inhabitant of the Earth. The surface soil of the world is being depleted at the rate of 0.7 percent per year. By the year 2000, the amount of topsoil available per person will decline by 32 percent.[13] The rate of soil degradation is accentuated by accelerated soil erosion—a severe hazard in humid, subhumid, and semiarid tropics. Soil erosion has supposedly caused irreversible degradation of some 430 million hectares of once fertile land since the dawn of settled agriculture.[14] Deforestation in regions of high potential for climatic-induced erosion is a reason for accelerated soil erosion. Accelerated erosion is particularly a severe problem in tropical Africa[15] and Latin America.[16] More than half of the estimated 11 million hectares of forests cleared annually can be attributed to the need to replace degraded agricultural soils.[17]

The available data on the extent of soil erosion in the regions of tropical rainforest are obsolete and often unreliable. Long-term, continuous, and reliable records of streamflow and sediment transport are not available for major rivers of the regions. Although the annual rate of conversion of tropical rainforest has increased over the past quarter of a century, the available records on sediment transport do not permit development of a meaningful cause–effect relationship. For example, Fournier's estimates of erosion rates made in 1960 are far lower than those estimated in mid-1980s (Table 9–4). Erosion rates have increased by several orders of magnitude over the quarter of a century ending in 1985. The economic costs of such erosion are hard to estimate. Officials in Nigeria estimate that controlling gully erosion would cost about 2.2 billion U.S. dollars.

TABLE 9–4   COMPARATIVE EROSION RATES IN THE HUMID TROPICS OVER 25 YEARS (FROM 1960 TO 1985)[a]

Region	Erosion Rate (t/ha/yr)	
	1960	1985
Amazon Basin	6–10	18–190
Congo Basin	0.5–1	1–10
West Africa	0.5–1	2–7
Madagascar	10–20	25–40
Southern India	10–20	40–100
Burma	20	139
Java	20	43

[a] Collated from Fournier (1960), Strakhov (1967), Walling (1964), and Pimental et. al. (1986). The 1960 data derived from Fournier (1960) are based on few measurements and may have over-estimated the sediment transport in African rivers.

Based on current and worldwide soil loss, erosion-caused degradation of arable land may depress cropland productivity by as much as 15 percent to 30 percent by the year 2000.[18] Principal reasons for this drastic decline in crop productivity are reduction in water availability to plants and frequent drought stress,[19] nutrient deficiency and depletion of soil organic matter content, and deterioration in soil structure.[20] Laterization, enhanced by the exposure of a hardened layer to the surface by accelerated erosion, is a major problem in semiarid and arid regions of tropical Africa. Once this soil is exposed to the surface, cropland productivity is virtually reduced to zero.

Prevention is better than attempting to control soil erosion. Erosion preventive techniques include those based on vegetation cover. A tropical forest decreases soil erosion through its influence on the following processes.

1. Preventing Raindrop Impact: The multistory canopy of a mature tropical rainforest comprises five strata at different heights above the soil surface. In addition, the soil surface is buffered against the raindrop impact by a thick layer of leaf litter and partly decomposed biomass.[21]

2. Increasing Infiltration Rate: Favorable soil structure and high activity of soil fauna are responsible for high infiltration rates commonly observed in forested soils. Consequently, surface runoff is generally lower from forested than cleared lands.[22]

3. Low Soil Erodibility: Highly structured soil with a relatively high proportion of macropores and a predominance of water-stable, large aggregates render the soil under forest canopy more resistant to raindrop splash and the abrasive forces of running water than cleared and cultivated soil.

4. Reduction in Runoff Velocity: Rough soil surfaces and a thick root mat just beneath the leaf litter offer resistance to runoff velocity and drastically reduce sediment transport capacity.

The forest canopy, therefore, prevents erosion by controlling both the origin and transport of sediments. Deforestation changes this harmonious balance and increases soil erosion by increasing raindrop impact, soil erodibility, and the amount and velocity of surface runoff. The data in Table 9–5 show drastic increases in runoff and erosion caused by deforestation. Methods of deforestation also have an important effect on the magnitude of soil erosion. In general, machine clearing causes more severe erosion than manual clearing.

TABLE 9–5   RUNOFF AND EROSION CAUSED BY LAND CLEARING FOR CASSAVA GROWING ON ALFISOL SOILS IN SOUTHERN NIGERIA[a]

Land Clearing	Runoff		Soil erosion
Methods	mm	% of Rainfall	(kg/ha)
Tree pusher-root rake	18.2	44.8	4474
Shear blade	5.1	12.5	74
Manual	0.04	1.0	Traces
Natural forest (uncleared)	0.00	0.00	0.00

[a] Rainfall of 24 September 1979 = 40.6 mm.

Careful consideration should be given to the use of heavy machinery for removing a tropical rainforest. Vast areas of tropical rainforests are presently being cleared with heavy machinery in west and central Africa, Sumatra and adjacent regions in southeast Asia, the Amazon Basin, and Central America. In view of the susceptibility of these soils to compaction, erosion, and erosion-induced degradation, it is important to realize that the bigger and faster the machine, the more adverse its effects on soil erosion and environments (Table 9–5). The objective should not be to clear land as quickly and as cheaply as possible, but to clear it in such a way as to preserve the ecological balance. It is short-sighted to use certain machines in favor of short-term gains at the expense of irreversible damages to the soil and environments. If the existing forest cover must be removed, clearing should be done in such a way that the delicate balance between soil and climate is not drastically disturbed. The indiscriminate use of heavy power equipment can cause havoc to soil and environments. Equally important is the consideration on the part of planners and researchers to develop technologies for restoring economic productivity from lands that have been degraded due to past mismanagement. Bringing degraded lands under production would reduce the need to convert forested lands to cropland.

Soil erosion control on deforested lands is also achieved more effectively by biological rather than engineering techniques. Biological measures of erosion control are based on the principle of minimizing runoff, sediment origin, and sediment transport. Engineering techniques, aimed at safe disposal of surface runoff, are expensive, require regular maintenance, and are often applied too late. Terraced farming is an ancient practice and has been used for millenia in steep land management in Asia and South America. It is a cultural tradition of many societies. The Incas terraced hill slopes for water management. Rice paddies have been developed on steep hill slopes in southeast Asia for thousands of years. Terracing decreases rill and gully erosion but does not effectively reduce splash and interrill erosion. Splash erosion is

TABLE 9–6   EFFECT OF CULTIVATION TREATMENTS ON RUNOFF AND EROSION WITH A TROPICAL ALFISOL SOIL[a]

Treatment	Runoff (mm)	Soil Erosion (t/ha)
Terraced-plowed	18.1	0.7
Unterraced–no-till	18.8	2.3

[a] Single rainstorm event recorded at IITA in July 1981, (Lal, 1984).

curtailed by the ground cover and crop residue mulch. The data in Table 9–6, for example, show the greater effectiveness of no-till agriculture than terracing in reducing runoff and soil erosion. Extensive gully erosion occurs whenever terraces fail.

The decline in soil productivity by accelerated erosion is a serious concern in Africa and elsewhere in the tropics, where the loss of productive topsoil cannot be compensated for by inputs (e.g., supplemental irrigation and chemical fertilizers). Reduction in plant-available water reserves is the most serious aspect of erosion-caused degradation. Even where it is useful, supplemental irrigation is often not available. High runoff losses cause water shortages even in years with good rainfall. For most soils with a shallow effective rooting depth and where the nutrients are confined to the surface soil horizons, the acceptable level of annual soil erosion is less than one ton per hectare (t/ha) and may be as low as 0.5 t/ha. The observed soil loss from arable lands often exceeds the tolerable level of soil erosion by several orders of magnitude. There is often an exponential decline in crop yields with increases in the depth of topsoil lost by erosion.[23] The paucity of research data and the lack of standardized terminology have led to the common belief that tropical rainforest ecosystems are fragile and that tropical soils are easily degraded. It is only solid scientific data based on well-designed field experiments that will replace myths with facts. It is important to understand processes and factors of soil degradation so that the trends can be curtailed and even reversed.

CLIMATE

Another major consequence of forest conversion is its effects on local, regional, and global climate. Locally, deforestation causes drastic alterations in air and soil temperature, relative humidity, and solar radiation received at the ground level. Of the total rainfall received at the canopy level, about 10 percent is intercepted/retained by the cover.[24]

Experiments currently in progress at the joint International Institute of Tropical Agriculture and United Nations University Project at Okomu, near Benin City, Nigeria, shows that 25 times as much sunlight reached the ground level in a cleared area as it did beneath an intact forest canopy. Consequently, soil and air temperatures for cleared and forested areas are drastically different. Data from the Nigeria Oil Palm Plantation in Okomu show that at the 5-cm depth the average soil temperature under the forest was 3 degrees centigrade lower than in the open (Figure 9–3),[25] and the maximum soil temperature under the forest canopy was generally lower by 6 degrees centigrade compared to

EFFECTS OF DEFORESTATION ON SOIL PROPERTIES

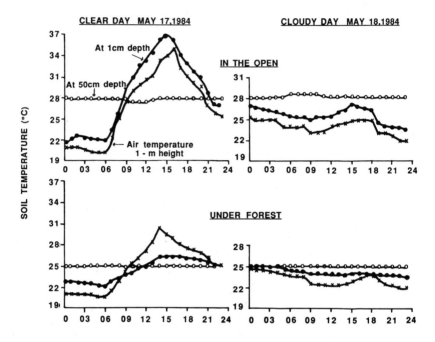

FIGURE 9–3 Diurnal fluctuations in soil and air temperature under forest and on cleared land. [Note: B.S. Ghuman and R. Lal, 1986, "Effects of Deforestation on Soil Properties and Microclimate of a High Rainforest in Southern Nigeria," *The Geophysiology of Amazonia*, R.E. Dickinson (ed.), John Wiley & Sons, New York, pp. 225–244.]

the cleared area. Similarly, the maximum air temperature measured at 1 m height above the ground level was lower under forest than in the open by about 3 to 6 degrees centigrade. Even during the dry season in November and December, the relative humidity was always higher under forest than in the open even a few meters away. Consequently the pan evaporation inside the forest was only 25 percent of that in the cleared area. There are, therefore, significant differences in the microclimate of the forested and cleared areas.

On a regional scale, deforestation influences water balance. Deforestation causes an increase in streamflow, and the total water yield from a watershed is dramatically increased. Lal[26] reported a steady increase in total water yield from the cleared watershed with time after deforestation (Figure 9–4). The interflow component increased from an unmeasurable trace in January during the dry season of 1978 before

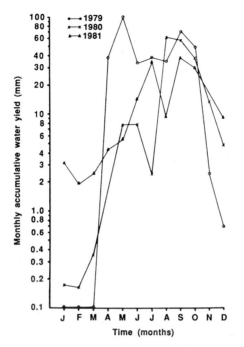

FIGURE 9–4   Effects of deforestation on total water yield from an agricultural watershed in Nigeria. [Note: R. Lal, 1986, "Conversion of Tropical Rainforest: Agronomic Potential and Ecological Problems," *Adv. Agron.*, Vol. 37.]

clearing to 0.1, 0.18, and 3.2 mm/month after clearing in January 1979, 1980, and 1981, respectively. The progressive increase in interflow is due to a gradual decrease in forest regrowth and in the lack of utilization of subsoil water by the shallow-rooted annuals that succeed the forest cover. Increase in streamflow following deforestation has also been reported from east Africa by Pereira,[27] from New South Wales, Australia,[28] and from the Amazon. Gentry and Lopez-Parodi[29] reported a pronounced and statistically significant increase in the Amazon flow measured at Iquitos between 1962 and 1978, indicating an increase in the runoff of water from the upper Amazon. This increase in flow has been linked to deforestation[30] and to possible irreversible changes in the Amazon water balance.

Important global effects of deforestation are related to the role of forests in recycling carbon and water vapors. Forests play a significant role in regulating the concentration of CO_2 in the atmosphere. An increase in the concentration of greenhouse gases in the atmosphere is an important global effect of deforestation. There has been an overall

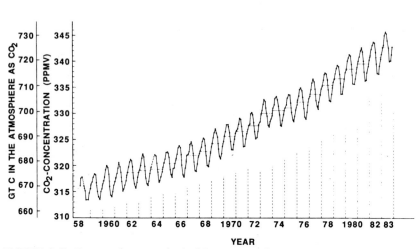

FIGURE 9–5   Increase in atmospheric CO_2 concentration at Mauna Loa, Hawaii, USA, 1958–1983. (GT C = gigatons of carbon) [Note: W. Bach, 1986, "Trace Gases and Their Influence on Climate," *Natural Resources and Development*, Vol. 24, pp. 90–124.]

and steady increase in CO_2 concentration in the atmosphere from about 316 ppm in 1958 to 346 ppm in 1983 (Figure 9–5).[31] This increase is primarily due to burning of fossil fuel, but also is the result of deforestation.[32] It is estimated that doubling the atmospheric concentration of carbon dioxide would raise the mean temperature of the global surface by 2.8 degrees centigrade.[33]

Tropical forests affect the regulation of the CO_2 concentration in the atmosphere by three mechanisms: (1) through major river systems draining tropical rainforests by which a considerable amount of carbon is transported to the oceans; (2) through resynthesis of carbon within the tropical biomass; and (3) through storage in soil as soil organic matter. These recycling mechanisms are disrupted by deforestation. Deforestation releases the vast quantity of carbon stored in the biota through burning and decomposition.[34] Deforestation also increases CO_2 release from the soil by increasing the rate of mineralization.

The most direct effect of deforestation on global climate is through increased albedo (which is the ratio of light received by a surface to the light energy reflected from it). Although deforestation increases albedo locally from 0.07 to 0.25,[35] it is difficult to estimate global change in albedo by deforestation in the tropics. Computer simulation models, based on strong assumptions, indicate drastic changes in global climate if these assumptions were to come true.[36] A considerable amount of well-planned basic research is needed to replace myths by facts regarding the warming trends presumably related to tropical deforestation.

GUIDELINES AND POLICY ON DEFORESTATION

While the need for additional forest conversion should be minimized by development and adaptation of improved cropping and land management systems for sustained and increased production on existing farmlands, some countries have no choice but to expand agricultural activities to forested tropical lands. If forest conversion is inevitable, it is important to adopt those methods of land clearing and postclearing land use that would least disrupt the biophysical, hydrological, and chemical balance of tropical soils. IITA and many other research institutions have made considerable progress in developing ecologically compatible land clearing and management technologies. The choice of appropriate technology of land clearing, however, depends on many factors, including the intended land use. If mechanized land clearing must be done, the choice of appropriate equipment is essential.

One of the severe constraints to provide guidelines for environmental management is the paucity of research data obtained from properly designed and adequately equipped field experiments where systematic observations have been made regarding the ecological impact of deforestation. For example, the effects of deforestation on hydrology, microclimate, soil, and biota have been measured and related to crop production for few locations. And yet, vast areas of primary and secondary forests are rapidly being converted for agricultural use. It is thus necessary to conduct field experiments on different soils and ecological environments to bridge the knowledge gap and validate existing data. Data from such studies could provide answers to the questions listed in Figure 9–6 in adopting an ecological land use in the tropics. Prior to implementing the change in land use through deforestation and subsequent management, it is important that baseline data be obtained for soil, hydrology, microclimate, vegetation, and other biotic factors. These factors should also be continuously monitored for at least five years during the period of changed land use. Simultaneous observations are required for agricultural output to relate the soil's productivity for a range of management systems to the environmental factors.

In view of the current food shortages in tropical Africa, the question most often asked is whether soil productivity in tropical environments can be sustained with intensive and continuous farming. The research data from IITA and other organizations indicate that most tropical soils can be intensively cultivated and produce high and sustained yields by adapting an ecological approach to agriculture.[37] In this connection, land clearing techniques play an important role. The effects of improper land clearing methods are observed even eight to ten years after the land

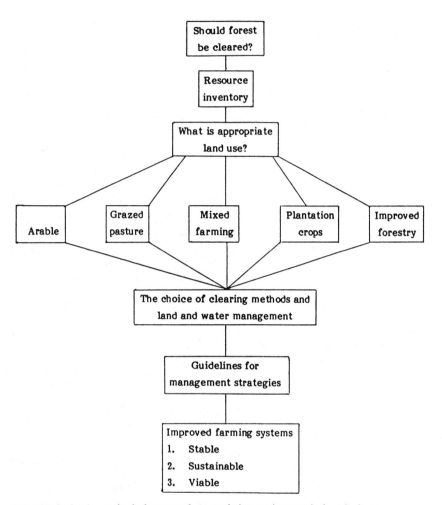

FIGURE 9–6   An ecological approach towards improving tropical agriculture.

has been cleared and especially when the overall soil fertility has drastically declined. Adapting a land use system that may produce 60 percent to 80 percent of maximum returns but without causing environmental degradation is a better choice than land use systems that bring high short-term returns but severely degrade the resource base.

NOTES

[1] P. Buringh, H.D.J. Van Heemst, and G.J. Staring, 1975, "Computation of the Absolute Maximum Food Production in the World," Afd. Bodemkunde en Geologie Publ. No. 598; M. Chou, D.P. Harmon, Jr., H. Kahn, and S.H. Wittwer, 1977, *World Food Prospects and*

Agricultural Potential, Praeger Publishers, New York; C.T. de Wit, 1967, "Photosynthesis: Its Relationship to Overpopulation," in *Harvesting the Sun—Photosynthesis in Plant Life,* A. San Pietro, F.A. Greer, and T.J. Army (eds.), Academic Press, New York; A.H. Kassam and J.M. Kowal, 1973, "Crop Productivity in Savanna and Rainforest Zones in Nigeria," *Savanna,* Vol. 2, pp. 39–49.

[2] P.R. Crosson and K.D. Frederick, 1977, *The World Food Situation,* Resources for the Future, Inc., Washington, DC, 230 pp.

[3] E. Salati and P.B. Vose, 1982, "Deforestation: Environment Research and Management Priorities for the 1980s," The Royal Swedish Academy of Sciences, Stockholm, November 23–26.

[4] M.T. El-Ashry, 1986, "Soil and Water Conservation in Relation to Food Production in Africa," *Proc. 41st Annual Meeting of the Soil Conservation Society of America,* Winston-Salem, North Carolina, August 3–6.

[5] J.F. Richards, 1984, "Global Patterns of Land Conversion," *Environment,* Vol. 26, pp. 6–13, 34–38.

[6] F.H. Bauer, 1977, "Cropping in North Australia: Anatomy of Success and Failure," *Proc. First NARU Seminar,* Darwin, Australia, pp. 24–27.

[7] E. El-Hinnawi and M. Hashmi, 1982, *Environmental Issues,* UNEP, Tycooly International Publishing Ltd., Dublin, Ireland, 236 pp.

[8] V.A. Kovda, 1983, "Loss of Productivity of Land Due to Salinization," *Ambio,* Vol. 12, pp. 91–93.

[9] R. Dudal, 1981, "An Evaluation of Conservation Needs," in *Soil Conservation Problems and Prospects,* R.P.C. Horgan (ed.), John Wiley & Sons, Chichester, UK.

[10] N.R. Hulugalle, R. Lal, and C.H.H. ter Kuile, 1984, "Soil Physical Changes and Crop Root Growth Following Different Methods of Land Clearing in Western Nigeria," *Soil Sci.,* Vol. 138, pp. 172–179.

[11] C.E. Seubert, P.A. Sanchez, and C. Valvarde, 1977, "Effects of Land Clearing Methods on Soil Properties of an Ultisol and Crop Performance in the Amazon Jungle of Peru," *Trop. Agric.,* Vol. 54, pp. 307–321.

[12] Seubert et al., 1977, *op. cit.*

[13] L.R. Brown and E.C. Wolf, 1984, "Soil Erosion: Quiet Crisis in the World Economy," Worldwatch Paper 60, Worldwatch Institute, Washington, DC.

[14] Kovda, 1983, *op. cit.*

[15] R. Lal, 1981b, "Soil Erosion Problems on an Alfisol in Western Nigeria. VI. Effects of Erosion on Experimental Plots," *Geoderma,* Vol. 25, pp. 215–230.

[16] R. Lal, 1984. "Soil Erosion from Tropical Arable Land and Its Control," *Adv. Agron.,* Vol. 37, pp. 183–248.

[17] D. Pimental, J. Allen, A. Beers, L. Guinand, R. Linder, P. McLaughlin, B. Meer, D. Musonda, D. Perdue, S. Poisson, S. Siebert, K. Stoner, R. Salazar, and A. Hawkins, 1987, "World Agriculture and Soil Erosion," *BioScience,* Vol. 37, pp. 277–283.

[18] R. Lal, 1987, "Effects of Soil Erosion on Crop Productivity," *CRC Critical Reviews in Plant Science,* Vol. 5, pp. 303–367; M.M. Shah, G. Fischer, G.M. Higgins, A.H. Kassam, and L. Naiken, 1985, "People, Land, and Food Production—Potentials in the Developing World," International Institute for Applied Systems Analysis, CP-85-11, Laxenburg, Austria; Pimental et al., 1987, *op. cit.*

[19] H.A. Elwell, 1985, "An Assessment of Soil Erosion in Zimbabwe," *Zimbabwe Science News,* Vol. 19, pp. 27–31; R. Lal, 1976, "Soil Erosion Problems on an Alfisol in Western Nigeria and Their Control," ITA Monograph 1, 208 pp.

[20] Lal, 1987, *op. cit.*

[21] R. Lal, 1986, "Conversion of Tropical Rainforest: Agronomic Potential and Ecological

Problems," *Adv. Agron.,* Vol. 37.

[22] R. Lal and B.T. Kang, 1982, "Management of Organic Matter in the Soils of Tropics and Subtropics," XII Congress of the ISSS, New Delhi, India.

[23] Lal, 1981b, *op. cit.*

[24] T.L. Lawson, R. Lal, and K. Oduro-Afriye, 1981, "Rainfall Distribution and Microclimatic Changes Over a Cleared Watershed," in *Tropical Agricultural Hydrology,* R. Lal and E.W. Russell (eds.), John Wiley & Sons, Chichester, UK, pp. 141–152.

[25] B.S. Ghuman and R. Lal, 1986, "Effects of Deforestation on Soil Properties and Microclimate of a High Rainforest in Southern Nigeria," *The Geophysiology of Amazonia,* R.E. Dickinson (ed.), John Wiley & Sons, New York, pp. 225–244.

[26] Lal, 1986, *op. cit.*

[27] H.C. Pereira, 1973, *Land Use and Water Resources,* Cambridge University Press, Cambridge, UK, 246 pp.

[28] J.A.N. Brown, 1972, "Hydrological Effects of a Bushfire in a Catchment in Southeastern NSW," *Journal of Hydrology,* Vol. 15, pp. 77–96.

[29] A.H. Gentry and J. Lopez-Parodi, 1980, "Deforestation and Increased Flooding of the Upper Amazon," *Science,* Vol. 210, pp. 1354–1356; A.H. Gentry and J. Lopez-Parodi, 1982, "Deforestation and Increased Flooding of the Upper Amazon," *Science,* Vol. 215, p. 426.

[30] H. Sidi, 1975, "Tropical Rivers as Expressions of Their Terrestrial Environments," in *Tropical Ecological Systems,* F.B. Golley and E. Medina (eds.), Springer-Verlag, Berlin, pp. 275–288.

[31] W. Bach, 1986, "Trace Gases and Their Influence on Climate," *Natural Resources and Development,* Vol. 24, pp. 90–124.

[32] G.M. Woodwell, R.A. Houghton, J.E. Hobbie, J.M. Melillo, B. Moore, B.J. Peterson, and G.R. Shaver, 1983, "Global Deforestation: Contribution to Atmospheric Carbon Dioxide," *Science,* Vol. 222, pp. 1081–1086.

[33] Bach, 1986, *op. cit.*

[34] G.M. Woodwell, 1978, "The Carbon Dioxide Question," *Sci. Amer.,* Vol. 238, pp. 34–43; G.M. Woodwell, J.E. Hobbie, R.A. Houghton, J.M. Melillo, B. Moore, C.A. Palm, B.J. Peterson, and G.R. Shaver, 1982, "Biotic Contributions to the Global Carbon Cycle," Woods Hole Conf., Massachusetts, February 10–12; G.M. Woodwell, R.H. Whittaker, W.A. Reiners, E.G. Likens, C.C. Delwiche, and D.B. Botkin, 1978, "The Biota and the World Carbon Budget," *Science,* Vol. 199, pp. 141–146.

[35] G.L. Potter, H.W. Ellraesser, M.C. MacCracken, and F.M. Luther, 1975, "Possible Climate Effects of Tropical Deforestation," *Nature,* Vol. 258, pp. 697–698.

[36] J. Hansen, A. Lacis, D. Rind, G. Russell, P. Stone, I. Fung, R. Ruedy, and J. Lerner, 1984, "Climate Sensitivity: Analysis of Feedback Mechanisms," in *Climate Processes and Climate Sensitivity,* Geo. Monogr. 29, American Geophysical Union, pp. 13–163.

[37] R. Lal and B.T. Kang, 1982, "Management of Organic Matter in the Soils of Tropics and Subtropics," XII Congress of the ISSS, New Delhi, India; P. Sanchez, D.E. Bandy, J.H. Villachica, and J.J. Nicholoides, 1982, "Amazon Basin Soils: Management for Continuous Crop Production," *Science,* Vol. 216.

EDITORS' INTRODUCTION TO:
G. CARRIER
Nuclear Winter, Current Understanding

One theme of this book is that there are many links between technology and the environment. Nowhere are the deleterious aspects of these links more dramatically illustrated than in the case of the nuclear winter hypothesis—the hypothesis that the detonation of many nuclear weapons would force dust and smoke into the stratosphere, darkening the skies to the extent that a human-caused winter would occur. This possibility which has been recognized only in the last several years, was impossible when George Perkins Marsh wrote *Man and Nature* in the nineteenth century, and was inconceivable at the time of writing of the 1955 conference, *Man's Role in Changing the Face of the Earth.* No discussion of man's role in changing the global environment can ignore this topic today. The events that might take place leading to a nuclear winter are at a scale similar to those introduced in Chapter 6 by Dr. Loucks, but they would occur far more quickly and the effects appear to be more far-reaching. In this chapter, Dr. Carrier explains the nuclear winter hypothesis, discusses the critical issues, and points out a change that has occurred in research arising from the development of the nuclear winter hypothesis.

Dr. Carrier, of Harvard University, has participated in a U. S. study of the effects of nuclear winter.

10 NUCLEAR WINTER, CURRENT UNDERSTANDING

GEORGE F. CARRIER
Harvard University
Cambridge, Massachusetts

INTRODUCTION

Among the most serious effects that we could impose on our environment are those that would follow a major exchange of nuclear weapons. Included among the widely known immediate consequences of such an event are an enormous loss of life; the devastation of cities; and the elimination of fuel supplies, transportation systems, communications systems, and, for all practical purposes, any access to medical aid, food supplies, or other badly needed resources. These particular environmental changes would be confined primarily to that portion of the globe occupied by the contending nations and their neighbors, presumably the northern temperate zone, and it is very unlikely that these environmental changes could be attenuated in any timely manner.

On a somewhat larger time scale, however, further degradation of the environment might occur. Of the many contaminants that nuclear explosions would inject into the atmosphere, there are three in particular that could be of grave concern. These are: 1) The nitric oxide that is formed wherever such an explosion occurs would be incorporated into the fireball and would be deposited high in the atmosphere. For yields typical of the weapons in present-day arsenals, much of that nitric oxide would penetrate into the stratosphere and might be depleted on a time scale of a year or two; 2) The dust generated by a ground-level explosion would also be incorporated into the fireball of the weapon, and those particles that were about one micron or less in size could also be lifted high into the atmosphere and could remain there for times on the order of a year; and 3) Weapons allocated to targets that are in or near large cities would, in all likelihood, initiate massive fires, and it is to be expected that large amounts of combustible material would be consumed and that large amounts of smoke would be carried high into the troposphere by the buoyant fire plume.

The amounts of these materials that would take part in the foregoing phenomena are uncertain indeed, and the time during which they would continue to reside in the atmosphere is also uncertain. Nevertheless, there is a clear possibility that the amounts, durations, and atmospheric responses would be so large as to be of great concern even in regions remote from the portion of the globe in which the weapons might be detonated. However, before turning the focus of our attention to the sources and implications of these contaminants and their spatial, temporal, and quantitative uncertainties, one observation deserves emphasis. Many of the undesirable impacts that humans exert on their environment are the by-products of otherwise supposedly beneficial activities. Unwanted emissions from factories, automobiles, and power production plants are obvious cases in point. Our task in each such case is a difficult engineering and organizational problem wherein one must retain the beneficial aspect of the activity while eliminating the undesirable impact. The environmental problem associated with nuclear weapon usage is not of this kind. There are no beneficial contributions to anyone that attend an exchange of nuclear weapons. As such, we have only to avoid the unwanted aspects of their use. But, of course, there will be no detrimental effects if nuclear weapons are not used. Our only requirement, difficult to ensure though it may be, is the avoidance of their use. For as long as nations successfully avoid such use, the environmental impact of nuclear weapons will remain as hypothetical as it is now.

Detailed accounts of the estimation of the amounts of dust, nitric oxide, and smoke that would accompany a major nuclear exchange have been recorded in the reports of several major studies.[1] Recent ongoing audits of the fuel that would be exposed in devastated parts of cities do suggest that the fuel amounts cited in earlier reports are too large. Note, however, that these audits are not yet comprehensive and that the uncertainty range of these estimates has not decreased significantly. The technology of remote sensing and information systems could be employed to improve these estimates. Finer details can be found in the references that make up the bibliographies of those reports. We confine ourselves here to a more concise and qualitative account of that contamination and its consequences. What follows, in part, is identical with much of an earlier account[2] except for minor modifications that permit some attention to the biological aspects of the potential environmental impact of a nuclear exchange.

Any assessment of the environmental threat posed by the use of nuclear weapons depends on: quantitative estimates of the numbers of weapons that might be used against the various types of targets; the yields of those weapons; the amounts of contaminant (dust, NO_x, and smoke) that would be produced; and their temporal, lateral, and vertical distribution in the atmosphere. The assessment also depends on calculations of the atmosphere's response to the presence of those contaminants—that is, the evolving temperature distributions, motions, and precipitation patterns. However, any attempt to make such calculations with today's knowledge and understanding of many of the pertinent phenomena is severely impeded by a large number of major uncertainties.

To understand the extent of those uncertainties and their role in attempts to estimate the degree of the atmospheric degradation that would follow a nuclear war, it may be useful to consider the ways in which uncertainties would be compounded in the events that accompany a major weapons exchange. There is an uncertainty associated with each of these events, in particular, the nuclear weapons scenario; the production of smoke and its injection into the atmosphere; and the atmospheric response to contaminants on the scale envisioned.

The first set of uncertainties cannot be removed. One cannot know in advance of the nuclear phase of the postulated hostilities, for example, the number of weapons that any combatant would actually use, the distributions of targets against which those weapons would be directed, or the number of those weapons that would reach their targets and detonate successfully. One can postulate, however, a plausible hypothetical exchange and the time of year at which it is to occur and then try to estimate the atmospheric degradation caused by that exchange.

In contrast, the second set of uncertainties can be estimated by a process illustrated in the following example. A moderate amount of observational data exists concerning large fires in irregularly littered solid fuel, such as would be found in a city in the aftermath of a nuclear explosion.[3] These data suggest that between 2 and 6 percent of the fuel actually burned would be converted to smoke. The data do not imply that the fraction converted to smoke cannot be larger; in fact, if the fuel largely consisted of synthetic organic materials, it is known that the smoke production could be much larger than 6 percent. Alternatively, distributions of fuel and air supply are possible for which smoke production can be much lower than 2 percent. Nevertheless, no competing arguments seem to refute the plausibility of the 2 to 6 percent range, which we will refer to as the uncertainty range. Furthermore, because the largest number in this range is three times the smallest, we will say that the uncertainty factor is three. Recent small-scale experiments indicate that, in generously ventilated configurations, structural wood is likely to produce less smoke than was cited in the earlier reports. One should not read too much into this finding, however, because, there is no authenticated methodology by which one can extrapolate this finding to configurations of large size; neither is there any set of criteria for quantifying the dependence of the degree of ventilation on configurational characteristics. Of greater concern is the observation that the fuel in devastated cities will be interspersed with noncombustible structural materials such as concrete, steel, and plaster. No effort has yet been initiated to gather information concerning the effects of such rubble on either the completeness of burning or the production of smoke in fires of any size. This observation implies that the uncertainties in smoke production now loom larger than they did in earlier assessments and that those uncertainties cannot be diminished until more knowledge and understanding of fires in heterogeneous mixes of fuel and rubble have been acquired.

The size of the smoke particles and the height to which they rise in the atmosphere are important because a given mass of larger particles will impede the passage of solar radiation less effectively than will the same mass of smaller particles. Furthermore, larger particles and those injected at lower altitudes will be more rapidly removed. To estimate the amount of submicron smoke that would rise above this altitude requires quantitative estimates for the amount of fuel in the regions where burning would occur (the fuel supply), the fraction of the fuel supply that would burn, and the fraction of the fuel burned that would emerge as smoke. It also requires estimates of the fraction of smoke particles that would remain at submicron size during their ascent in the smoke plume, despite their coagulation and incorporation into moisture condensation droplets that would form at higher altitudes. I would assert that the uncertainty factor for the fuel supply is not less than two, that the uncertainty factor in the fraction burned is not less than two, that the uncertainty factor in the fraction of fuel burned that becomes smoke is not less than three, and that the uncertainty factor in the nonagglomerated fraction of the total smoke is not less than three. Thus, the composite uncertainty factor associated with this second set of uncertainties is not less than 36. Still other uncertainties are not included in this estimate, the height of the smoke plume (and hence the height at which the smoke is injected); the optical properties of the smoke (the more opaque the smoke, the more it obscures sunlight); and changes in the smoke's optical properties over a period of time.

In the National Research Council's recent report, *The Effects on the Atmosphere of a Major Nuclear Exchange*,[4] a particular scenario for a nuclear exchange in which somewhat less than half (6,500 megatons) of the world's arsenal is expended was adopted as a baseline case. In other words, this scenario was used to illustrate the process of estimating the atmospheric effects of a nuclear exchange. No pretense is made that this is a "most likely enchange." It is merely a plausible assumption whose estimated consequences can give some guidance regarding possible atmospheric degradation. For this assumed nuclear exchange, the amount of submicron smoke that would survive the ascent in the fire plume is between 20 million tons and 650 million tons. Those numbers are generally consistent with the uncertainty factors given earlier. (Some small and unimportant discrepancies arise, however, because this discussion is a highly simplified recasting of the National Research Council's report.) In that report, for purposes of inquiry, the investigators chose to assume that 180 million tons of submicron smoke were injected at altitude (4 to 9 kilometers) in the atmosphere.

The third set of uncertainties—those dealing with the atmosphere's response—complicates the continuing analysis. Atmospheric scientists have at their disposal a variety of computational procedures designed to reproduce some of the large-scale features of the atmosphere's response to various conditions. These mathematical models are designed to deal with relatively small variations in normal atmospheric behavior. The modeling of small-scale processes (such as precipitation, particle removal, the mixing effects of turbulence, to name a few) are chosen and refined so that they satisfactorily represent the large-scale consequences of those small-scale processes. They are satisfactory because they are designed, insofar as possible, to conform to the observed behavior of the normal atmosphere. In the phenomena of interest here, however, the conditions include strong and abnormal temperature gradients and millions of tons of smoke particles at an altitude of several kilometers, yet there are no observations of an atmosphere in such a severely modified state that could be used to validate the models.

Accordingly, it is especially difficult to assess quantitatively the inaccuracies that may result when making calculations with existing mathematical models. Clearly, it is eminently sensible to use these models to estimate the order of magnitude of the temperature change caused by smoke, but the results can only be regarded as suggestive. They are definitely not predictions. There is an ongoing evolution of General Circulation Models (GCMs) used to estimate atmospheric response to the presence of smoke. Much of that evolution has rested on the incorporation of parameterization of processes that take place on scales smaller than the grid size of the overall model, and, it is important to establish the credibility of those parameterizations. It is apparent to the writer that this establishment of credibility is a central part of few ongoing GCM studies. Some changes in the central foci of the research arising from the nuclear winter hypothesis are needed.

The fire phenomenology associated with heterogeneous distributions of solid fuel (with and without rubble) must be quantified. I think this can be done in a satisfactory way only if a sequence of experimental studies at laboratory scale is conducted. The sequence should proceed from relatively inexpensive small-scale fundamental studies toward large-scale experiments and possibly culminate in a small number of rather large burns. All of this should be interwoven with attempts to evolve a model for the fire dynamics that is consistent with fundamental principles, mechanistic hypotheses, and observational results. In particular, the models must include mechanisms that allow the determination of the fuel consumption rate and completeness of burning instead of (currently used) postulated values for these aspects of the phenomenon.

Another indispensable set of results is that associated with the smoke removal processes. There must be an experimental effort to provide a quantitative characterization of the effect of candidate processes (dry coagulation, interaction with moisture, others?) on the smoke distribution and morphology, and there must be an incorporation of such results into the GCM parameterizations.

A variety of computational models have been applied to the baseline war scenario described earlier and to some variations on that case.[5] The results must be interpreted with care, but they boil down to the suggestion that the atmospheric response to smoke injection on the order of 180 million tons, as estimated using currently available computational models, would include temperature changes that could be of serious concern. In particular, the results suggest that for an exchange occurring in the summer, with all of the foregoing quantitative uncertainty, intermittent temperature drops in the northern temperature zone could be on the order of 20°C and might continue for a few weeks. Although it is even more uncertain, smaller temperature drops might occur in the tropics of the northern hemisphere. It is even possible that areas in the southern hemisphere could experience long-lasting temperature drops of several degrees.

Finally, one must correlate possible biological responses with the foregoing atmospheric modifications, uncertain as they are, both in magnitude and detail.

Recorded attempts to correlate quantitatively the modifications of crop yields and other ecological productivities with changes in temperature, precipitation, and light levels are few. Further, they rest on data consistent with naturally occurring climatic variability, which, with some certainty, does not usefully resemble the variability at issue here. Accordingly, the inferences attempted must be regarded, once again, not as predictions but only as coarsely approximate indications of the sensitivity of the biological environment to the atmospheric changes discussed earlier. With that caveat firmly in mind, we note that a model developed to characterize Canadian wheat production[6] suggests that a reduction of 3 or 4 degrees from normal in the average temperature during the growing season in the Canadian wheat-growing regions could virtually eliminate the entire crop. Another study that deals with rice production in Japan includes a model that suggests that a 3 or 4 degree reduction in that average growing season temperature could have a comparable effect.

Clearly, in view of the uncertainties emphasized here and in view of the variabilities that are so familiar to workers in atmospheric science, fire phenomena, and biology, no firm predictive conclusions can be drawn. Uncertainties, such as those discussed here, usually have two sources. One of those sources lies in the inherent variability associated with the configurations at issue. There is a broad variability in the distribution (both in size and location) of the fuel and rubble. In principle, one could know the smoke production that would accompany a given particular realization of those distributions but, in fact, one cannot predict that production because of the second source of uncertainty (i.e., the limited understanding and knowledge regarding such phenomena). Again in principle, one might systematically remove a part of the second category but not the first and, most importantly, the extent of the inherent variability cannot be known until much of the ignorance underlying the second category has been removed! One of the broad uncertainties is related to the role of moisture condensation in the fire plume on the coagulation or removal of smoke. Research on these processes has been initiated, but timely progress requires a more intense effort.

But no recent refinements of knowledge and understanding have diminished the truth of the qualitative conclusion of the National Research Council study that says, in effect: there is a clear possibility that the atmospheric modification accompanying a major nuclear exchange could be of serious concern, even in regions somewhat remote from those in which detonations would occur.

I will close this discussion with a repetition of an important part. The environmental problems associated with an exchange of nuclear weapons are, today, merely hypothetical. The best attack on that particular environmental problem is to keep them hypothetical.

NOTES

[1] P. J. Crutzen and J. W. Birks, 1982, "The Atmosphere After a Nuclear War: Twilight at Noon," *Ambio*, Vol. 11, pp. 114–125; R. P. Turco, O. B. Toon, T. P. Ackerman, J. B. Pollak, and C. Sagan, 1983, "Nuclear Winter: Global Consequences of Multiple Nuclear Explosions," *Science* (December 23), pp. 1283–1292; National Research Council, 1985, *The Effects on the Atmosphere of a Major Nuclear Exchange*, National Academy Press, Washington, DC; A. B. Pittock, T. P. Ackerman, P. J. Crutzen, M. C. MacCracken, C. S. Shapiro, and R. P. Turco, 1985, "Environmental Consequences of Nuclear War," *Scope* 28, Vol. I, *Physical and Atmospheric Effects*, John Wiley and Sons; 1985, Vol. II, *Environmental Consequences of Nuclear War, Ecological and Agricultural Effects*, M.A. Harwell and T.C. Hutchinson (eds.), John Wiley and Sons, Great Britain.

[2] G.F. Carrier, 1985, "Nuclear Winter: The State of the Science," *Issues in Science and Technology*, Vol. I, No. 2, pp. 114–117.

[3] National Research Council, 1985, *op. cit.*

[4] *Ibid*

[5] Crutzer and Berks, 1982, *op. cit.*; Turco et al., 1983, *op. cit.*; National Research Council, 1985, *op. cit.*; Pittock et al., 1985, *op. cit.*

[6] Pittock et al., 1985, *op. cit.*

SECTION II
SOME OF THE WAYS WE
CAN LEARN ABOUT
OUR GLOBAL
ENVIRONMENT

Over the past several decades, society has become increasingly aware of a variety of serious environmental problems. These problems are large in terms of geographic scale, complex, and increasingly hard to solve. As we have seen in the first section of this book, these problems include depletion of the ozone layer and the resulting potential for decline in vegetation production and a significant increase in human skin cancer; acid rain and the death of lakes in the northeastern United States, Canada, and Europe; the introduction of toxic materials into our environment, as with chemical accidents in Bhopal, India, and in the Rhine River from Switzerland to the sea; desertification and the problems of African drought and famine in Ethiopia. All are significant within a human context, but they are part of a broader picture of environmental disruption.

The need for maintaining the quality of our global environment is eloquently spelled out in the preceding section. To gain a clear picture of the state of the Earth and the dynamic fluxes that affect it, information is required of a nature and type that until the last decade was essentially unavailable. The purpose of Section II is to describe some new technologies that allow the Earth to be studied from scales ranging from cubic centimeters to entire continents.

At one extreme, in the first chapter Dr. A. Orio discusses the revolution that is taking place in the procedures for analyzing and disposing of environmental pollutants—tiny devices measuring the concentration of chemicals at the site of occurrence with precision never possible before. At the other extreme, several chapters describe the application of satellite remote sensing to deal with environmental issues at national and regional scales. The chapters by Paul et al. and by Brooner show how remote sensing has been applied to problems important to developing nations; the chapters by Taranik and by Tinney and Lackey show how remote sensing is used to locate nonrenewable resources and to help us deal with the consequences of their extraction and use. Dyer and Crossley discuss the application of remote sensing to increase our understanding of natural ecosystems; finally, Tuyahov et al. discuss some aspects of the future of these techniques.

Computers are a crucial component of these new techniques, from the microprocessors that make miniaturized chemistry possible to the large mainframe systems that analyze millions of bits of information received from satellites. One of the newest applications of computers is in geographic information systems. These are computer-based techniques that have evolved as means for gathering, organizing, managing, manipulation, analyzing, and presenting information with spatial components. Maps can be made in a digital form that allow statistical analyses of the spatial patterns, rapid methods of categorization of material, and the study of correlations among several different sets of information. The chapter by Risser and Iverson discusses the application of a geographic information system to a region, the state of Illinois; the chapter by Gwynne and Mooneyhan describes the new development of such a system with global applications.

These chapters suggest that we are making impressive strides in increasing our understanding of the biosphere. These chapters describe the current status of the technologies that make it possible to know about our global environment. To realize their potential, it is necessary that knowledge about these technical advances reaches the suite of disciplines whose studies contribute to our fundamental understanding of the complex interactions that sustain life on Earth. The material in this section is one step in that direction.

EDITORS' INTRODUCTION TO:
DR. A. A. ORIO
Modern Chemical Technologies for Assessment and Solution of Environmental Problems.

Our ability to create pollution is well known and publicized, but our ability to both detect and deal with these pollutants has received much less attention. Perhaps this is because, as Chapter 11 makes clear, most of the techniques to detect and deal with pollutants are new, many having been developed since 1980. Dr. Orio, of the University of Venice, is one of Italy's leading experts in environmental chemistry, and in this chapter he illustrates what we can do with modern chemical technologies to lessen the impact of pollutants, monitor the state of the environment, and assure that pollutants introduced remain below harmful levels. The array of sophisticated chemical techniques include fiber optic devices that measure minute quantities of chemical pollutants directly in place in groundwater and wells. Interestingly, at the heart of many of these techniques are modern computers that provide speed and access to great quantities of data. The methods discussed here were not available for environmental analysis at the time of the 1955 conference, *Man's Role in Changing the Face of the Earth;* their development represents one of the fundamental changes that is occurring in the management of the environment.

Dr. Orio describes a variety of environmental issues that could be addressed using new chemical technologies. He concludes that the solution for most of these problems are already available. Solutions could be achieved provided that society is willing to accept the costs of implementation, a theme that is taken up again in Section III of this volume. With these techniques and those of remote sensing and geographic information systems, which will be discussed in the chapters by Taranik and others we seem to be entering a new and hopefully constructive stage in our relationship with our environment.

11 MODERN CHEMICAL TECHNOLOGIES FOR ASSESSMENT AND SOLUTION OF ENVIRONMENTAL PROBLEMS

ANGELO A. ORIO
University of Venice
Venice, Italy

INTRODUCTION

With the publication of Rachael Carlson's *Silent Spring,* it has become commonplace to acknowledge that the development of our modern civilization has led to a great amount of serious pollution of our environment. However, what is less well known is that in this decade we have begun to develop new methods to detect and to deal with pollutants. Although there has been considerable concern about pollutants and their environmental effects, we often know little about their transport and the fate of these substances. Therefore, we often find ourselves in a position where we are unable to evaluate their actual or potential effects. These new technologies allow us to measure and trace these substances with incredible precision. In addition, there are new approaches to the control of pollutants. These approaches make significant use of recently developed technologies. The purpose of this chapter is to review and evaluate the most promising of these new technologies. I hope to show that there is much that we can do to lessen the impact of pollutants on our environment and to monitor the state of the environment to assure that pollutants introduced remain below harmful levels.

We can categorize the production of pollutants into three groups: 1) internationally produced toxic substances, such as pesticides and chemicals used in certain industrial process; 2) pollutants that are by-products of energy production and use, such as emission of ashes containing toxic elements including mercury, beryllium, cadmium, arsenic, lead, and radioactive elements, and of gases such as sulfur, nitrogen, and carbon oxides, with resulting acid rain, as well as thermal pollution; and 3) pollutants associated with waste disposal. The atmosphere and water on the Earth's surface are used to disperse, either directly or some time after degradation such as incineration, the wastes resulting from a large variety of human activities. These wastes are diluted to low concentration, sometimes below the level of detection.

It used to be said that "dilution is the solution to pollution"; but there are disadvantages to this older approach. First, since mixing is not an instantaneous process, local concentrations may be quite high and have undesirable effects that are frequently evident in densely populated regions or on local ecosystems. Attempts simply to increase the area over which wastes are dispersed—increases the dilution—also have not always been successful. For example, older smelter operations and coal-burning plants used shorter smokestacks, which led to local deposition of toxic ashes. As a result, new, much taller smokestacks were built, which have resulted in the spread of acid rain over larger areas. In this case, the attempt to dilute the ashes further merely spread a new problem over a larger area. Furthermore, many modern chemical products, such as some pesticides, plastics, and other chlorinated compounds, are only slowly broken down to normal environmental components or rendered inaccessible by being trapped in sediments. As a result, the rate of removal of these substances from a given sector of the environment may be slower than their rate of input, and concentrations may build up to undesirable levels.

For several reasons, it is not always possible to establish what concentration of a pollutant is necessary before harmful effects are produced. In fact, the effects may be slow to develop and may be noted only statistically in a large number of samples. Because of synergistic effects, substances that separately present no problems may be much more deleterious when found in combination. In addition, various natural processes may reconcentrate materials to levels well above those expected on the basis of uniform mixing. Bioconcentration along the food web of mercury and chlorinated hydrocarbons are two examples of this mechanism.

The only processes that do not depend on dilution of wastes are disposal by burying on the land and dumping of insoluble materials in the oceans. These processes also are not free of disadvantages since leaching of radioactive materials, with consequent contamination of groundwater, and corrosion of barrels or other containers by sea water, with resulting contamination of the biological environment, is a well-known phenomenon.

As a consequence of these processes, chemical reactions of all kinds occur continuously in the atmosphere, in oceans, lakes, rivers, in all living things, and even underneath the Earth's crust. Chemistry, together with other disciplines, has proved a useful discipline for providing the basis for a better understanding of these processes and in helping to find proper solutions for most of them.

The solution of any environmental problem involves two major technological aspects: 1) the assessment of the problem based on quantitatively precise and reliable experimental data, which must be converted into the information needed for any subsequent decision; and 2) the application of proper technologies for abatement and control of pollutants and, therefore, avoidance of their emission into the environment.

ASSESSMENT OF ENVIRONMENTAL PROBLEMS

The first aspect, usually referred to as monitoring, encompasses the identification and measurement of samples in all environmental media. The large variety of environmental issues of deep concern to scientists, technologists, administrators, and to the public in general, has stimulated research for new analytical methods and for the production of new instruments to generate raw, intermediate, and finished data for the management of the environment.

In some cases, well-tested analytical techniques have been modified to meet the high selectivity and sensitivity needed for detecting the extremely low concentrations of noxious compounds normally present in the environment. Thus, the use of atomic absorption spectrometry (AAS), formerly restricted to biomedical analyses, has been widened to include environmental analyses as a consequence of the introduction of hollow cathode lamps, graphite furnaces, and the application of the Zeeman effect, which allows the determination of heavy metal concentrations at the level of one-billionth of a gram per milliliter. In other cases, it has been necessary to work out completely new technologies to satisfy the same needs. High pressure liquid chromatography (HPLC) has been recently developed to analyze highly boiling and labile compounds. Some advanced analytical technologies applied in environmental studies will be discussed in this chapter.

TABLE 11–1   CHARACTERISTICS OF SOME TECHNIQUES FOR ANALYSIS OF POLLUTANTS

Type of Pollutant	Analytical Technique	Detection Limits	Analysis Time*	Some Applications
Volatile organic compounds	GC/MS with MSD	10 μg/L 50–200 μg/kg	50 min	Waters Sediments
Phenols and nitrocompounds	GC/MS with MSD	25 μg/L	1 h	Waters and wastewaters
Amines	LC/MS	100 ng	30 min	Soil and water samples
Pesticides	GC/MS or GC/ECD	1–10 ng/L 2–10 μg/kg	30 min	Liquids Plants, animals
Polychlorophenols	HPLC	25 μg/L	35 min	Wastewaters
Polyaromatic hydrocarbons	HPLC	10–25 μg/L 0.2–0.5 mg/kg	30 min	Waters Soil samples
Halocarbons	GC with headspace sampler	25 μg/L	20 min	Waters and wastewaters
PCBs	GC/ECD or GC/MS	2 μg/kg 10 ng/L	30 min	Waters Soil samples
Heavy metals	DPP	0.1–5 μg/L	45 min	Waters
Heavy metals	AAS	10–500 μg/kg	10 min	Solids

*Only for the instrumental step of the analysis.

GC: gas chromatography; MS: mass spectrometry; MSD: mass selective detector; LC: liquid chromatography; HPLC: high pressure liquid chromatography; ECD: electron capture detector; DPP: differential pulse polarography; AAS: atomic absorption spectrophotometry; PCBs: polychlorinated biphenyls.

A major feature of most modern instruments applying the analytical methodologies indicated in Table 11–1 is the integration of detectors with microprocessors capable of controlling the instrument and displaying the results in digital form. The most versatile and powerful instruments, such as gas chromatography mass spectrometry (GC/MS) systems, may be purchased with powerful minicomputers, large capacity disk drives, multiple workstations, fast printers, industrial standard magnetic tape drives, and multiuser software capable of doing several tasks simultaneously. These systems can produce vast quantities of data in short periods of time. Some systems have software that can provide the desired final results. More often, however, additional external calculations are required to convert raw and partially processed or intermediate data into usable information in the final form. Although the number of target samples has increased along with quality controls requirements and cost per sample, the cost for analysis for each analyte is far lower than it was in the mid-1970s because of advancements in instruments technology.

The GC/MS technology is usefully applied to the analysis of trace level organic contaminants, such as polychlorinated dibenzophenyls (PCBs), polychlorinated dibenzo–p–dioxines (PCDDs), and polychlorinated dibenzofurans (PCDFs), present in complex mixtures. The number of chlorine atoms in these compounds can vary between one and ten to produce up to 209 PCBs, 75 PCDDs, and 135 PCDFs positional isomers.

By 1970, PCBs were recognized as persistent, bioaccumulative and possibly hazardous contaminants. Although little, if any, data have been able to link them directly to serious human health effects, during the 1970s national and international regulations restricted their use. Animal studies and in vitro experiments have indicated that the toxicity and biologic effects among the different PCDD and PCDF isomers can vary by a factor of 1,000.[1] The isomers with the highest activity are those having four to six atoms of chlorine.[2]

Since even small concentrations of PCBs can be detrimental to the environment, the U. S. Environmental Protection Agency (EPA) has recommended a marine water quality standard of 0.03 parts per billion [(ppb) (μg/L)] to protect marine life. Marine and fresh water organisms have been found to concentrate PCB in their bodies to levels 100,000 to one million times higher than those levels present in ambient water. The U. S. Food and Drug Administration (FDA) has established regulatory limits for PCB in various foods. The limit for fish and shellfish is 5 parts per million (mg/kg wet weight) in the edible portions of foodstuffs. At this time, there are no United States federal standards for PCB in drinking water or ambient air. The National Institute for Occupational Safety and Health (NIOSH) recommended a workplace standard of 1 μg/m^3 total PCB.

PCDD and PCDF are present in 0.1 to 500 μg/g concentration[3] in commercial products such as chlorophenols used as herbicides and PCB.[4] In dust and smoke emitted from municipal incinerators, heating facilities, and thermal power plants, concentrations up to 1 μg/g have been found.[5] In soil and vegetation samples from Seveso, Italy, immediately after an accident at a chemical plant producing 2, 4, 5-trichlorophenol, up to 50 μg/g of 2, 4, 7, 8-tetraPCDD were observed.[6]

Because of the extreme toxicity and the very low concentration in natural samples of some PCDD and PCDF isomers, high sensitivity and selectivity coupled with specific analytical techniques are required for the measurements. A detection limit of 1 picogram (10^{-12} g) or less might be required to find the most toxic isomers in 1 g sample and GC/MS is the only analytical technique that can satisfy so highly demanding performances.

Such a high sensitivity is normally obtained by extracting the compounds to be analyzed from the sample with an organic solvent. The solution is cleaned to eliminate possible interfering chemicals and concentrated to low volume (1 to 0.1 ml) before the injection into a capillary column where the gas chromatographic separation into single analytes is achieved. The selectivity of the entire analytical procedure mainly depends on the stationary liquid phase coating the inner walls of the column and on the possibility of tuning the mass spectrometer on ions characteristic of single molecule fragmentations. Up to 50 m long capillary columns (with 0.2–0.3 mm inside diameter) can be used, enabling the separation of a large number of compounds in a single run. Each compound is then broken down (by 70 eV energy electrons in the most common instrument configuration) into several ionic subunits whose mass/charge ratio and intensity are converted from an analog signal to digital and recorded by a computer. The mass spectrum thus obtained can be compared with those of the compounds already in the computer data bank. This can be done by a computer-assisted procedure, which spares the operator time-consuming and tedious steps. Up to several thousand different substances can be identified at the picogram level or lower. Using the special technique of the single ion monitoring the sensitivity of the mass spectrometer can be enhanced to reach the nanogram per kilogram level needed for the most sophisticated analyses of trace organic compounds.

When water bodies are of interest, a prerequisite for a good analysis is to be sure that the sample analyzed really represents the sampled water body. Chemical and biological reactions may take place in the sample during transportation to the laboratory and conservation before the execution of the analysis. Changes in the analyte concentration within the sample may occur during this time. Ideally, one should always carry out even the most sophisticated analysis in the field immediately after collecting the sample. Until recently, this was not possible.

For a number of pollutants this is now possible by applying electrochemical methods such as voltammetric striping or ion-selective electrodes. These methods have the necessary characteristics that make them suitable for environmental analysis: accuracy, precision, sensitivity (in the parts per billion range), reliability for automation, and low cost, which is important for monitoring purposes in field applications where many monitoring stations must be operated.

FIGURE 11–1   Automatic instruments for in-field continuous monitoring of pollutants and nutrients (in figure are shown the analyzers of Phenols, Total Organic Carbon, Ammonia, and Heavy Metals).

In soil or sediments, inorganic pollutants such as mercury, cadmium, lead, zinc, copper, nickel, cobalt, arsenic, and selenium are present in the 0.1 to 1,000 μg/g range, and they can be determined conveniently with other analytical techniques, such as atomic absorption. However, for natural waters, where the concentration is in the 0.1 to 1,000 μg/L range, voltammetric methods, such as differential pulse striping and differential voltammetry, are more reliable and practical analytical techniques.[7] Automatic systems, such as the one shown in Figure 11–1, can analyze the elements mentioned earlier in the field together with chemicophysical parameters (pH, redox potential, temperature, conductivity, dissolved oxygen, turbidity) and other important chemical species, such as ammonia, nitrite, nitrate, phosphate, chloride, fluoride, sulfides, phenols, and hydrocarbons, became available in the early 1980s for continuous monitoring of lakes, rivers, estuaries, lagoons, and wet deposition of atmospheric precipitates.[8]

An advanced and promising technique for the in situ detection and quantification of low levels of contaminants is remote fiber fluorometry in which chemical sensors attached at the distal ends of optical fibers can

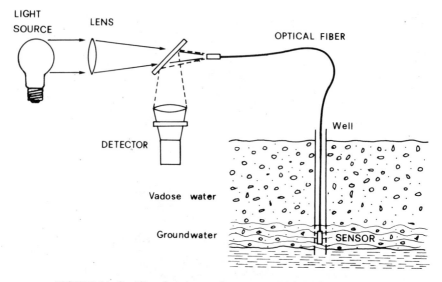

FIGURE 11–2   Fiber fluorimetry for remote sensing of groundwater.

be inserted in the sampling region to be monitored. This can be done in place for groundwater or wells (Figure 11–2). Excitation light of the wavelength appropriate for the specific compound to be analyzed is focused into a single-strand optical fiber transmitter to the sensor in the sampling region, causing a fluorescence in the target molecules. A small amount of this fluorescent light is collected by the optical fiber and returned to the optical coupler that separates the excitation light from the returning fluorescent light. Finally, the fluorescence can be analyzed by a spectrometer. Although this technique has only been tested for the analysis of organic chlorides,[9] it seems to be of a more general use and applicable to the detection and monitoring of many organic and inorganic compounds directly in the field, thus solving the difficulties associated with sampling, transporting, and properly monitoring water samples containing trace quantities of pollutants and broadening the number of toxic species that can be analyzed in situ.

ABATEMENT AND CONTROL TECHNOLOGIES

For the abatement and control of pollutants, well-known biological, chemical, and physical principles are normally applied. However, in most cases the solution of environmental problems has been made possible through the development of totally new techniques and advanced technologies.

FIGURE 11–3 Incineration plant for chlorinated hydrocarbons.

One example is the thermal destruction of hazardous wastes involving the exposure of the wastes, mostly organic, to high temperature (usually 900°C or greater). Industrial boilers, cement kilns, and industrial furnaces are traditionally used for oxidizing or incinerating the wastes. The use of high temperature processes such as molten salts and plasma or electric furnaces represents the most advanced technologies in this field.

Properly designed and operated thermal destruction systems offer the prospect of destroying hazardous organic components of wastes streams, reducing the volume, and in some cases recovering energy and valuable materials such as sulfuric or hydrochloric acid. A typical example of this technology is the incineration of chlorinated hydrocarbons, which are one of the largest volume products manufactured by the chemical industry (Figure 11–3). In the production of chlorinated solvents or monomers, the yield of waste by-products is often as much as 5 percent of the main product. An average size production of 300,000 tons/year of chlorinated solvents may result in 4,000 to 15,000 tons/year of various wastes containing more than 60 to 70 percent of chlorine by weight. Wastes are also generated in the manufacture of vinyl chloride monomer and epichlorohydrin. In some countries, such as the United States and Germany, burning of these wastes aboard incinerator ships at sea is practiced. This practice, however, does not allow the recovery of valuable components or incineration products from the wastes for subsequent recycling. Burning at sea also does not permit any of the potentially considerable caloric value to be recovered.

Processes that effectively destroy chlorinated hydrocarbons while recovering up to 75 percent of the wastes' heat value as high-quality steam and up to 99.5 percent of chlorine as 33 percent HCl solution or 100 percent HCl gas were developed in France and Italy in the late 1970s and are now licensed and built in many countries, including the United States and Russia. However, thermal destruction is currently used to dispose of only 2 percent of the 264 metric tons of hazardous wastes produced annually in the United States.[10] Although data are not available for other countries, it is reasonable to assume that similarly low percentages are applicable.

Prospects for an increased use of this technology are good when incineration is proven, commercially available, and demonstrated to be suitable for handling a wide variety of wastes. Further, these high temperature processes are flexible since they can be modified to destroy different hazardous material including PCBs, dioxines, and dibenzofurans. They can be operated in self-contained transportable units to mitigate the public opposition that frequently plagues the construction of centralized permanent incineration facilities. However, there is a public concern that, because of technical and human failings under real operating conditions, incineration plants may lead to atmospheric contamination with dioxines and possibly other toxic halogenated compounds.

Many other new technologies can deal with the reuse of products or materials. Probably the best example of reclamation using the most advanced technologies is represented by the water factory 21 (WF21) conceived and constructed about eight years ago by the Orange County Water District in California.[11] The reclaimed water is injected into the ground through a series of 23 wells to create a barrier to push back the seawater infiltrating from the sea and preventing the loss of groundwater supplies.[12] Prior to injection, unchlorinated, municipally treated activated sludge effluent (0.66 m³/s) undergoes lime clarifaction, air striping, recarbonation, perchlorination, mixed media filtration, granular activated carbon absorption, final chlorination, and reverse osmosis (RO) demineralization (Figure 11–4).

Although well-proven technologies are applied in these processes, some interesting new results have been obtained. For instance, lime clarification has been found effective not only in reducing suspended

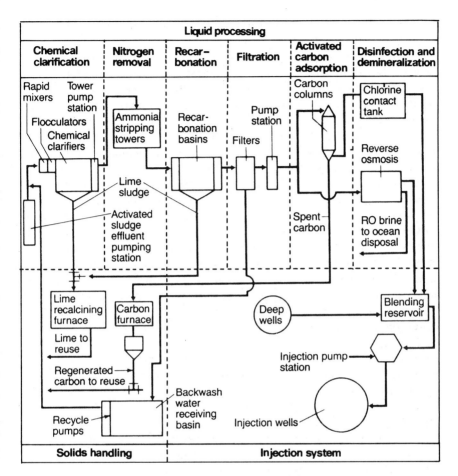

FIGURE 11–4   Flow chart of water factory 21.

solids, turbidity, and phosphates, but also in decreasing the concentration of heavy metals, organic compounds, bacteria, and viruses present in the effluent. Similarly, the carbon absorption reduces total organic carbon from more than 100 mg/L to approximately 5.0 to 10.0 mg/L, which compares pretty well with the 30 mg/L specified for injection water as maximum level or carbon oxygen demand by the state of California. Total carbon is further reduced to 1 mg/L or less by the reverse osmosis process, which represents the most advanced technology used in waste factory 21. This part of the plant, designed to remove a maximum of 90 percent of all the dissolved salts and to achieve at least 85 percent recovery of product water, allows water factory 21 to meet all the EPA secondary criteria for drinking water.

When the cost of providing the advanced treatments applied at water factory 21 is compared with that of alternative supplies available in the region, the data indicate that reclaiming wastewater in southern California is competitive with alternative methods in terms of both energy consumption and money. For example, the energy consumption for delivering water from northern California to southern California is approximately 2,600 kilowatt-hours per thousand cubic meters (kWh/10^3m^3), that for furnishing Colorado River water is approximately 1,622 kWh/10^3m^3.[13] In comparison, the energy consumption for water reclamation at water factory 21 is 1,782 kWh/10^3m^3. In financial terms, it has been estimated[14] that the cost for delivering new sources of water from northern California range from $325 to $725 per thousand cubic meters. By comparison, the most expensive wastewater treatment system operated at water factory 21 costs $370 per thousand cubic meters. These results are interesting since shortage of water is becoming an important issue for water supply planners in the United States as well as in many other countries, including those in Europe and the Middle East. Water recycling is today available as a technological option. Although not all the answers to the issues regarding the suitability and feasibility of wastewater reclamation are provided by this example, no evidence has been found that reclaimed wastewater would pose a significant health risk if used as a source of domestic water.

The reverse osmosis demineralization cost in water factory 21 represents 70 percent of the total treatment cost. Therefore, any advance in the membrane technology that makes reverse osmosis possible may decrease energy requirements and thereby decrease the costs for filtering or demineralizing wastewater. Important research efforts are now being carried out in the filtration field to develop thin-film composite membranes, which are effective in removing a wide range of pollutants, including many trace organics. For these purposes, new technologies such as crossflow microfiltration or ultrafiltration are now applicable.[14] Microfiltration membranes have pores that can range from 50 to 1.5 nm in diameter, and this technique can be used to remove carbon black, proteins, viruses and other pathogens, silica, and particles in colloidal form. One problem common to these filtration systems is that membrane pore size cannot be made uniform. If pores could be made the same size, such systems would be much easier to operate and control. Because they cannot, actual filtration must be based on average pore size. Moreover, after a time of use, clogging or fouling of most of the membrane pores occurs, so backflushing and chemical cleaning of the membranes is necessary when operating these systems. Of course, some provision for proper land site disposal or reuse of rejected materials must be made.

Most of the efforts of the filtration industry are now devoted to the development of stronger membranes and hardware materials able to withstand more rigorous conditions, such as higher pressures and temperatures, which enhance the efficiency of filtration processes. Devising membranes that have a more uniform pore size and can withstand a greater variety of chemical environments are other important objectives of the research in this field.

Finally, a recent and promising development in environment control technology is the use of supercritical fluids for separation or detoxification processes. A supercritical fluid is a substance that has been heated above and compressed beyond its critical temperature and pressure (i.e., the temperature and pressure at which the vapor and the gas phase of a substance in equilibrium with each other become identical, forming one phase). The variety of interesting and useful characteristics of fluids near critical point makes them ideal for mass transfer processes. Their high density gives them high capacity as solvents, since the solubility varies exponentially with the solvent density. They are highly compressible, which means that, within the critical region, their properties undergo large changes with relatively small changes in the operating conditions. Molecular diffusivity of supercritical fluids is one or two orders of magnitude higher than that of normal liquids, while the viscosity is almost as low as that of gases, facilitating both pumping and natural convection.

The several common substances with critical temperatures near ambient, that are therefore interesting as process fluids include ethylene (10°C), carbon dioxide (30°C), ethane (32°C), fluoroform (45°C), and sulfur exafluoride (46°C). Ammonia, alcohols, and water can be considered for uses at higher temperatures. Carbon dioxide is particularly suitable for use in food, pharmaceutical, and environmental control processes because it is hazardless and nontoxic.

A number of rapidly developing possibilities for using this technology in various aspects of environmental control are highly interesting. A process that uses supercritical carbon dioxide to obtain pyrethrin—a natural insecticide highly toxic to insects but totally harmless to warm-blooded animals—from pyretrum flowers has been patented by Botanical Resources.[15] Pyrethrin decomposes with time; therefore, it does not accumulate in the environment and the insects do not develop a resistance to it. Another example is the decontamination of soils containing DDT and chlorinated aromatic hydrocarbons such as chlorophenols or PCBs.[16] Atrazine and dinitrobutylphenol, which are contaminants present in the waste stream of a pesticide factory, are successfully removed from activated carbon at an estimated cost of 14 to 29 cents per pound of carbon, which compares favorably with the thermal regeneration cost of 29 to 36 cents per pound.[17] The same technology can also be applied to the regeneration of resinous absorbents and to the removal and concentration of trace contaminants from gas streams. Finally, a modular process for the direct destruction of organic wastes has been proposed using supercritical water.[18] In this process, which takes advantage of the high critical temperature of water (374°C), the organic compounds are rapidly oxidized to light gases (carbon dioxide, carbon monoxide, and molecular nitrogen) and the halides are precipitated as salts. Contaminants that have been destroyed with this process include DDTs, biphenyl, trichloroethane, ethylene dichloride, and two PCBs. The conversion of these compounds to light gases has been reported as greater than 99.99 percent.

SUMMARY

Most of the environmental problems facing mankind today result from a development model of industrialized societies in which technology has been asked to exploit natural resources and to provide a constantly increasing living standard. In the past, it was felt that "dilution is the solution to pollution," but this is no longer acceptable or necessary. New techniques provide alternative solutions. Technology is being asked to play a major role in the solution of a large variety of environmental problems:

- improvement of sensitivity, selectivity, and accuracy in monitoring environmental contaminants for a better evaluation of their health effects
- increased performance of urban and industrial wastewater treatment plants
- development of low or zero wastes technologies in the production of goods
- better control, disposal, and destruction of polluting species, thus avoiding problems for future generations

- better assessment of the risks associated with the use of new technologies for energy production
- development of new information systems for the management of the environment
- use of natural resources in ways compatible with the protection of the environment
- development of new methods to reclaim contaminated ecosystems and to recycle used materials
- production of new harmless materials to be used instead of the toxic ones
- development of safer and environmentally harmless pest controlling systems.

The solutions for most of these problems are already available or can be obtained using the most advanced chemical technologies, provided there is the will to implement them properly and that society is ready to accept the relative costs of such implementation.

NOTES

[1] J. A. Bradlaw and J. L. Casterline, 1976, *J. Assoc. Off. Anal.*, Vol. 62, p. 904.

[2] A. Poland, E. Glover, and A. S. J. Kende, 1976, *Chem. Biolog.*, Vol. 251, p. 4926; E. E. McConnell, J. A. Moore, J. K. Haseman, and M. W. Harris, 1976, *Toxicol. Appl. Pharmacol.*, Vol. 36, p. 65; J. A. Moore, E. E. McConnell, D. W. Dalgard, and M. W. Harris, 1979, *Ann. N. Y. Acad. Sci.*, Vol. 320, p. 151; J. E. Huff, J. A. Moore, R. Saracci, and L. Tomatis, 1980, *Environ. Health Perspect.*, Vol. 36, p. 221.

[3] C. Rappe, H. R. Buser, and H. P. Bosshardt, 1979, *Ann. N. Y. Acad. Sci.*, Vol. 320, p. 1.

[4] G. W. Bower, M. J. Mulvihill, and R. W. Simoneit, 1975, *Nature*, Vol. 256, p. 305.

[5] H. R. Busersand and H. P. Busshardt, 1978, *Mitt. Geb. Lebensmittelunteres. Hyg.*, Vol. 69, p. 191.

[6] D. Firestone, 1978, *Ecoll. Bull.* (Stockholm), Vol. 27, p. 39.

[7] H. W. Nurnberg, 1979, *Sci. Total Environ.*, Vol. 12, p. 35; H. W. Nurnberg, 1982, *Pure Appl. Chem.*, Vol. 54, p. 853.

[8] L. Mart, H. W. Nurnberg, and D. Dyrssen, 1982, in *Trace Metals in Sea Water*, C. S. Wong and K. Bruland (eds.), Plenum Press, NY; H. W. Nurnberg, P. Valenta, and V. D. Nguyen, 1982, in *Deposition of Atmospheric Pollutants*, H. W. Georgy and J. Pankfrath (eds.), D. Reidel, Dordrect and Boston.

[9] F. P. Milanovich, 1986, *Environ. Sci. Technol.*, Vol. 20, p. 441.

[10] U. S. EPA, "Assessment of Incineration as a Treatment Method for Liquid Organic Hazardous Wastes—Background Report III: Assessment of the Hazardous Wastes Incineration Market," U. S. Government Printing Office Order NO. 1985–526–778/30376.

[11] D. G. Argo, 1985, *Environ. Sci. Technol.*, Vol. 19, p. 208.

[12] Metropolitan Water District of Southern California, 1983, "The Need for Major Addition to Metropolitan's Distribution System," Report 979, Los Angeles, California.

[13] D. G. Argo, 1984, *J. Water Pollut. Control Fed.*, Vol. 56, p. 1213.

[14] J. Josephson, 1984, *Environ. Sci. Technol.*, Vol. 18, p. 375A.

[15] M. Sims, 1982, *Chemical Engineering*, Jan, 25, Vol. 50.

[16] F. C. Knopf, B. Brady, and F. R. Groves, 1985, *CRC Crit. Rev. Environ. Control*, Vol. 15, p. 273.

[17] R. P. deFilippi and R. J. Robey, 1983, "Supercritical Fluid Regeneration of Adsorbents," Project Summary, EPA-600/52-83-038, EPA, Washington, DC.

[18] M. Modell, 1979, "Destruction of Hazardous Wastes Using Supercritical Water," Modar, Inc., and H. P. Bosshardt, *Ann. N. Y. Acad. Sci.*, Vol. 320, p. 17; G. W. Bower, M. J. Mulvihill, and R. W. Simoneit, 1975, *Nature*, p. 256.

EDITORS' INTRODUCTION TO:

EDITORS' INTRODUCTION TO:
DR. J. V. TARANIK
The Search for Nonrenewable Resources in the Next Twenty Years

This chapter deals with two issues: the impacts of the search for hydrocarbon and mineral resources on our global environment and the use of remote sensing to locate new sources of minerals. Dr. Taranik is President of the Desert Research Institute of the University of Nevada, Nevada System, an organization that conducts environmental research in atmospheric, biologic, hydrologic, energy, and social sciences. He was Dean of the Mackay School of Mines of the University of Nevada, and has long been active in the application of remote sensing to the utilization of mineral resources. In this chapter, Dr. Taranik focuses on the application of advanced aerospace remote sensing technologies to the search for and the development and reclamation of mineral and hydrocarbon deposits. Satellite remote sensing systems from Landsat to SPOT are discussed, as are imaging radars and spectrometers. The history of the use of these systems for exploration and research are recounted and schedules for future missions presented.

12 THE SEARCH FOR NONRENEWABLE RESOURCES IN THE NEXT TWENTY YEARS

JAMES V. TARANIK,
University of Nevada
Nevada System,
Reno, Nevada

INTRODUCTION

The U. S. Department of Interior predicts that the global demand for nonfuel minerals will increase. This raises two issues: What will be the impact of increasing demand on the global environment? and How will we find new sources of these minerals? This chapter discusses both issues.

THE IMPACT OF MINERAL EXTRACTION ON THE GLOBAL ENVIRONMENT

Land directly utilized for nonfuel mineral production currently amounts to approximately 5,000 square miles and is expected to increase to approximately 8,000 square miles by the year 2000. Lands indirectly affected by nonfuel mineral production are estimated to be 94,000 square miles, or about 0.2 percent of the total surface of the Earth.[1] In the United States, 0.16 percent of the total land area is currently used for nonfuel mineral extraction and 40 percent of some previously mined land has been reclaimed.[2]

The majority of the nonfuel mineral production in the United States is directly related to the production of construction materials (sand, gravel, clay, stone, cement, and other nonmetals), which are used primarily for construction of highways and buildings in urban-industrial areas. In 1983 the value of construction materials produced in California (population 38 million) was over $1.5 billion, while metals only accounted for $15 million in production.[3] Today most of the new precious metal production in the United States is in Nevada (population 900,000), where it is estimated that some $20 billion in gold and silver will be produced by the year 2000. Currently, 0.64 percent of Nevada's land area is involved in nonfuel mineral extraction and that percentage is not expected to increase significantly because, under environmental regulations, mined land must be reclaimed for other possible multiple uses.

In the Global 2000 report, prepared at the request of the President of the United States in 1981, production and use of nonfuel minerals were judged to have no projected impact on global, regional, and urban terrestrial and atmospheric environments.[4] However, in rural areas, mining of bauxite, sand, gravel, and limestone, and to a limited degree, metallic minerals, may result in long-term and sometimes "permanent" local land disturbance and loss. In the United States, environmental regulations require mined land to be returned to other multiple uses through reclamation and, thus, the impact of mineral production and nonfuel mineral use is not expected to increase. Also in the United States, metal refining plants now have to comply with strict air pollution guidelines. The closure of such plants because of low metal prices, plus the implementation of new equipment for scrubbing stack gases of unwanted pollutants, has now demonstrated that metal-producing operations are not the major causes of the acid rain problem on the North American continent. The major contribution to acid rain appears to come from industrial plants and automobile emissions in urban-industrial areas.

Global supplies of hydrocarbons are expected to become increasingly scarce by the end of this century. Global demand for coal and oil shales will begin to increase rapidly in the next century. Likewise, there may be an increasing demand for uranium minerals if atomic power is developed as an alternative energy source beyond the year 2000. Reclamation of mined lands in the United States will return much of the land utilized for coal and uranium mining to other and multiple uses. Therefore, the process of extraction of energy resources in the United States is not expected to result in a significant impact on the global environment. In addition, increasing emissions standards in the United States will insure a minimal increase in the impact of energy on the environment. However, increased energy *use* worldwide and industrialization in developing countries may have a significantly increasing adverse effect on the global atmosphere, unless there are international and national incentives to regulate emissions from industrial plants and automobiles.

In the United States, new regulations require that those exploring for minerals post bonds for reclamation of lands before access permits are granted. Changes to the environment that result from the exploration activity must be corrected prior to departure from the area and prior to approval of permits for resource extraction. Resource extraction activities in the United States are now strictly controlled by state and federal regulations that require protection of groundwater sources, control of effluents, control of dust, and, ultimately, return of mined land to other multiple uses.

The recent worldwide decline in energy prices and reduction in the demand for base metals has caused a severe contraction in exploration activity in the United States and throughout the developed areas of the Western world. This decline is expected to continue to the end of this decade. Many large metals manufacturing firms in the United States and Europe have closed their exploration offices. Presently, some 71 percent of the potentially employable exploration geologists in the United States are unemployed, and many are retraining for other professions. Most global exploration activity is occurring outside the United States where highly economic deposits of base metals (copper, tin, zinc, lead, and mercury), ferro-alloy metals (manganese, chromium, nickel, tungsten, molybdenum, vanadium, cobalt, columbium, tantalum, and rare earth metals) and platinum-group metals are being sought. Major new deposits of these minerals have been discovered in Latin America, Indonesia, and Africa, where there is at this time little incentive to reclaim lands for other multiple uses or to control emissions from metal manufacturing plants.

NEW TECHNOLOGIES TO SEARCH FOR MINERAL DEPOSITS

Many companies involved in the search for and extraction of nonrenewable resources are using new advanced aerospace technology to establish an environmental baseline prior to exploration, and they are designing exploration activities to minimize impacts on the environment while maximizing the results of ground-based investigations. Major companies are utilizing aerospace data in poorly mapped, lesser developed countries that do not permit acquisition of aircraft data, except by their own military forces. Aerospace remote sensing data are now utilized to document environmental conditions prior to resource extraction, and they are used to design resource extraction plants to take maximum advantage of the landscape. This avoids local conditions that may be hazardous to plant development while insuring that landscape restoration can be accomplished efficiently and with minimal long-term impact. In addition, many companies are using aerospace remote sensing data to design reclamation plans and to monitor the effectiveness of reclamation over time. This chapter discusses new developments in aerospace remote sensing, which have great potential applications in global mineral and energy resource assessment, exploration, development, and production, and which can help to minimize the negative impacts on our environment from extractive industries.

THE LANDSAT SYSTEM

The first land observations satellite was launched into orbit in 1972. That satellite, initially called the Earth Resources Technology Satellite (ERTS), was renamed Landsat when its essentially identical twin, Landsat-2 was launched in 1975. A total of three Landsats were launched into polar 914-kilometer orbits between 1972 and 1978. From that altitude, their Multispectral Scanner (MSS) instrument detected solar radiation in four wavelength bands from 80-meter ground areas. A backup instrument, the Return Beam Videcon (RBV), produced limited data. Multispectral Scanner Images are formatted so entire images cover areas 185-by-185-km. Photographic data were initially planned as the major data product. However, digital computer compatible tape data are now in widespread use. Landsat MSS data exist for most of the Earth's land areas between 87 degrees north and south latitude. These data can provide an essential environmental baseline for monitoring subtle changes in global landscape conditions that may be occurring because of changes in global atmospheric conditions.

In July 1982, a fourth Landsat was launched into a 705-km orbit with an advanced seven-wavelength band sensor called the Thematic Mapper. This multispectral scanning sensor covers the same ground areas as the MSS, but at higher spatial resolution—30 meter ground instantaneous field of view. Landsat-4 experienced difficulties in power and data transmission that prevented acquisition of Thematic Mapper (TM) data soon after launch. Therefore, Landsat-5 was launched in 1984 with an essentially identical complement of sensors. Landsat-5 performed well until 1985, when the direct downlink to Earth receiving stations was lost. Currently, data external to the continental United States could only be acquired through the Tracking and Data Relay Satellite System (TDRSS) in geosynchronous orbit. Although much of the United States has been covered one time with Thematic Mapper data, there is still little seasonal multispectral data coverage. Thematic Mapper coverage of foreign areas is limited and noncontiguous.

On September 26, 1985, the U. S. government signed an agreement to transfer management of the Landsat program to industry. The program will now be operated by EOSAT. EOSAT's subcontractors include Hughes Aircraft Corporation, RCA, and Computer Sciences Corporation. The Earth Satellite Corporation (EARTHSAT) will provide value-added data processing and will be involved in marketing the data. The Earth Resources Observation System (EROS) Data Center operated by the U.S. Department of Interior will archive and distribute Landsat data for some time to come. EOSAT is building a ground station at Norman, Oklahoma, and eventually all Landsat data will be acquired and distributed from that location. EOSAT's proposal to the U.S. government includes an advanced Thematic Mapper scheduled to be launched in 1991 with a 15-m panchromatic band in addition to the existing six solar reflectance bands and one thermal band. In 1991, EOSAT may launch an instrument with the same wavelength bands as Landsat-6, but also with multiple bands in the thermal infrared.

APPLICATIONS OF LANDSAT DATA TO MINERAL AND ENERGY INVESTIGATIONS

Landsat-1, -2, and -3 data have now been used for ten years in energy and mineral exploration. Many larger petroleum companies have a worldwide data base of Landsat images, and interpretation of these data is now a routine procedure used in identifying exploration targets. Both mineral and energy exploration companies have found that large area coverage of Landsat data can be used to great advantage in delineating new structural features and identifying abundances of key mineral indicators and anomalous vegetation patterns, which are indicators of mineralization and potential areas for hydrocarbon accumulations.

When Landsat data were first distributed to the geological community, geoscientists discovered linear features displayed by Landsat imagery that extended to tens, if not hundreds, of kilometers. The extent and pervasive nature of these linear features had not been previously recognized by geologists, and no geological models existed that could be used to explain them satisfactorily. Lineament mapping became common practice. Intersections of lineaments were postulated to control mineralization or fracturing in the subsurface. Little geological expertise was needed to map and interpret the interconnectability of these features on imagery. Skilled photogeologists found that they could not duplicate the linear features delineated by others from imagery. Skepticism set in, and conservative geologists rejected lineament analysis as a useful analytic technique. Although the geological significance of these landscape features is still not completely understood, many of them have now been recognized as the surface expressions of zones of fracturing in the Earth's crust that have been active for long periods of geological time. Often the displacements on these features are slight, and they are difficult to recognize on the basis of subsurface indicators alone. Repeated movements throughout geologic time cause these fundamental fractures to be maintained. As a geological basin evolves, they often control the alignment of streams and the distribution of reservoir rocks, and create fracture porosity in sedimentary sequences. Explorationists have determined that many of these surface features display trends in the subsurface that are also substantiated by gravity and magnetic data, but are not well delineated by seismic data. Systematic linear features mapping from Landsat data is now accepted practice when used with geophysical or subsurface data to define exploration targets.[5]

For the past 15 years, the U.S. Geological Survey (USGS) and the California Institute of Technology's Jet Propulsion Laboratory (JPL) have cooperatively investigated the applicability of Landsat data to geologic mapping for mineral resource investigations. This research has shown that iron oxide coatings can be detected on rocks and soils using computer processing of Landsat Multispectral Scanner data. The USGS and industry geologists now routinely use iron oxide abundance maps, derived from Landsat data, to guide their field investigations.[6]

In 1976, a group of industry geologists formed a committee to advise the U.S. government on their needs for global remote sensing data. This organization, called the Geosat Committee, now includes over 60 energy and minerals firms that contribute more than 25 percent of the gross national product of the United States. That same year the Geosat Committee and USGS geologists organized a workshop in Flagstaff, Arizona, to document their remote sensing data requirements for the next decade. A similar Flagstaff workshop was held in fall 1985. At the 1976 workshop, geologists realized that NASA (National Aeronautics and Space Administration) planned to launch Landsat-4 primarily for agricultural inventory, and they were concerned that the needs of geoscientists would be overlooked. Research by USGS and JPL scientists in 1975 had shown that if another infrared band was added to the new Landsat-4 sensor, the Thematic Mapper, it would be possible to discriminate not only iron oxides but clays as well.

In 1977, NASA decided to include the additional channel on the Landsat-4 sensor. To further understand the requirements of the geological community, NASA offered to acquire data using aircraft and satellites over eight industry test sites in the United States selected by the Geosat Committee. In return for the data, industry was asked to evaluate the usefulness of the data for geological investigations. This project, called the Joint NASA/Geosat Test Case Project, has now been completed, and its results have been published by the American Association of Petroleum Geologists. Three sites were evaluated in Arizona to determine the applicability of remote sensing data for targeting porphyry copper mineralization, two sites in Utah and Wyoming evaluated deposits of uranium, and three sites in Texas, West Virginia, and Wyoming were oil/gas sites.

One of the early and most striking results of the joint industry/NASA investigations was in Silver Bell, Arizona. Aircraft scanner data, simulating those to be acquired by the new sensor on Landsat-4, clearly delineated the alteration zone characterized by iron oxide stain and clay alteration. Simulated Landsat-2 Multispectral Scanner data were acquired at 12-m resolution, three times better than those acquired by the satellite scanner. These aircraft data were processed to determine if additional information could be extracted with higher resolution data, but no additional rock materials were delineated. A most important conclusion of this study was that increased spectral resolution (narrower wavelength band) was significantly more important than increased spatial resolution. ASARCO geologists who have spent many years mapping and analyzing the Silver Bell deposit were surprised at the correlation between the alteration mapped by ground techniques and laboratory analyses, and the clay and iron oxide abundances detected and mapped through processing and analysis of aircraft scanner data. Another important conclusion of this research was the development of methods for mapping mineral abundances that geologists do not normally observe in the field. In fact, clay abundances are usually determined using a petrographic microscope.[7]

In many areas of the world, vegetative cover obscures rock and soil materials, and remote sensing techniques must use variations in vegetation patterns as indicators of the underlying geologic materials. In some areas, such as Australia, rock types are not well exposed, and even outcrops do not produce characteristic weathering patterns. However, in Australia, plant species are selective in associating with particular rock materials, and geological mapping can be almost completed without actually observing the rock materials themselves. With repetitive Landsat data, the optimal times of year for "geobotanical" mapping can be selected and the data can be computer-enhanced to bring out subtle variations in plant cover.

Anomalous concentrations of elements may cause vegetation to be stressed and stunted, and these areas can be readily identified from Landsat data. The U.S. Geological Survey and U.S. Steel studied an area in Indonesia to determine if sites having nickel laterite duricrusts could be identified. Not only could the duricrusts be detected and mapped in known areas on the basis of the vegetation patterns, but the analysis could also be extended to unknown areas to delineate potential new sites for exploration. In this study, the areas analyzed were covered with dense, triple-canopy vegetation.[8]

AIRBORNE AND SPACEBORNE IMAGING SPECTROMETRY

In 1976, NASA asked for proposals for instruments to be flown as a part of the Shuttle engineering test program. Eventually six Earth-viewing experiments were selected, including two geological remote sensing experiments—Shuttle Multispectral Infrared Radiometer and Shuttle Imaging Radar.[9] The flight of both instruments in November 1982 resulted in spectacular breakthroughs in geological remote sensing.[10]

Analysis of the radiometer data demonstrated that direct mineralogic identification of carbonates and specific clay minerals was possible by narrow (20-nanometer) wavelength band measurements from Earth orbit. On an orbital pass over Baja California, analysis of the data showed that clay and iron oxides were abundant. Subsequent field work documented that a zone of hydrothermal alteration had been detected. When the Mexican government analyzed samples, they confirmed that a previously unknown deposit of silver and molybdenum had been discovered.[11]

On the basis of the results from the analysis of the Shuttle Multispectral Infrared Radiometer data, the Jet Propulsion Laboratory has proposed that NASA fund the development of a Shuttle Imaging Spectrometer. This instrument would view an 11-km ground swath with 30-m ground resolution in 196 different spectral wavelength bands, 10 nanometers wide. This instrument has been approved for flight on the Shuttle for some time in the late 1980s or early 1990s.

To prepare the scientific community for analyzing Shuttle Imaging Spectrometer Experiment (SISEX) data, NASA is supporting investigators through an aircraft airborne imaging spectrometer program. In 1983, NASA began flying an Airborne Imaging Spectrometer (AIS) in a C-130 aircraft based at NASA Ames Research Center. The AIS instrument utilizes a 32-by-32-element area array that is scanned with four different grating positions to provide 128 spectral bands having 9.3-nm bandwidths. The instrument is usually flown 6,000 meters above terrain and the 32-element area array covers a 400-meter swath at 12-meter resolution. In the future, NASA plans to begin flying an advanced version of the AIS in its high-altitude ER-2 aircraft. This instrument, called the Advanced Visible and Infrared Imaging Spectrometer (AVIRIS), will be flown at 75,000 feet and at that altitude will image an 11-km swath at 20-meter ground resolution. AVIRIS will have 224 bands with 9.6-micrometer bandwidths. Using three instruments, iron oxide and clay abundances can be mapped and specific minerals can be identified. In the Hot Creek Range of Nevada, montmorillonite and kaolinite associated with rock alteration were identified from AIS spectra.[12] In addition, the feasibility of direct identification of limonite, jarosite, geothite, alunite, calcite, and gypsum has been demonstrated.

Ultimately, an imaging spectrometer will be flown on the polar orbiting space platform—on Instrument Orbital Configuration-2 (IOC-2). Current plans for the NASA Space Station call for a manned platform in a 28.5-degree inclination orbit approximately 400-km above the Earth and two unmanned polar orbiting platforms with inclinations of 98 degrees.[13]

Another important multispectral instrument will complement High Resolution Imaging Spectrometer (HIRIS). That instrument is called the Moderate Resolution Imaging Spectrometer (MODIS). MODIS will acquire global coverage every two days at a spatial resolution of 0.5 kilometers with an imaging swath of 1,500 kilometers. The 35 spectral bands of the instrument will include wavelength bands for discriminating soils from vegetation, assessing vegetation properties, suspended sediments, atmospheric effects on solar radiation, cloud altitudes, snow/cloud discrimination, cloud and surface temperatures, and stratospheric aerosols.

AIRBORNE AND SPACEBORNE MULTIBAND THERMAL MAPPING

Research in the thermal infrared also shows great promise for geological applications. When thermal mappers were first made available to geologists, research indicated that the manner in which geological materials heated up in response to the sun and cooled down when it was not illuminating the surface, was related to the body properties of the materials. These body properties are heat capacity, density, and thermal conductivity. These three body properties combine to determine the thermal inertia of the material. Thermal inertia images have been used to locate faults in unconsolidated alluvium, map structures in areas having thin alluvial cover, and differentiate dolomitic rocks from limestones.

Recently, NASA developed a multiband thermal mapper, which the agency has been flying in a Lear jet over sites thoughout the United States. Data collected over Donner Pass, California, were recently analyzed and the feasibility of separating diorites from granodiorites in the Sierra Nevada batholith was demonstrated.[14] Thermal Infrared Imaging Spectrometer (TIMS) data collected over Virginia City, Nevada, have been evaluated and preliminary findings indicate that bulk silica composition of volcanic rocks can be assessed and that hydrothermally altered areas can be separated from deposits of diatomaceous earth.[15] The ability to map the identity and abundances of iron oxides and clay minerals, and to differentiate carbonates and sulfates from these minerals with solar reflectance data, when combined with the ability to evaluate silicon-oxygen bonding and free-silica abundance with TIMS data, represents a technological breakthrough for global geological investigations. NASA has proposed development of Shuttle TIMS and plans are underway for adding a multiband thermal capability to the HIRIS instrument planned flight on the space platform in 1996. Plans are also underway to include four thermal bands on Landsat-7 scheduled to be flown in the 1991 time frame.

AIRBORNE AND SPACEBORNE RADAR

The Shuttle Imaging Radar provided remote sensing image data not previously available to geoscientists. The Shuttle Imaging Radar-A (SIR-A) flown in 1981 actually penetrated dry sand cover of the Sahara Desert and detected bedrock that in field verification work was determined to be more than 3-m below the surface. The L-band (23-cm wavelength) radar images from this mission revealed previously unknown buried valleys, geologic structures, and possible Stone Age archaeological sites. Because the radar is sensitive to variations in roughness and soil moisture, and its illumination geometry is independent of the sun, analysis of its data provides information not available from Landsat data. Scientists from the U. S. Geological Survey have concluded that the radar data collected on the second flight of the Shuttle represent a major breakthrough in geological remote sensing.[16] SIR-B, the second version of this radar instrument, was flown in 1984 at a higher inclination. Digital L-band radar data was acquired at multiple incidence angles over diverse global topography. SIR-B data have been analyzed over the Candelaria region of Nevada and several significant structural features were identified.[17] Problems with data acquisition and transmission greatly limited the utility of the radar data on SIR-B. Therefore, NASA agreed in 1985 to refly the radar into a polar orbit. At the time of this writing, the SIR-B reflight is scheduled to be launched in the late 1980s, to be followed by the SIR-C radar system. SIR-C will be a two-frequency radar system (C-band, 8-cm wavelength, and L-band), and it will have different polarizations (HH, VV, HV, VH). SIR-D is planned for launch in 1993 and that mission will probably be an engineering model of the radar system to be flown on the EOS polar orbiting space platform in 1995 or 1996. The polar orbiting platform will be most likely flown at an altitude near 800-km.

NASA was flying an airborne imaging radar system on a Convair 990 aircraft until 1985. An unfortunate accident destroyed the aircraft on the ground, and a replacement aircraft is currently being acquired. The 990 aircraft was used to fly the Jet Propulsion Laboratory radar system consisting of phased array L-band and C-band systems having a quad-polarization capability. The quad-polarization, multifrequency airborne radar system has demonstrated great potential for significantly expanding the application of radar data to a wide variety of earth resource problems. Multifrequency data can be used to evaluate variations in surface roughness and dielectric constant, two parameters that are difficult to separate with single frequency radar. Presently, the greatest applications of multifrequency radar data appear to be analysis of vegetation and soil moisture, although C-band and L-band, and perhaps P-band (70-cm wavelength) radar could have major utility in discriminating geological units on the basis of their roughness. Although most natural surfaces will be smooth at P-band, that frequency could have great potential for subsurface detection of bedrock that is buried below dry, unconsolidated alluvium.

GLOBAL STEREOSCOPIC ANALYSIS OF THE EARTH

In early 1986 the French launched Systeme Probatoir d'Observation de la Terre (SPOT). SPOT as currently configured has two modes of imaging: panchromatic and multispectral. The panchromatic mode of imaging produces 10-meter spatial resolution and the multispectral mode provides 20-meter resolution. The satellite images a 60-km swath and is capable of pointing its sensors for viewing of the same area from five different perspective angles. As presently planned, SPOT-2 will fly the same sensors as SPOT-1, but SPOT-3 will have additional wavelength bands. SPOT simulator data (data acquired by an airborne scanner with characteristics similar to those of SPOT) were evaluated over Goldfield, Nevada, and were found to be of far greater utility than Landsat MSS data because of the increased spatial resolution. Ratios of SPOT simulator data were utilized to map iron oxide abundances and the image data could be easily used in the field.[18] The SPOT satellite is able to acquire stereoscopic data with four angles other than nadir. The stereoscopic analysis permitted with SPOT panchromatic data represents another major breakthrough in the analysis of terrain. For the past decade, satellite data acquired about the Earth's surface has been collected to investigate spectral reflectance landscape cover. Geoscientists develop geological information by analyzing both landscape cover and topography. SPOT represents the first analytical tool for the systematic analysis of topographic patterns (landform and drainage patterns).

The NASA Large Format Camera (LFC) was flown for the first time on the Shuttle in October 1984. The camera is a 12-inch focal length, 9-by-18-inch precision camera that provides 10-meter ground resolution. The data from the LFC will permit 1:50,000 scale topographic maps to be constructed with 20-meter contour accuracy. However, present U. S. government policy will not allow experimental data, like that acquired by the Large Format Camera, to be utilized for commercial purposes. NASA does not have additional plans to refly the LFC because it fulfilled most of its mission objectives during the STS41-G mission.

OTHER ENVIRONMENTAL SATELLITE SYSTEMS

The Canadian government has approved development of a radar satellite (Radarsat), a C-band imaging system that will provide coverage of northern Canadian lands and waterways. The European Space Agency is also conducting design reviews for a C-band imaging radar (ERS-1) to fly in the 1990s. Japan has plans for an L-band imaging radar system similar to SEASAT. In addition, Japan is conducting feasibility studies for an Earth Resources Satellite, also called ERS-1, that would provide approximately the same imaging capability as the Landsat Thematic Mapper. With the assistance of the Soviet Union, India has been developing an earth resources satellite system. Brazil has been studying the possible development of an earth resources satellite as well.

GLOBAL GEOPHYSICAL INVESTIGATIONS WITH SATELLITE DATA

For almost twenty years, NASA has been studying the Earth's magnetic and gravity fields using satellite techniques. Since 1971, scientists have discovered that variations in these fields at satellite altitudes can be measured and inferences can be made regarding the Earth's crustal composition and structure on a global basis. NASA has proposed repeatedly, but unsuccessfully, a mission called the geopotential field mapping mission, which would fly vector and scalar magnetometers for mapping the Earth's magnetic field to an accuracy of 1/50,000 of the total field (1 gamma) and would map the Earth's gravity field to 1 milligal in A-by-A-degree blocks of latitude and longitude. Detection of compositional changes (density and magnetic susceptibility differences in rocks) will be possible because the platform will be flown at an altitude of 160-km and this will provide 80-km spatial resolution. Such data will be extremely valuable for recognition of regional trends in airborne and ground-based potential field data. These regional trends can then be systematically removed to identify local anomalies that may be related to structures and compositional differences on a local scale.[19]

Subtle variations in the ocean surface may also be detected and mapped with sensitive altimeters to determine the ocean geoid—the equipotential gravitational surface that the Earth's oceans would adopt if they were solely under the influence of gravitational and rotational forces. NASA has unsuccessfully proposed a topography experiment (TOPEX) designed to measure repeatedly the sea surface to +2 cm over 30-to-40-km distances and 10 cm over 3,000- to 5,000-km distances. At this level of accuracy, it may be possible to detect subtle variations in the marine geoid that are related to subsurface continental shelf structures, such as salt domes.[20]

Several new space-related techniques have been developed for determining distances between points and elevations on the surface of the Earth and permitting dynamic movements of crustal plates to be evaluated on a global basis. Very Long Baseline Inferometry (VLBI) using extragalactic radio sources and laser ranging to satellites like Lageos (5,000-km altitude orbit) and the moon have now made it possible to measure distances over thousands of km to the +2-cm accuracy level. This capability will allow systematic establishment and maintenance of global geodetic grids. The new global positioning satellite (GPS) system will allow location of points on the ground through inferometry to an accuracy of 1 m with a few simple in-the-field calculations. These capabilities will revolutionize global geodesy and cartography.[21]

DEVELOPMENT OF AEROSPACE TECHNOLOGY FOR GLOBAL NONRENEWABLE RESOURCES EXPLORATION AND PRODUCTION

In the 1990s the world may witness rapid development of advanced aerospace remote sensing systems capable of globally collecting data about most of the Earth's continental areas. Repetitive coverage with low-resolution systems, like MODIS, will allow identification of key seasonal conditions for optimal sensing of landscape parameters of geological importance. Higher spatial resolution sensors like Landsat TM and SPOT will be able to provide regional coverage of 3,600-to-12,500-km areas at the optimal seasons for investigation of landscapes. Stereoscopic imaging systems like SPOT and multiple incidence angle radar will provide for systematic analysis of drainage and landforms on a worldwide basis.

By the mid-1990s it may be possible to acquire global digital topographic data that will permit the construction of 1:50,000 scale topographic maps with 20-m contour accuracy for any area on the Earth's continental surfaces. Site-specific areas will be defined for collection of high spectral and spatial resolution data based on regional analysis of landforms, drainage, and cover types. These data sets will permit the definition of targets for ground-based exploration, and they will document the physiography of landscapes for purposes of minimizing the impact of ground exploration on the environment. Global gravity and magnetic data will be utilized along with digital topographic data to correct airborne and ground-based high-resolution potential field data. Seismic exploration plans will be developed to minimize impact on the environment, while insuring that geological structures of interest are properly investigated. The same data sets will permit efficient design of postexploration reclamation programs. Once drilling targets have been identified, access to the areas to be drilled, availability of water, and sensitive environmental areas can all be defined with the aerospace remote sensing data sets.

If economic deposits are found as a result of the exploratory drilling, then the aerospace remote sensing data can be utilized to evaluate local site conditions for the resource extraction plant; to locate haulage roads, pipelines, and power lines, and to identify locations for shipping facilities. All this can now be done with the idea of minimizing the impact of the development of these facilities on the environment and developing efficient reclamation plans that will permit other multiple uses of the same landscape.

Currently, aerospace remote sensing data is being utilized to monitor the effectiveness of reclamation of lands previously used for resource extraction in the United States. With these new global tools, the effectiveness of land reclamation can be monitored on a global basis.

NOTES

[1] G. O. Barney, 1980, *The Global Report to the President: A Report Prepared by the Council on Environmental Quality and the U. S. Department of State,* Volume II, U. S. Government Printing Office, 382 pp.

[2] J. Paone et al., 1976, *Land Utilization and Reclamation in the Mining Industry, 1930–1971,* U. S. Department of the Interior, U. S. Bureau of Mines Circular 8642.

[3] J. L. Burnett, 1984, "1983 Mining Review," in *California Geology,* October, California Division of Mines and Geology, Sacramento, California, pp. 215–220.

[4] Barney, 1980, *op. cit.*

[5] J. V. Taranik and C. M. Trautwein, 1977, "Integration of Geological Remote-Sensing Techniques in Subsurface Analysis," in *Subsurface Geology, Petroleum and Mining Construction,* LeRoy and LeRoy, eds., Colorado School of Mines, pp. 564–586.

[6] J. V. Taranik, 1985, "Characteristics of the Landsat Multispectral Data System," in *The Surveillant Science, Remote Sensing of the Environment,* 2nd ed., R. K. Holz, ed., John Wiley and Sons, New York, pp. 328–351.

[7] J. V. Taranik, 1982, "Geological Remote Sensing and Space Shuttle: A Major Breakthrough in Mineral Exploration Technology," *Mining Congress Journal,* Vol. 68, No. 7, American Mining Congress, July, pp. 18–25.

[8] J. V. Taranik et al., 1978, "Targeting Exploration for Nickel Laterites in Indonesia with Landsat Data," in *Proceedings 12th International Symposium Remote Sensing of Environment,* Volume II, Manila, Philippines, Environmental Research Institute of Michigan, Ann Arbor, Michigan, pp. 1037–1051.

[9] J. V. Taranik and M. Settle, 1981, "Space Shuttle: A New Era in Terrestrial Remote Sensing," *Science,* Vol. 214, November 6, American Association for the Advancement of Science, Washington, DC, pp. 619–626.

[10] J. V. Taranik, 1981, "Advanced Technology for Global Resource Applications," in *15th International Symposium Remote Sensing of Environment,* Vol. I, Environmental Research Institute of Michigan, Ann Arbor, Michigan, pp. 1–15.

[11] A. F. H. Goetz, L. C. Rowan, and M. J. Kingston, 1982, "Mineral Identification from Orbit: Initial Results from the Shuttle Multispectral Infrared Radiometer," *Science,* Vol. 218, December 3, American Association for the Advancement of Science, Washington, DC, pp. 1020–1024.

[12] S. C. Feldman, J. V. Taranik, and D. A. Mouat, 1985, "Airborne Imaging Spectrometer (AIS) Analysis of the Southern Hot Creek Range, South Central, Nevada," JPL Publication 85–41, AIS Workshop, April 7–9, Pasadena, California, pp. 56–61.

[13] IOC-2 will probably fly at 705-km altitude. The spectrometer, called High Resolution Imaging Spectrometer (HIRIS), will probably fly for over 15 years and have a 50-kilometer swath width, 30-meter ground resolution, and 196 spectral bands with 11-nm bandwidths.

[14] J. V. Taranik, D. Davis, and M. X. Borengasser, 1985, "Application of Thermal Infrared Multispectral Scanner (TIMS) Data to Mapping of Plutonic and Stratified Rock Assemblages in Accreted Terrains of the Northern Sierra, California," in *Proceedings, First TIMS Conference*, National Space Technology Laboratories, June 18–20, Jet Propulsion Laboratory, Pasadena, California (Abstract).

[15] J. V. Taranik, A. Hutsinpiller, and M. X. Borengasser, 1985, "Detection and Mapping of Volcanic Rock Assemblages and Associated Hydrothermal Alteration with Thermal Infrared Multispectral Scanner (TIMS) Data, Comstock Lode Mining District, Virginia City, Nevada," in *Proceedings, First TIMS Conference*, National Space Technology Laboratories, June 18–20, Jet Propulsion Laboratory, Pasadena, California (Abstract).

[16] J. F. McCauley et al., 1982, "Subsurface Valleys and Geoarcheology of the Eastern Sahara Revealed by Shuttle Radar," *Science*, Vol. 218, December 3, American Association for the Advancement of Science, Washington, DC, pp. 1004–1020.

[17] M. X. Borengasser and J. V. Taranik, 1985, "Application of Shuttle Imaging Radar-B (SIR-B) Data to Tectonic Analysis of the Candelaria Region, Nevada," in *Proceedings 4th Thematic Conference on Remote Sensing for Exploration Geology*, International Symposium on Remote Sensing of Environment, Environmental Research Institute of Michigan, Ann Arbor, Michigan, April.

[18] M. X. Borengasser and J. V. Taranik, 1985, "Comparison of SPOT Simulator Data with TMS and MSS Imagery for Detection of Alteration, Goldfield/Cuprite, Nevada," *Photogrammetric Engineering and Remote Sensing*, Vol. 51, No. 8, August, pp. 1109–1114.

[19] J. V. Taranik and P. G. Thome, 1980, "Development of Space Technology for Resource Applications in the 1980s," 14th Congress, International Society of Photogrammetry, Hamburg, Germany, 22 pp.

[20] M. Settle and J. V. Taranik, 1982, "Mapping the Earth's Magnetic and Gravity Fields from Space: Current Status and Future Prospects," in *Proceedings Symposium on Remote Sensing and Mineral Exploration*, COSPAR XXIV, Ottawa, Canada, May.

[21] Taranik and Thome, 1980, *op. cit.*

EDITORS' INTRODUCTION TO:
C. K. PAUL, M. L. IMHOFF, D. G. MOORE AND A. N. SELLMAN
Remote Sensing of Environmental Change in the Developing World

This chapter carries forward the discussion of advanced technology begun by Drs. Orio and Taranik. The authors of this chapter, Paul, Imhoff, Moore, and Sellman, have worked for organizations that have used remote sensing to study problems of the developing world. Here they focus on the need to monitor environmental changes and the linkages between such change and economic development.

Three case studies are presented to document the integral role remote sensing is playing along with other planning tools in guiding development of resources: the use of remote sensing as an input to geographic information systems for natural resource management in Peru; the monitoring of forests that serve as buffers against the impacts of severe storms in Bangladesh; and the monitoring of drought in Africa. The unifying theme in these examples is the basic need to develop monitoring systems that can be applied to large-scale problems. While the chapter by Drs. Gwynne and Mooneyhan discussed a single system that could be used to investigate problems that were inherently global in scale, the discussion in the current chapter suggests that remote sensing can be employed at a more local level in many areas of the globe to provide a deeper understanding of the complex linkages between deteriorating resources, economic poverty, and social unrest.

13 REMOTE SENSING OF ENVIRONMENTAL CHANGE IN THE DEVELOPING WORLD

C. K. PAUL
U. S. Agency for International Development, Washington, DC

D. G. MOORE
EROS Data Center, U. S. Geological Survey Sioux Falls, South Dakota

M. L. IMHOFF
Goddard Space Flight Center NASA, Greenbelt, Maryland

A. N. SELLMAN
Environmental Research Institute of Michigan, Ann Arbor, Michigan

INTRODUCTION

There is much concern that our natural resource base is deteriorating around the world, but the reality of such a deterioration has been questioned because so little data are available. This is especially true in regard to deforestation in the humid tropics, the desertification of once fertile soils, and the consequent loss of biological diversity as habitats are destroyed. Global monitoring by remote sensing purports to document these losses, but making an inventory of natural resources at large scales (larger than a map scale of 1:250,000) for specific development projects puts the technology of remote sensing to the test, especially in the developing countries where data are generally lacking. In local development projects, remote sensing must go farther than measuring trends; timber volumes must be estimated, crop production must be forecast, the soil resource must be mapped, and water must be found. Lack of such information impedes development and can create the potential for environmental degradation should development proceed on the foundation of an inadequate knowledge base.

Three case studies are documented in which remote sensing technology is playing an integral role with other planning tools in guiding development. The first study concerns the management of natural

205

resources in highland forests in Peru, the second is the mapping of lowland humid forests in Bangladesh, and the third is a plan to gauge the effect of the African drought on crop and range production. While these studies focus on developing nations, the problems transcend these areas and point to the use of remote sensing and geographic information systems as tools for monitoring the environment in the industrialized world as well.

NATURAL RESOURCE MANAGEMENT IN PERU

Modern data organization and analysis tools, such as computerized geographic information systems, can help assess the agricultural development potential of a region. An excellent example of the use of a geographic information system for assisting resource management occurred recently during a two-year U.S. Agency for International Development (A.I.D.) funded program with Peru's National Office for Natural Resource Evaluations (ONERN). One of three training and demonstration sites used in this program covered the town of Aucayacu in the Huallaga Valley in east central Peru. For this demonstration, a 100-square-kilometer area was selected where, among other things, new cropping systems are being considered in order to improve farm income.

Soil and topographic maps, and color-infrared aerial photography (1:10,000 scale) collected by ONERN were assembled. They provided complete and uniform coverage of the demonstration area, which is depicted in Figure 13–1 (see color insert). Information about roads and rivers was entered into the computing system directly from topographic maps, but road network information was updated with aerial photography (see the upper left and center of Figure 13–1). Elevation contours were also included and then converted into slope classes using specialized software. The resulting slope classes are depicted in the center of Figure 13–1. The soil classes (center left), forest classes (upper right), and crop types (center right) are also illustrated. Forest and agricultural land use classes were derived by manual interpretation of the aerial photography. With these basic features, precisely registered with one another as digital files within the geographic information system data base, a variety of analyses can be performed as illustrated by the bottom row of images in Figure 13–1.

The lower left image shows intensively cultivated row crops that are grown on a specific soil type (Tocache) with virtually no slope (no more than 2 percent). The lower center image shows the location and distribution of a specific crop on four different slope classes. In this case the intent was to discover what percent of the crops grown in this area was found on the two steepest slope classes (10 to 25 percent; greater than 25 percent). Soils on these slopes would, in all likelihood, suffer the most serious erosion problems given existing agricultural practice. The lower right image was generated to depict the suitability (best to worst depending upon color) of the entire region for intensive cultivation.

This particular demonstration of an information-system-based approach to resource analysis is significant for several reasons. First, the geographic information system is operated and maintained in Peru by ONERN staff, and these personnel performed the data collection and analyses as well—demonstrating that people can be trained readily to use the sophisticated geographic information system technology. Second, many iterations of the analysis can be performed, with the relative influence of parameters changed in each case, until acceptable alternatives are derived that meet specific resource management needs. These alternatives can often be viewed and assessed in a matter of minutes or a few hours, owing to the inherent processing speed of computerized systems. Also, all factors may be treated in a consistent manner, which can help to eliminate much of the bias and errors associated with conventional analysis, since it is limited only by the availability of appropriate data and the insight of the analyst using the system.

FORESTS AS BUFFERS TO SEVERE STORMS IN BANGLADESH

Bangladesh suffers continually from floods. Mangrove forests serve as a buffer, which can lessen the areal extent and severity of flooding of coastal areas. In these regions, deforestation of the critical mangrove buffer allows tidal waves to surge more deeply into the densely inhabited interior. The proper planning of development and conservation activities in this region is currently hampered by a lack of hydrographic, climatic, and standard survey data. The acquisition of conventional synoptic imagery that could provide much of this information has been limited in the past due to obscuring cloud cover during periods of inclement weather. It is the possible acquisition of cloud-penetrating, synoptic image data of the Earth's surface by the L-band (23.5 cm) Shuttle Imaging Radar (SIR) during periods of severe weather that can revolutionize flood control and disaster prevention efforts.

A shuttle-based radar flight in October 1984 permitted penetration of the cloud cover and produced detailed images showing surface conditions during the preceding monsoon period. Radar images at incidence angles of 26, 46, and 57 degrees were collected to analyze the effects of the important mangrove forests on controlling inundation by flood waters.

Mangrove forests in Bangladesh represent an important economic resource to the country. Revenue from the licensed harvesting of tree crops and fish can exceed 10 million U.S. dollars in any one year, and the presence of these coastal forests protects the interior from tidal surges. The health and productivity of these forests depend upon the environmental conditions of the soil and carefully balanced cycle of tidal inundation, freshwater flooding, and drying. The ability of the radar to penetrate the vegetation canopy and portray the shape of the forest floor could provide the means for mapping the carrying capacity of mangrove soils and for monitoring changes in the soil environment.

Images from SIR-B radar using all three incidence angles were evaluated for their ability to penetrate vegetation and for their usefulness in mapping the density, extent, and health of mangrove vegetation. Analysis showed that all three incidence angles were penetrating the 12.5-meter-high mangrove canopy. The resulting images actually showed the topographic contour of the forest floor as it was being differentially flooded by the tidal cycle. Little difference in information content was found as a function of radar incidence angle for mangrove areas. The ability to map and monitor the contour of the mangrove forest floor will aid foresters and ecologists in their determinations of stand development and health.

The SIR-B radar at the three incidence angles provided a valuable set of image data over the flood plains of eastern and central Bangladesh. The somewhat intermediate angle of 46 degrees proved most useful for separating important features, such as agricultural areas and floodwaters. Area classes corresponding to flood boundaries, village embankments, and agricultural crops were derived and merged with Landsat Multispectral Scanner (MSS) land use/cover classification of the same area taken weeks earlier. The merged Radar/MSS data set showed the extent of flood inundation and provided a real measure of the temporal change that had taken place between the collection of the MSS and Radar data sets. The change in flood area was then correlated to river gauge measurements and an index was derived describing inundated land area as a function of river level. This application of remotely sensed data demonstrates the utility that radar-based measurements can have on activities that must be carried out in flood-prone environments.

MONITORING THE AFRICAN DROUGHT

The recent famine in east Africa and portions of the Sahel has reminded the world's food suppliers that an adequate early warning system is among the highest priorities of the food-donor and the African nations. Although several technologies provide pieces of the total information system, these are not yet fully integrated into a global early warning "alert" system. Even with the El Niño/Southern Oscillation phenomena and its suspected climatological effects being reported almost daily, the recent African drought seems to have caught a number of the industrialized food-producing donor nations by surprise. Abandoned sites and spot price rises in food, indicators of drought years before the full brunt of famine is evident, tend to be ignored as researchers pursue the elusive panacean informational solutions offered by a limited set of data collection methodologies. Early warning and the integration of several advanced information processing and analysis technologies are so important that the subject was highlighted at the Economic Summit in Bonn, Germany, in 1985.

A variety of data collection and analysis methods are used or proposed for estimating food and fiber supply and utilization for global programs of famine relief. The approaches vary from country to country but the common goals are to provide early warning, reliable supply, and utilization statistics for food balance sheets, and to prepare and implement food distribution programs.

The technical and administrative challenges of implementing these bilateral, yet global, programs are complex. General guidelines are being developed by the Economic Summit states for the design of a technically appropriate and cost-effective automated system that can take maximum advantage of existing cultural and resource data. The system must provide early warning but must balance this with reliability. The procedures must be sensitive to costs, must be amenable to implementation by developing countries, and must create time and cost efficiencies for skilled administrators and technicians, especially during times of famine crisis. Errors both of overestimation and underestimation can result in human suffering and population instability. The system must be capable of accepting data that are being collected and used in population, development, and environmental programs. The system must have both an assessment and a predictive capability. These and the constraints normally associated with such systems are being examined to make certain that the system will truly respond to famine relief information needs of both the donor and recipient countries and not just to the convenience of modern technology.

Both supply and use of food must be considered in any proposed system. Supply data must be gleaned for current production figures, stored food, imports, exports, and food aid. Use of food can be optimized by considering various data sources such as land potential, population statistics, socioeconomic data, current land use, agricultural practices, and historic consumption patterns. Satellite, aircraft, and field data are collected and, through computer automation, formatted, archived, retrieved, analyzed, and synthesized for reports. Data are best collected at the country level, with a regional approach to data analysis and reporting.

Figure 13–2 shows a qualitative comparison of reporting accuracy versus various data collection approaches. Accuracy of food estimates increases as one proceeds from strictly climatic and vegetation indices through other approaches such as agroclimatic models and land potential stratification to field-level area frames. Costs go up as one moves to higher resolution data collection. As shown at the right side of the accuracy curve, when food policy analysts attempt to use these current approaches and data to expand to regional estimates and make predictions, accuracy quickly falls off in a discontinuous manner—a condition that foreign assistance agencies now face in dealing with the African famine.

A computerized geographic information system incorporating these data approaches could assist in determining the next level of approach for continuing observations to improve food supplies. By factoring in costs, the optimal combination of approaches can theoretically be determined. The eight Economic Summit states are now collaborating in the development of such a system.

An additional advantage of a geographic information system approach to the problems caused by drought and food and fodder supply shortages is the ease of incorporation of information concerning spatial distributions of both problem areas and the infrastructure for supply relief. It is not only important to understand that a given problem exists and where it is located, but relief planners must also know the locations of roads, rail lines, shipping and port facilities in order to get supplies to affected areas. This data must also be incorporated into a geographic information system.

SUMMARY

The three case studies discussed describe ways in which remote sensing and associated technologies can contribute to an understanding of environmental change and its link to economic development. The

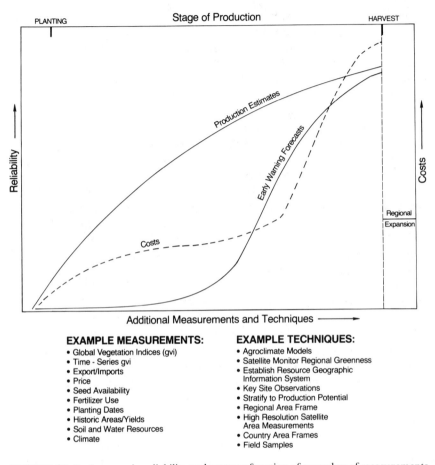

FIGURE 13–2 Increase in reliability and cost as a function of a number of measurements and techniques.

African drought is a continentwide phenomenon; the suggested integration of various measurement approaches has significance for global science and monitoring programs that purport to provide status reports on the "health" of the planet. However, the long-range goal of the integrated approach is to alleviate poverty, feed the hungry, and hopefully restore the land base of Africa. For this reason, likely resources to support the development of an integrated approach will probably be orders of magnitude larger than studies to push the frontiers of knowledge about our planet, even though methodologies of providing data may be similar. An approach must also be advanced that understands the complex linkages between environmental conditions and socioeconomic conditions. It is important, then, that we are sensitive to the variety of needs of developing nations. Transportation, effective organizations, and a knowledge of culture are also critical as we approach these problems in the developing world.

In the cases of Bangladesh and Peru, the studies presented cover small areas, but are representative of problems generic to lowland tropical moist rainforests (in the case of Bangladesh) and highland development (in the case of Peru). Bangladesh highlights the dramatic destruction of mangrove forests due to severe storms—because of cloud conditions in the tropics, radar systems probably represent the only long-term feasible solution to remote sensing data collection, especially during the rainy season. Peru typifies the environmental problems associated with forest clearings on fragile upland slopes—a geographic information system with data inputs from remote sensing was essential to look at natural resource problems in the region.

The unifying theme in all three case studies is the link of a basically global monitoring system (including earth resources satellites) to local development problems. The United Nations Environment Programme (UNEP) Global Resources Information Database (GRID), discussed in the chapter by Gwynne and Mooneyhan in this volume, tabulate and geocode information on a global scale that sets the physical basis on which all development and quality of life depend. The anthropomorphic impacts on the natural resource base can only be truly assessed at local scales—within areas on the scale of a single or a local village. The understanding of physical resource use, misuse, enhancement, or degradation requires local data collection at high resolutions (30 meters or better). It requires the active participation of local indigenes. Geographic information systems at this level must include social, economic, political, and other forms of data that capture the hopes and trials of human endeavor in order to be relevant to planning and other development and survival processes. The merging of the global types of data bases (such as GRID) and local ones such as those discussed in this chapter and that of Risser and Iverson, should prove of great importance in the coming years as more of the natural state of the planet is challenged to cope with the effects of intensive development.

EDITORS' INTRODUCTION TO:
W.G. BROONER
The Application of Remote Sensing in South America:
Perspectives for the Future Based on Recent Experiences

This chapter carries forward the theme of the use of remote sensing to monitor environmental conditions in the developing world discussed by Dr. Paul and his colleagues in the previous chapter. Here Mr. Brooner focuses on the use of remote sensing in South America. Drawing on more than 15 years of experience with Earth Satellite Corporation and more than 10 years work with remote sensing on developmental issues in South America, Mr. Brooner makes the case that remote sensing can provide information to guide field studies and aid in planning and monitoring, even though a number of unanswered questions exist with respect to the sources of analysis error in the use of these data. Brooner uses a number of examples to illustrate his point: the impact of Vinal, a woody invader plant in the Chaco; a multitemporal analysis of the riverine environment of the Rio Pilcomayo; production of base maps and improved cartographic materials in Peru; and monitoring of agricultural development in Amazonia. In the first section of this volume, Dr. Lovejoy described the ecological affects of deforestation in the South American tropics. In the current chapter, Mr. Brooner shows that we can measure other effects of this deforestation using remote sensing.

Mr. Brooner presents a compelling case for the application of satellite remote sensing in South America, stressing the need for coordinated satellite, airborne, and ground data collection. He also makes the point that we must keep abreast of new technological developments such as those presented by Taranik and Tuyahov, Star, and Estes later in this volume. Indeed the more focused work by Tinney and Lackey (Chapter 17) carries this theme forward as well, presenting the use of remote sensing for monitoring energy facilities in the United States.

14 THE APPLICATION OF REMOTE SENSING IN SOUTH AMERICA: PERSPECTIVES FOR THE FUTURE BASED ON RECENT EXPERIENCES

WILLIAM G. BROONER
Earth Satellite Corporation
Bethesda, Maryland

INTRODUCTION

Following the launch of Landsat-1 in July 1972, many remote sensing training and technology transfer programs have been developed in South America. During the 1980s, programs have progressed beyond earlier training stages and institutional development to the level of ongoing resource and environmental analysis and management programs. These operational remote sensing applications and programs may be quite sophisticated, but they are often constrained by inadequate funding.

In 1980, a professional scientific society, the Society of Latin American Specialists in Remote Sensing (SELPER), was created to promote professional improvement and affect regional cooperation. After six years, SELPER has over 450 members, 15 national chapters, has organized annual symposia and workshops, and begun a scientific journal (*Revista SELPER*). In addition, international training programs (e.g., CIAF in Bogota and IAGS/Carto School in Panama) have modernized their remote sensing and geographic information system curriculums and facilities to accommodate recent technology and applications developments.

Successful programs have usually employed several types of remote sensing systems, including aerial cameras, Landsat and meteorological satellites, and have provided information in both analog and digital form. Experience gained in these programs by the Earth Satellite Corporation has demonstrated that computer-enhanced Landsat imagery has been especially useful for resource inventories and evaluations in remote, sparsely settled areas, and over large areas where basic maps and other current data on resource use are lacking or inadequate.

Over the past ten years, the author has had the opportunity to observe and participate in a wide variety of remote sensing applications and to work closely with organizations applying advanced remote sensing technologies throughout South America. Several examples are briefly reviewed in the following discussion. These examples provide an overview of current applications of remote sensing and information systems technology and their relationship to issues of regional environment and development in South America.

VINAL: A WOODY INVADER PLANT IMPACTING THE CHACO REGION

Vinal (*Prosopis ruscifolia*) is a tree found in the Chaco region of northern Argentina. The Vinal tree grows rapidly, and at three to five years of age may achieve heights of 3 to 5 meters. An important characteristic of this tree is the presence of sharp thorns of up to 15 cm in length.

Vinal is an invader species, aggressively colonizing different environments, and it grows in association with a wide variety of vegetation. Massive colonization of Vinal over extensive areas has created serious problems for pasture and range management. Typically pastureland is lost, but in addition, soil erosion can increase when Vinal replaces other natural vegetation.

The explosive diffusion of Vinal has affected large areas in the Argentinean provinces of Formosa and Chaco, as well as Santiago del Estero and Salta. This invasion of Vinal has been aggravated since 1935 by the interaction of natural as well as cultural factors. Cultural factors that have contributed to the spread of Vinal include: land clearing for cultivation, pastures, and roads; drainage of marshes; and inappropriate use of fire. Periodic flooding has also produced conditions suitable for Vinal colonization (Figure 14–1). As a result, there are presently areas with densities of Vinal as high as 2,000 to 3,000 plants per hectare. Once established, Vinal plants produce seeds that are transported to new sites by several mechanisms, including cattle and water.

Soil type appears to not be an important factor for the establishment and growth of Vinal. Rather, Vinal appears to be able to establish itself

FIGURE 14–1   Behind the fence, the low forest is Vinal, 20 to 25 years old and of mature height. The wide grass right-of-way between the fence and a road has been excavated for roadbed fill; following heavy rains, these low areas fill with water. Vinal, in the form of low bushes, typically invades such grass areas along roadsides.

and thrive on any soil, limited only by inundation of water. In fact, research conducted in recent years has demonstrated that flooding of Vinal sites for a period of six months is considered to be one of the most effective control mechanisms.

Investigations conducted in 1980–1981 demonstrated the application of Landsat analysis, complemented by aerial reconnaissance and ground surveys, for mapping the distribution of Vinal, its occurrence in different ecosystems, and its temporal dynamics.[1] Specific objectives included analyses to: 1) distinguish areas occupied by Vinal from other principal forest species; 2) distinguish associations of Vinal and other classes of *Prosopis* species; 3) detect areas of instability or potential invasion by Vinal; and 4) assess changes and rates of expansion by Vinal.

Successful results were demonstrated for a 7,500 km^2 project area located in Formosa Province, Argentina. Multidate Landsat imagery, multispectral photography (both aerial and terrestrial), field survey data, meteorological records, and available literature were integrated into the study.

Landsat data for two dates were processed by computer to produce images used in the analysis: October 8, 1980 and June 13, 1975. The 1980 scene was one of the first Landsat scenes acquired by the Argentinean ground receiving station and was the first scene to be processed digitally and analyzed outside of that station.

Landsat analyses were conducted at 1 : 100,000 scale. For field surveys, prints were enlarged to 1 : 60,000 scale, a scale found particularly useful for detailed field observations and for image-to-ground feature correlations for subsequent image classification and analysis (see Figure 14–2).

Multispectral classification of the 1980 Landsat scene used a unique statistical approach (in technical terms, combining unsupervised and supervised techniques) to "map" nine categories of land cover in a 294,400-hectare area. Each land-cover category was related to ecological characteristics of the environments. The probable occurrence of Vinal in each category, as opposed to other vegetation species or land-cover types, was estimated. Vinal occupied an estimated 82,270 hectares, or 28 percent, of the classified image area in October 1980. The study concluded that: 1) It is possible to differentiate and stratify Vinal occupied areas from other principal species in some but not all ecosystems. It is, however, possible to differentiate and stratify ecosystems, and Vinal occurrence may then be estimated when its probable density in

FIGURE 14–2 During ground surveys, locations of observations were correlated with 1 : 60,000 and 1 : 100,000 scale Landsat GEOPIC™ images. They also provide a "map" to guide navigation (both ground and aerial) in project study areas.

each ecosystem is known. 2) Areas of instability or high potential for invasion of aggressive woody plants may be identified by correlation between land cover and ground observations. These include grassy covered areas bordering esteros, along roads and in pastures, as well as areas of low forest with open canopies (which are also typically grazed).[2]

Subsequently, vegetation change was analyzed using Landsat imagery for an area of approximately 7,900 km^2. The June 13, 1975, Landsat scene was digitally combined, processed, and analyzed with the October 8, 1980, Landsat scene to produce a three-category change image that showed: 1) areas of increased vegetation that commonly corresponded to increased occupancy of Vinal; 2) areas of decreased vegetation that are generally in esteros and cultivated areas, interpreted as residual effects of images acquired in different seasons; and 3) areas of no significant change.

The category that included changes in Vinal area represented 976 km^2, or about 14.2 percent of the temporal change image. This analysis suggested that Vinal invaded over 9,000 hectares per year, an annual rate of increase of 7 to 10 percent.

Rapid proliferation of Vinal continues to affect both the economy and ecology of Formosa Province. This use of Landsat data represented original and unique applications. In this study, objectives were successfully met, new areas of research were identified, and new and additional information on Vinal was provided to provincial and national scientists and planning officials in a cost-effective and timely manner.

RIO PILCOMAYO: LANDSAT TEMPORAL ANALYSIS OF A RIVERINE ENVIRONMENT

The Pilcomayo River flows out of the Bolivian Andes across the Gran Chaco of Argentina and Paraguay. In northwestern Formosa Province of Argentina, the river forms part of the international boundary between Argentina and Paraguay, flowing northwest to southeast over a slight gradient of only about 0.2 meters per km. There the river deposits sediments, frequently in excess of 100 million tons annually.

As noted in other studies, multitemporal characteristics have increased Landsat data's value for many resource management applications. In a program to monitor temporal hydrological variations on the Pilcomayo River, the nature and impact of changes in hydrologic features were analyzed. As part of this program, both natural and man-made changes in the Pilcomayo River floodplain that occurred between 1972 and 1981 were analyzed using Landsat data.[3]

Interventions in the Pilcomayo River's course, through both mechanized channelization and by seasonal overbank flow, and varying depositional processes continually alter the landscape. The most significant events in the study area during the nine-year period of Landsat coverage was the construction of canals to divert flow from the Pilcomayo River into adjacent basins in both Argentina and Paraguay; these diversions constructed in the mid-1970s are upstream from normal overflow points. A computer-enhanced temporal sequence of Landsat images showed clearly precanal and postcanal hydrological patterns.

Physiographic and hydrological characteristics of an adjacent basin in Formosa Province, Banado La Estrella, result from the dynamics of overland and shallow channel flow of seasonal floodwaters originating from the Rio Pilcomayo. During the past decade, additional waters, fed into the Banado through several canals from the river, expanded the areal extent of inundation in the Banado and lengthened the period of inundation. Because of increased water levels, several settlements had to be relocated away from flooded areas. The flooding also destroyed substantial areas of forest within the Banado that had served as winter habitat for migratory fowl.

Seven Landsat scenes covering the 17,500 km^2 Pilcomayo River study area were acquired and processed at 1 : 100,000 scale by computer. These images covered the period between September 1972 and March 1981, and their acquisition involved three different Landsat satellites and three ground receiving stations (one each in Argentina, Brazil, and the United States).

A series of maps were produced of surface hydrology from Landsat acquisitions of September 1972, March 1976, February 1978, March 1979, and November 1980. Each map contains probable limits of maximum inundation; active canals; abandoned canals (or those that transport water only part of the year); primary inundation surfaces and areas of active sediment deposition; areas covered by water on a corresponding image date; areas generally covered by water during periods of peak inundation; and areas generally not subjected to inundation.

Comparisons of imagery taken in temporal sequence illustrate where the river has shifted its channel by normal down-valley sweep and migration of meanders and bends. Within the single Landsat scene study area, between 1972 and 1980 the length of the Pilcomayo River was considerably shortened, from approximately 184 km in 1972 to only about 76 km in 1980, as a result of diversions of its discharge into both Argentina and Paraguay through the series of man-made canals.

The Pilcomayo River is now divided into two branches in the vicinity of Puerto Irigoyen, and its new branches are in the process of carving new channels.

A variety of additional Landsat analyses and products from the study included:

- Landsat photomaps, which provide a geographical overview of the physical resources and infrastructure of northwest Formosa Province for regional mapping, geographical and natural resources studies, and planning applications
- Temporal analyses of water-level change showing the spatial impacts of human-caused diversions of the Rio Pilcomayo. These led to extensive changes in the Banado La Estrella system, which were documented in color at 1 : 100,000 scale.
- A land-cover classification map, produced using the 1980 Landsat scene, shows twelve categories of land cover in an area of approximately 1,530,000 hectares. The land-cover categories generally relate to ecological aspects of the environments, incorporating soil characteristics, landform, and climatic variations. Results were shown both cartographically in a color classification image (1 : 100,000 scale) and in tabular form.

Within the project area, gross variations in vegetation appear to be related more closely to yearly ranges in near-surface soil moisture content than to soil texture or age. Changes in soil moisture caused by stream divagation and frequency can affect markedly the existing vegetation. The time required to establish dry-adapted hardwood vegetation on abandoned levees appears to be hundreds of years after the ground is left bare by the disappearance of hydrophytic vegetation. Invasion rates of preferred woody forest species, adapted to semiarid conditions characteristic of older uplands, are slow. In the meantime, barren areas are susceptible to rapid invasion of undesirable species, such as Vinal.

The information developed through the use of Landsat imagery on the hydrologic dynamics and morphological processes of the Rio Pilcomayo–Banado La Estrella system continues to be used to design and assess resource management in the Formosa Province.

BASE MAPS AND IMPROVED CARTOGRAPHY

Many areas of the world continue to be poorly mapped. Available cartographic products are commonly of small scale and contain significant errors, many of which the map user may not be aware of. Since the first launch in 1972, Landsat has been promoted for its cartographic value, and numerous examples have documented cartographic updating applications.

Landsat maps provide a broad geographical overview of physical resources and diverse environments for regional mapping and multiple resource applications. Landsat maps have provided a unique cartographic base map, particularly in areas where base maps of similar scale are unavailable. These provide an image of the landscape for interpretation, analysis, and planning, and a common reference for future maps.

Noteworthy among the lengthy list of Landsat maps produced by many organizations is the recent program to produce planimetric Landsat maps, at 1 : 250,000 scale, of the Republic of Peru. Institute for Applied Geosciences, West Germany, was selected in an international competition in 1983, and recently successfully completed the project, which includes over 90 maps sheets. The Landsat images were processed by computer to include rigid specification for geometric registration to ground control points. On each sheet, latitude and longitude as well as other geographic coordinates were accurately plotted. Many Peruvian and international agencies participated, coordinated by the Office of Natural Resources Evaluation (Officina Nacional de Evaluacion de Recursos Naturales) in Peru. Previously, topographic maps of Peru were confined principally to coastal regions, covering less than 40 percent of the country, and these maps were commonly many years out of date. Already, the new Peru Photomap Series is proving to be an invaluable planning and resource management tool. Its cartographic accuracy provides controlled map coverage for vast areas of the nation for planning development.

Most Landsat maps and mosaics prepared to date have used the Multispectral Scanner (MSS) imagery. Increasingly, Thematic Mapper imagery from Landsat-4 and Landsat-5 (and most recently, SPOT imagery) are being used because they provide significantly better resolution and geometric fidelity than the MSS imagery. Welch et al. have recently reported that:

> the geodetic accuracy of the Thematic Mapper data is compatible with relatively large-scale map products, and the completeness of planimetric data are best suited for the production of image maps of 1 : 100,000 scale as demonstrated by the USGS maps of Dyersburg (1983), Washington, DC (1984), and Great Salt Lake (1985). Another potential use of the TM data is for the revision of existing maps, particularly for changes in the boundaries of urban areas and water bodies, or the relocation of major transportation features.[4]

Thematic Mapper data for large areas of South America are acquired through the Brazilian ground receiving station at Cuiaba. In September 1985, four scenes were processed by computer and put together into a single map covering approximately 120,000 km^2 of northern Bolivia by Earth Satellite Corporation. The region is generally void of accurate maps or geodetic control; river courses and position of roads and water bodies are charted only at 1:1,000,000 or smaller scales. The TM imagery revealed significant misrepresentation of these features on existing maps. The TM mosaic of 1:250,000 scale and individual scenes at 1:100,000 and 1:50,000 scales have provided accurate base maps, sources for cartographic revision and compilation, and a pictorial base for resource analysis and reconnaissance.

MONITORING AGRICULTURAL DEVELOPMENTS

Colonization and agricultural expansion into the tropical forests of the Amazon Basin have both positive and negative connotations. For some the result is economic development, expanding food and fiber production, settlement, and livelihood for the Earth's expanding population. For others, deforestation has meant loss of species (documented in Chapter 6 by Dr. Lovejoy) and an increase in carbon dioxide in the atmosphere.

The Amazon Basin includes about 66 percent of the remaining tropical forest in the world and is estimated to contain about 14 percent of all the carbon stored in the Earth's vegetation. In addition, the Amazon is thought to contain the world's greatest remaining pool of biotic diversity. Taxonomists have identified one-half million of the several million species living in this biome.[5] The Brazilian government has supported programs of colonization and exploitation of the Brazilian portion of the Amazon Basin over the past two decades in attempts to integrate the "isolated" and "undeveloped" Amazon Basin into the national economy. Large forest areas along the Transamazonian Highway and in the states of Rondonia, Mato Grosso, Para, and Maranhao have been cleared rapidly for agricultural settlement.

Estimates of the tropical forest area cleared in Brazilian Amazonia over recent years vary by 100 percent, while estimates of the annual rate of clearing vary by 1,000 percent, from as low as 10,000 km^2 to 100,000 km^2 per year. As stated by Tucker et al.,[6] "The lack of any systematic data on the subject has hampered discussion of what many scientists feel to be a major ecological crisis with serious biological, climatic and political ramifications." Application of remote sensing technology can help narrow these differences by providing an unbiased source of data from which accurate information on forest clearing can be obtained.

The population of Rondonia grew 16 percent per year during the last decade.[7] The doubling time for this rate of growth is about five years. Fearnside in 1982[8] projected that Rondonia may be completely cleared of forests by 1988 if trends observed continued to that date.

Landsat imagery over central Rondonia was analyzed to determine the expansion of colonization activities and analyze changes in forest land clearing. This work involved personnel of a number of organizations, working collectively and independently.[9] Work by Tucker et al.,[10] for example, has reported the acquisition and analysis of Advanced Very High Resolution Radiometer (AVHRR) imagery from United States National Oceanic and Atmospheric Administration (NOAA) satellites to identify an area of about 100-by-400 km in Rondonia where massive forest clearing is occurring. Information gained from an analysis of the AVHRR data were subsequently verified by field studies, which observed large-scale colonization programs in the areas identified on the images.

Landsat imagery reveals a series of linear features extending from highway BR-364 in central Rondonia (Figure 14–3). These linear features are typical of organized colonization programs involving large-scale, systematic forest clearing. Similar linear patterns can be found in other colonizing areas of the Amazon Basin, in Brazil as well as in neighboring Bolivia, Peru, and Colombia. The central Rondonia example is unique, however, because of the magnitude of recent change occurring and monitored by Landsat.

Within a single Landsat scene acquired in June 1976, linear features interpreted as associated with colonization along a 220-km stretch of highway BR-364 totaled some 1,300 km in length. This area is immediately west of the town of Rondonia.

Comparison with a September 1981, Landsat image of the same area reveals massive expansion of land clearing and settlement; the characteristic linear features then totaled more than 3,300 km in length, a 5-year increase of 2,000 km, or approximately 250 percent.

Field surveys conducted in September 1982 and reported by Tucker et al.[11] noted that the "linear forest-disturbance features were found to be forest-clearing swaths some more than 80-km long, about 4 km apart and 600–800-m wide. The large-scale primary forest disturbance is a government-planned colonization project in which immigrants to Rondonia are sold 100-ha lots of forest to use for agriculture or pasture. Each lot has about 500 m of road frontage. Clearing on a lot begins nearest to the access road then gradually moves away as the farmer

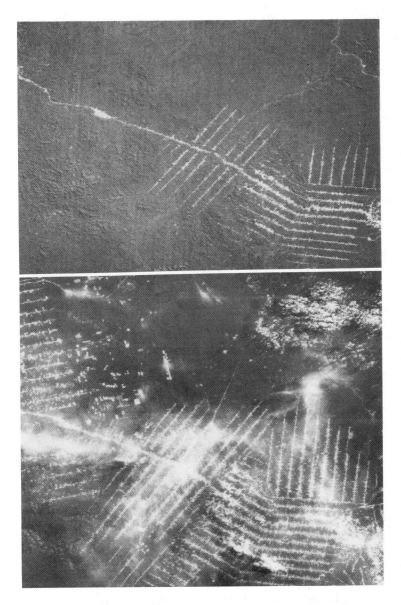

FIGURE 14–3 *Top:* Landsat-GEOPIC image (June 21, 1976) shows arterial highway BR-364 extending west of Rondonia, Brazil. Linear features show large-scale forest clearing for agricultural colonization (Path 248/Row 067, Scene No. 76173-12586, band 5). *Bottom:* Landsat GEOPIC image (September 20, 1981) showing an estimated 250 percent expansion in forest clearing since the 1976 image. Smoke from ongoing forest clearing is evident, as well as some clouds (Path 248/Row 067, Scene No. 81263, band 5).

clears more land to maintain or increase agricultural production and to reach soil not yet depleted of nutrients."

Between 1975 and 1978, the extent of primary forest clearing in Rondonia was 1,200 km^2 and 4,200 km^2, respectively, a three-year increase of 3,000 km^2. In 1982, researchers estimated about 9,200 km^2 was cleared in areas along BR-364.[12] All estimates are conservative. Forest clearing and colonization continue, further accelerated by improvements and paving of the arterial highways and new government-sponsored colonization programs.

The Rondonia example is ideal for testing remote sensing techniques to monitor and analyze tropical deforestation and colonization, and such work is being continued by several Brazilian and international organizations. Finer spatial resolutions, however, such as those currently being acquired by SPOT Image and EOSAT Corporation, are needed to refine the accuracy of tropical deforestation rates and to improve the quality and quantity of analyses and applications within the colonizing agricultural areas.

Similar studies, both recent and ongoing, are assessing the extent and location of agricultural land utilization in frontier regions of Brazil, Colombia, Peru, and Bolivia, among others. From both a technological and cost-efficiency viewpoint, data acquired from Landsat and SPOT are well suited to this task (i.e., providing locational and distributional information at mapping scales up to 1 : 50,000).

Detailed agricultural surveys often require analyses of small parcels involving ground studies and large-scale aerial photographs. High-altitude aerial photography (1 : 50,000–1 : 80,000 scale) has been used to define the agricultural zones within larger regions followed by large-scale photography over selected agricultural zones and field surveys. In a land-use inventory project conducted by the Servicio Nacional Aerofotogramatria de Bolivia in 1980, for example, 1 : 80,000 scale aerial photography for an area of 15,000 km^2 as analyzed for the initial stratification of areas of settlement and agricultural activity. Subsequent data were acquired for small parcel surveys.[13]

In a later study, Landsat imagery was evaluated as a potential substitute for high-altitude aerial photography. A single Landsat scene covers an area equivalent to over 100 nonoverlapping 1 : 80,000 scale aerial photographs. Comparative acquisition and processing costs for the 15,000 km^2 area, in 1980 U.S. dollars, was $2.55/km^2 for aerial photography and $40.12/km^2 for a computer-processed and computer-enhanced Landsat MSS image (for a more current comparison, costs for Landsat TM data would be less than $0.21/km^2).

The comparative analysis that was conducted involved the production of maps of areas of active cultivation based on interpretation of Landsat MSS imagery and 1 : 80,000 photos, which were composited and rephotographed at 1 : 100,000 scale. The resulting areas were 1,270.6 km^2 from Landsat and 1,256.3 km^2 from aerial photos, a difference of less than 1.12 percent. Estimated total costs (acquisition, processing, and analysis) were \$0.44/km^2 for Landsat and \$2.63/km^2 for the aerial photography. Hence, to locate and stratify areas of cultivation, Landsat images at 1 : 100,000 scale yield results comparable to high-altitude aerial photography results, both more rapidly and considerably more economically. Based on current research, Thematic Mapper imagery further improves the quality and accuracy of results, while the significant economic advantage of Landsat over aerial photography for such broad region analyses remains. It is noted, however, that in either case, state-of-the-art quality Landsat digital image processing (MSS or TM) is critical.

SUMMARY

The examples discussed here as well as those discussed in other chapters in this volume are only samples of the diverse range of current applications of Landsat and other advanced aerospace remote sensing technology to issues and programs of regional environmental and development concern. Analyses conducted in the course of these and many other programs have shown that a combination of existing published information, effective ground-survey techniques, remote sensing systems, state-of-the-art processing techniques, and analyses by qualified scientists and engineers can provide both new and additional information in a cost-effective and timely manner. Such information is essential for the conduct of sound resource development planning and monitoring efforts around the globe.

Collection of field data, or "ground truth," on soils, vegetation, landforms, cultural features, etc., is common practice and an important project component in the extraction of accurate information from Landsat imagery. Interpretation of Landsat images without benefit of ground truth information normally results in generation of many questions regarding the natural and man-made features within a project area. Even an experienced image interpreter who has some knowledge of the area is not certain to identify all Landsat images features or to interpret correctly subtle tonal differences whose significance must be resolved by an on-the-ground visit or low-altitude aircraft overflight of the area in question. Unanswered, these questions represent a potential source of analysis error that can adversely affect the management process. Yet the value of Landsat as a tool in planning field surveys cannot be overstated.

The studies discussed here have yielded significant findings and results. They can certainly be considered successful, but our measures of success must also account for the rapid improvements in technology and techniques over only a few years. To be successful this year or next, our approaches and techniques must stay abreast of the types of rapid technological improvements discussed in the chapters by Taranik and Tuyahov et al. If this is accomplished, the benefits will continue to accelerate rapidly.

The success of programs for resource inventory, change analysis, monitoring, development, etc., using remotely sensed data requires a variety of professional skills and experience if the program is to result in useful information for the planning, management, and decision-making processes. Principal skills and professional experience are required in three areas: resource inventory and mapping; resource development economics, planning, and implementation; and remote sensing and geographic information system technologies.

Throughout South America, there are today many qualified individuals and groups conducting ongoing operational resource management programs as well as applications research. These groups and their programs also have continuing technological and institutional needs in order to further the full development of resource analyses and information systems. These needs often involve attention to a variety of current natural, political, and economic complexities. Experience has shown that to address simultaneously real and immediate needs for management information while continuing ongoing programs effectively and successfully requires a number of essential elements found in a project team approach. These elements include:

- Resource scientists and engineers, both local and expatriate, experienced in resource inventory, mapping, and development projects in similar natural and cultural environments.
- A clear understanding of the types and complexities of information products (maps and reports) that can be effectively accommodated by existing institutions and systems.
- Practical knowledge of the application of satellite remote sensing technology to resource inventory/monitoring, and resource development programs.

- A management capability that can assure the meaningful integration of a wide variety of existing resource data required to meet program objectives and analysis, and synthesis of these data for the development of appropriate resource information.
- Project implementation experience in similar physical and cultural environments.

Clearly, the extent to which these elements are present in a given project, the greater its success potential. Based on my experiences as consultant and contractor on projects involving program design, implementation, and evaluation of applied remote sensing technologies to the information needs of developing nations, I am encouraged by the expanded and continuing applications of new technologies. Yet we must also temper our expectations of these technology applications with an understanding of socioeconomic, cultural, institutional, and political realities. Only then can the results of our analyses and products be expected to bear fruit.

NOTES

[1] W.G. Brooner and C.M. Viola et al., 1982, "Analysis of Multidate Landsat-GEOPIC™ for Mapping Vinal (*Prosopis ruscifolia*) and its Temporal Expansion," in *Proceedings, 16th International Symposium on Remote Sensing of Environment*, Buenos Aires, Argentina; and *Vinal: Investigacion Basada en la Teledeteccion Satelitaria*, 1981, Informe Final, Comision Nacional dal Vinal, Subsecretaria Recursos Naturales y Ecologia (Prov. de Formosa), Aeroterra, S.A. y Earth Satellite Corporation, Buenos Aires.

[2] *Ibid.*

[3] W.G. Brooner and C.M. Viola, 1983, "Landsat Monitoring of Temporal Hydrological Variations of the Pilcomayo River, 1972–1981," in *Proceedings, 17th International Symposium on Remote Sensing of Environment*, Ann Arbor, Michigan; and Secretaria de Planeamiento y Desarrallo, 1981, *Rio Pilcomayo-Banado La Estrella: Estudio Basado en la Teledeteccion Sateliteria Programa Landsat y Procesamient GEOPIC*, Informa Final, Prov. de Formosa, Aeroterra, S.A. y Earth Satellite Corporation, Buenos Aires.

[4] R. Welsh, T.R. Jordan, and M. Ehlers, 1985, "Comparative Evaluations of the Geodetic Accuracy and Cartographic Potential of Landsat-4 and Landsat-5 Thematic Mapper Image Data," *Photogrammetic Engineering and Remote Sensing*, Vol. 51, pp. 1239–1262.

[5] N. Meyers, 1980, *Conversion of Tropical Moist Forests*, National Academy of Science, Washington, DC, 20 pp.

[6] C.J. Tucker, B.N. Holben, and T.E. Goff, 1984, "Extensive Forest Clearing in Rondonia, Brazil, as Detected by Satellite Remote Sensing," *Remote Sensing of the Environment*, Vol. 15, pp. 255–261.

[7] World Bank, 1981, *Integrated Development of the Northwest Frontier*, Washington, DC, 101 pp.

[8] P.M. Fearnside, 1982, "Deforestation in the Brazilian Amazon: How Fast Is It Occurring?" *Interriencia*, Vol. 7, pp. 82–88.

[9] *Ibid.;* G.M. Woodwell et al., 1983, "Global Deforestation: Contribution to Atmospheric Carbon Dioxide," *Science,* December 9, Vol. 222, pp. 1081–1086.

[10] Tucker et al., 1984, *op. cit.*

[11] *Ibid.*

[12] *Ibid.*

[13] SNA, 1980, "Estudio a Nivel de Detalle del USO Agricola de la Tierra del Area del Chapare–Yapacami–Puerto Grether," Servicio National del Aerofotogrametria (FAB), La Paz, Bolivia.

EDITORS' INTRODUCTION TO:
P.G. RISSER AND L.R. IVERSON
Geographic Information Systems and Natural Resource Issues at the State Level

Previous chapters in this section have discussed new ways to obtain data about the environment, but the question remains, once one has the data, how does one use it? With the vast quantities of information becoming available about our global environment, especially through remote sensing, the problem is a major one. Ironically, we may think we know little about our global environment, but in fact in some ways we have much more data than we can use. Two important steps in the utilization of data are discussed in the next two chapters. One step is technical: the use of advanced computer techniques for the processing of spatial data. These are known as "geographic information systems," which have been mentioned but not elaborated on in several of the previous chapters. The other step is political and societal: developing the programs by which people collect data in a consistent form and agree to share and use it. The next two chapters illustrate these steps, but they do so at different spatial scales. The first chapter, by Drs. Risser and Iverson, deals with the application of the geographic information system to the state of Illinois in the United States, a spatial scale relevant to many nations, and many environmental problems that have a cumulative impact. In this chapter, we see that an efficient and unusual state organization, the Illinois Natural History Survey, has the social and political mechanisms to collect and use data; the new development is the implementation of a geographic information system.

Dr. Risser and Dr. Iverson worked on the development of the Geographic Information System at the Illinois Natural History Survey, where Dr. Risser was director and Dr. Iverson was a staff member. Now Vice President for Research at the University of New Mexico, Dr. Risser has been president of the Ecological Society of America; his research has focused on grasslands of North America.

15 GEOGRAPHIC INFORMATION SYSTEMS AND NATURAL RESOURCE ISSUES AT THE STATE LEVEL

P.G. RISSER
University of New Mexico
Albuquerque, New Mexico

L.R. IVERSON
Illinois Natural History Survey
Champaign, Illinois

INTRODUCTION

The state of Illinois has an area of about 146,500 sq km and a population of approximately 11.5 million. Although deciduous and pine forests are found in the southern portion and along rivers and streams, most of the land area was originally tall grass prairie that has been converted to such agricultural crops as corn, soybeans, orchards, and pastures. In certain regions, especially near Chicago, east St. Louis, and Peoria, substantial commercial and industrial development exists. Much of the state is underlain by coal, and significant localized petroleum reserves have been identified. Thus, the state is populous, with a complicated commercial-industrial system superimposed on a matrix of largely human-altered ecosystems. Under these circumstances, it is not surprising that severe pressures have been placed on the natural resources of the state and that decisions about the utilization and conservation of these resources always involve diverse points of view.

To facilitate the understanding of the state's natural resources and to assist in providing a rational framework for making decisions, the Illinois Natural History Survey has led the development of a statewide geographic information system (GIS). This system has been developed in concert with the Illinois Department of Energy and Natural Resources, which includes the following divisions: State Geological Survey, State Water Survey, Illinois State Museum, and the Office of Energy and Environmental Affairs. These organizations in combination acquire, archive, and make available a broad array of data sets and bring a high level of expertise to the complex issues of natural resources in the state. This chapter focuses on the biological aspects of this system, describing data sets and specific applications. Information about the hardware and software is given at the end of the chapter.[1]

DATA BASES

The data bases upon which these systems operate are designed to meet the needs of different types of users. As an operating principle, files of general data are shared among all system users; more specific or sensitive data sets are available to fewer users. There are essentially four components that make up the data base: 1) a statewide map data base; 2) a more detailed map data base for parts of the state underlain by coal; 3) site-specific data bases developed for special purposes or projects; and 4) a relational data base consisting of tabular data. Since all map data bases use the same projection and coordinate bases, appropriate files can be combined. Similarly, since tabular data files use the same codes as map data files, the two types of data can be manipulated as coherent files.[2]

The statewide map file consists of about 85 parameters that have been mapped and coordinates digitized (i.e., transferred into computer-compatible form) at the scale of 1 : 500,000 with a minimum polygon resolution of 2.6 sq km. The data for these files originated primarily from published maps at scales of 1 : 250,000 or 1 : 500,000. In all cases, these data files were checked by experts, usually on the basis of experience rather than by detailed interpretation of other sources of data or through additional collection of field data. These statewide maps include topics such as:

Biology
　　Natural Divisions of the State
　　Ecoregions
　　Potential Natural Vegetation
　　Pre–European Settlement Vegetation
　　Soil Associations
　　Land Use/Land Cover
Geology
　　Bedrock Geology
　　Quaternary Geology

Glacial Boundaries
Thickness of Loess
Coal Reserves
Hydrology
Stream Low Flows
River Mileages
High Groundwater Yield Areas
Sediment Survey Stations
Stream Channelization and Canalization
Administrative Units
Various Political Subdivisions
Dedicated Natural Areas
Soil and Water Conservation Districts
Surface and Underground Mines
Electric Utility Service Areas
Infrastructure and Special Features
Existing Roads and Railroads
Oil and Gas Pipelines
Airports and Air Navigation Facilities
Electricity Transmission Lines
Incorporated Settlements

These general data sets are used for regional analyses that do not require information mapped at a scale of finer resolution.

The first fully integrated geographic information system, the Canadian Geographic Information System (CGIS) developed in the early 1960s, was designed to provide information on marginal agricultural lands and lands suitable for agricultural production in Canada.

The state of Illinois geographic information system was originally developed to satisfy a similar purpose, the needs of the Illinois Lands Unsuitable for Mining Program. Under this program, petitions can be submitted to the Illinois Department of Mines and Minerals requesting that specific parcels of land be designated as unsuitable for some or all types of mining. A number of mandatory and discretionary criteria are invoked in this decision-making process. To assist in these decisions, the federal government requires that a data base be constructed to assist those who wish to submit petitions and to help those who will evaluate those petitions.

Because of the emphasis on coal resources, a coal-area data base has been developed. This map file incorporates most of the parameters included in the statewide map file but is mapped at the scale of 1 : 40,000 with a minimum polygon resolution of 1 hectare. Further, the data have been interpreted thoroughly, often with comparisons among several sources of information. Production of an integrated terrain unit map, for example, involves reconciling on the same map such sources as aerial photographs, soils maps, vegetation, and geological substrate. Integrated terrain unit maps have been constructed for relatively few areas, and the coal-area data base has been completed for eighty 12-minute U.S. Geological Survey topographic quadrangle maps.

The site-specific map data base consists of a large number of maps developed for specific purposes. All files employ a common coordinate and projection system to permit integration whenever appropriate. Examples of these files include habitat for deer and pheasant populations, erosion rates within watersheds, and streams likely to be vulnerable to acid deposition.

Many tabular data sets are also included in the system, and these files are linked through the coordinate system to the map files. Investigators can, on the one hand, merely manipulate the tabular data files; on the other hand, they can use the information in the tabular files as attribute information with the map files. For example, investigators can manipulate the tabular data and then plot the results, or they can select tabular data and manipulate it on the basis of the cartographic coordinates in the map file.

A typical tabular data set is the Illinois Plant Information Network (ILPIN), a file on the 3,300 plants that occur in Illinois. Entries for each species include: taxonomic information; life history characteristics; habitat requirements; geographic distribution; and information concerning how individual species respond to management and environmental insults. ILPIN allows rapid summarization of plant attributes by any number of selected criteria. For example, it is possible to identify in a matter of minutes which species grow in Cook County in bogs that have red flowers that bloom in June. It also allows generation of potential species lists for persons investigating any particular natural community, county, or natural division.

A data set entitled the Illinois Fish and Wildlife Information System (IFWIS) contains parallel information about the animals found in Illinois. Data sets such as these require an enormous amount of original field work, considerable biological expertise and judgment, and careful attention to ensure that the information represented by the data set is current and accurate. Thus, developing and maintaining such a data base should be undertaken only by a staff with scientific and technical expertise in the plants and animals of the region and in the hardware and software systems being employed.

APPLICATIONS

The geographic information system has had numerous applications assisting in the analysis and resolution of a variety of natural resource issues in Illinois. In addition, it is used routinely by the Natural History Survey to study plant and animal populations and the ecosystems of the state. For the purposes of this discussion, however, applications will be considered along a continuum from merely retrieving data from the system and applying it to using the cartographic and tabular data in descriptive and predictive mathematical manipulations. If geographic information systems are to reach their potential in assisting with natural resource issues, the administration, structure, and use of the system must incorporate the most advanced mathematical capabilities as well as the most powerful ecological concepts.

A geographic information system permits the reproduction of digitized maps and maps displaying data contained in the tabular data files. This enormously useful capability has been used consistently to depict the resources of Illinois, especially in the decision-making arena where participants are relatively unaccustomed to reviewing these types of information. In addition, interesting biological conditions of the state have become apparent by manipulating and displaying information. For example, the distribution of known historical locations of state threatened or endangered plants can be plotted along with the locations of protected nature preserves established to preserve natural communities including rare species (Figure 15–1, see color insert).

The overlay and buffering capabilities of the system have been employed in several projects. In one case, the vulnerability to soil erosion for part of Illinois was predicted using the overlay of five GIS layers. The K factor (erodibility) for the soils ranged from .23 to .47, with the highest values receiving ten points in an additive model and the lowest erodibility values receiving zero points. Similarly, slope in the area ranged from zero to 35 degrees; the points for these polygons ranged from one to nine points. Fifty-meter buffers were placed around the escarpment faces, and these were given three points. Landforms of various types were treated similarly. With the overlay of these four layers, it was possible to create five classes of vulnerability to erosion if permanent vegetation were removed (Figure 15–2, see color insert). Fortunately, much of the area is under permanent cover; when the vegetation map is added to the combined overlay, a map depicting probable areas currently undergoing erosion can be generated (Figure 15–3, see color insert).

In the same fashion, 14 GIS layers were used to predict suitable habitat locations for endangered plant species for the same area of southwestern Illinois. The ILPIN data base was accessed to find the most common habitat types for endangered species in Jackson County— wetlands, bluffs, and closed deciduous forests. The GIS was invoked to determine all areas under disturbance (i.e., agriculture, urban) or adjacent to disturbance (i.e., buffers around roads, pipelines, railroads, agriculture, and urban). These were scored in the negative direction. The favorable locations were then scored in the positive direction according to their closeness to the desired habitats. The resulting map (Figure 15–4, see color insert) depicts five classes of suitability to provide suitable habitat for endangered plant species.

Siting issues have also been approached by overlaying several environmental parameters and then evaluating the suitability of alternative sites, and several soil and vegetation characteristics have been evaluated by using overlay techniques to determine vegetation vulnerability to air pollutants. As one example, forest productivity for a portion of Illinois was predicted using point, polygons, and tabular data from the GIS and Landsat Thematic Mapper (TM) data coupled with a regression model. The U.S. Forest Service data on the Illinois Continuous Forest Inventory (CFI) plots, collected in 1985, included data pertaining to forest productivity at plot locations scattered throughout the state. Additional 1 hectare resolution data from the coal area data base used for the model included a woodland soil productivity index as derived from the soil layer. Landsat TM was then used in combination with the other data to develop a regression model aimed at predicting forest productivity. The resulting map depicts the variation of deciduous forest productivity classes based on the best regression fit for each 30-by-30-m pixel of northern Pope County, Illinois (Figure 15–5, see color insert). These analyses have obvious implications, not only for increasing our understanding of the ecological systems in Illinois, but also for improving our ability to manage these natural resources.

Current applications of the geographic information system emphasize the combination of spatial-tabular data sets with simulation models.[3] This approach retains the power of the geographic information system, but adds the specificity of mechanistically describing how natural resource entities or processes operate in the field. For example, the data system includes a large data base on the distribution of forest types in Illinois as well as descriptive data about forest stands, including species composition, stem sizes, and site quality. These data are now being coupled with a simulation model that projects forest growth and the economic returns that can be expected with alternative timber management techniques. In another example, simulation models of the population dynamics of agricultural insect pests have been developed that combine life history characteristics of the insects with climatic data and crop conditions. Coupling these models within the context of a geographic information system allows us to predict how various areas of the state will be affected by agricultural pests, thus increasing the likelihood the most appropriate management strategies will be recommended. The precision of these predictions enhances the efficacy of insect pest management treatments while reducing the hazards to the environment from needless applications of chemicals.

SUMMARY

Geographic information systems are powerful tools for addressing natural resource issues. Illinois is a diverse state with many natural resources and constant pressure to develop and utilize them. Only with the assistance of an organized data base that can be manipulated is it possible for the state to make wise decisions about these natural resources. Experience has shown, however, that biological data require considerable expertise, both when the data are collected and when the results are interpreted. Therefore, a geographic information system that focuses on natural resource issues can only be developed and used in conjunction with biological scientists.

Currently, most geographic information systems used for evaluating natural resource topics employ a number of powerful capabilities, including map overlays, buffering displays, and algebraic calculations. Future applications will combine the spatial and tabular data with simulation models that capture the mechanistic behavior of biological populations and ecological processes. The incorporation of these mechanisms will reinforce the requirement that the data system and the scientists operate in tandem; it will also enhance the predictive and prescriptive capabilities of the geographic information system at the state level.

NOTES

[1] The hardware system is located at the Illinois Natural History Survey and consists of linked Prime 750 and 9955 minicomputers with 24-mb main memory, two tape drives, 80 ports, and 3300-mg disk storage on six disk drives. A complete graphics workstation is located at each of the five divisions of the Department of Energy and Natural Resources.

Each consists of a digitizer and graphics terminal; three workstations have 86-cm drum plotters. In addition, approximately 40 graphics or alphanumeric terminals are connected to the system. Access to remote stations is provided by hard-wired connections, through a coax cable on the University of Illinois campus, through dial-up ports, or through dedicated telephone lines connected to multiplexers and high-speed modems.

Although a number of software packages are available on the system, ARC/INFO is the primary package for managing spatial and tabular data. ARC (marketed by Environmental Systems Research Institute [ESRI] in Redlands, California) is a series of geographic information processing routines for encoding, manipulating, and displaying spatial data (i.e., data that can be categorized into points, lines, or polygons, and that are stored in x, y coordinate form). Analytic capabilities include algebraic manipulations, overlays, and buffering. INFO (Marketed by Henco Software in Waltham, Massachusetts) is a relational data base management package. A major advantage of the ARC/INFO system is its capability for combining spatial and tabular data. Recently we have acquired the ERDAS software (marketed by Erdas, Inc., of Atlanta, Georgia), which provides image processing and raster GIS processing. This system has been integrated into the ARC/INFO software so that vector GIS files can be integrated with ERDAS for further processing. Other major software packages available to state of Illinois personnel include GRID, GRID/TOPO, TEXT, Minitab, ELAS, EMHASP, IGL, PRIMOS, Surface II, TERMS, FORTRAN 77, NETWORK, and MIDASPLUS.

[2]C.G. Treworgy, 1984, "Design and Development of a Geograhic Information System for Illinois," Proceedings of Pecora IX, Spatial Information Technologies for Remote Sensing Today and Tomorrow, Silver Springs, Maryland, pp. 29–32.

[3]P.G. Risser and C.G. Treworgy, 1986, "Overview of Ecological Data Management," in *Research Data Management, and the Ecological Sciences*, W.K. Michner (ed.), University of South Carolina Press, Columbia, pp. 9–22.

EDITORS' INTRODUCTION TO:
M. D. GWYNNE AND D. W. MOONEYHAN
The Global Environment Monitoring System and the Need for a Global Resource Data Base

Techniques to study our global environment cannot be put to good use without the social and political mechanisms that allow nations to work together. The United Nations Environment Programme (UNEP) developed out of the 1972 United Nations Conference in Stockholm on the environment and is a major step in this direction. One product of UNEP is the Global Environment Monitoring System now directed by Mr. Gwynne. The cooperative project among nations is a constructive and positive development, important in our role in changing the global environment. A new part of this activity is the application of geographic information systems at a global level through a pilot program called the Global Resource Information Database, which is under the direction of Mr. Mooneyhan. These programs are described in this chapter.

16 THE GLOBAL ENVIRONMENT MONITORING SYSTEM AND THE NEED FOR A GLOBAL RESOURCE DATA BASE

MICHAEL D. GWYNNE AND
D. WAYNE MOONEYHAN
United Nations Environment Programme
New York, New York

INTRODUCTION

The Global Environment Monitoring System (GEMS) is a collective effort of the world community to acquire data needed for rational management of the environment. Following recommendations of the United Nations (UN) Conference on the Human Environment held in Stockholm, Sweden, in 1972, the United Nations Environment Programme (UNEP) was given the task of coordinating and stimulating international environment monitoring activities, especially at regional and global levels. It was not until 1975 that UNEP moved into the field of monitoring in a deliberate and systematic manner. This was accomplished with the establishment of a Program Activity Centre (PAC) for GEMS, located at UNEP headquarters in Nairobi, Kenya. The role given to the GEMS PAC by governments participating in UNEP programs is to coordinate disparate international monitoring activities. The monitoring activities are conducted throughout the world, particularly within the UN system. In addition, GEMS PAC advises the Environment Fund of UNEP on how best to support and stimulate the initiation of new activities or the expansion of ongoing ones through allocation of financial resources to these activities.

In considering GEMS, it is essential to bear in mind that some activities fall under GEMS because of their international nature (e.g., those of the

Organization for Economic Co-operation and Development in the field of transboundary pollution and of contamination of wildlife by pollutants, those of the International Council for the Exploration of the Sea in marine environment monitoring, or a large portion of the World Weather Watch), while others are supported by the Environment Fund. Clearly the role of the GEMS PAC is much more direct in the latter, and it is these activities that will be discussed here as a prelude to a more detailed discussion of the Global Resource Information Database (GRID).

CLIMATE SYSTEM MONITORING

Climate system monitoring of the World Meteorological Organization/United Nations Environment Programme (WMO/UNEP as part of the World Climate Program) looks at world climate data in terms of climatic anomalies. These anomalies are shown in a monthly bulletin that indicates areas where it is too hot, too cold, too wet, or too dry, based on analysis and comparison with expected norms. Fluctuations in the position of the Inter-Tropical Convergence Zone and in the extent of polar ice are also shown. An annual report summarizes these climatic patterns on both global and regional scales as appropriate.

LONG-RANGE TRANSPORT OF POLLUTANTS

The program on long-range transport of pollutants over Europe was conducted in partnership with the UN Economic Commission for Europe and WMO as a followup to a preliminary study by the Organization for Economic Co-operation and Development on the transboundary movement of pollutants over western Europe. Data were provided on deposition of pollutants (particularly sulphur oxides and their transformation products, which are largely responsible for acidic deposition) in relation to the movement of air masses. The data, in essence, traced the path from pollutant sources to areas of deposition. This was done so that models of transport, especially over national boundaries, could be tested. As an integral part of this program, a network of 50 stations in 18 countries collects and analyzes samples of air, rain, and airborne particulates.

In 1977, the UN Economic Commission for Europe, in cooperation with UNEP and WMO, formed these activities into a cooperative program for the monitoring and evaluation of the long-range transmission of air pollutants in Europe, or EMEP as the program is known

(European Monitoring Evaluation Program). These activities have since been taken over by the signatory states to the Convention on Long-range Transboundary Air Pollution. GEMS now supports work on methods to inventory and monitor the effects on temperate forests of acidic deposition.

The UNEP/World Health Organization (WHO)/United Nations Educational Scientific and Cultural Organization (UNESCO)/WMO water quality monitoring program consists of a network of some 300 stations in approximately 70 countries. Emphasis in this program is given to water bodies (rivers, lakes, and groundwater) that are major supply sources for municipal areas, irrigation, livestock, and some industries. A number of stations for monitoring international rivers and lakes, rivers discharging into oceans and seas, and water bodies as yet largely untouched by the activities of humans (baseline stations) are included in the network. Variables being monitored fall into three categories—1) basic, 2) optional, and 3) of global significance. The first group includes variables that are important for the general assessment of water quality and are measured at all stations. The second contains those substances or factors that may be of importance in some areas but not in others and whose measurement, therefore, depends on circumstances. The third group of variables are those toxic substances that are of long-term significance. Harmonization of methods and quality control are an essential component of the program.

MONITORING OF FOOD CONTAMINANTS

Monitoring of contaminants in food in the framework of GEMS was begun in 1976 with WHO as the executing agency in cooperation with the Food and Agriculture Organization (FAO) of the UN. Collaborating Centres in over 20 countries have been designated to collect information on a number of contaminants in selected foods. Collection of existing national data gives information on distribution of the contaminants and comparison of data obtained yearly under standardized conditions provides information on trends and current levels for assisting governments in the management and control of this problem.

Analytical quality assurance was initiated in 1979 as part of this program to allow comparison of data from the different areas of the world as well as to assist the Collaborating Centres in improving their own techniques. Under coordination by WHO, the responsibility for quality control measurements has been undertaken within the program for aflotoxins by the International Agency for Research on Cancer, for metals (lead and cadmium) by a British laboratory, and for organochlorine compounds by a Swedish laboratory.

The health-related air quality monitoring network includes five pilot studies to be carried out in different parts of the world in cities with different climates, different life-styles, and different pollution profiles. Each pilot project includes indoor/outdoor monitoring, personal monitoring, and ambient monitoring components. Thus far, there are some 180 monitoring stations within the network. These stations are located in 60 cities around the world. The main purpose of this effort is to evaluate whether, under different conditions in cities, exposure of people to air pollution can be reasonably estimated from ambient monitoring alone. Alternatively, this monitoring will determine whether other complementary information should or must be gathered in order to permit an accurate assessment of exposure to air pollution. A first assessment of these data was issued recently by WHO and UNEP.

Owing to the natural variability of humans as subjects exposed to various forms of pollution, it will always be difficult to make precise estimates of exposure. In this connection it should be noted that no matter how good a monitoring program is, it will not be possible to obtain precise estimates of short- and long-term exposure conditions. The implicit difficulty is that humans tend to absorb, modify, accumulate, or dispose of environmental pollutants at rates that are not well known or understood and which, in any case, may vary enormously. To provide insight into these questions, a pilot project on the monitoring of biological fluids and tissues was implemented, involving the measurement of selected heavy metals and organochlorine compounds.

RENEWABLE NATURAL RESOURCE MONITORING

Although the activities of terrestrial renewable resource monitoring stem to a large extent from the recommendations of the 1972 United Nations Conference on the Human Environment held in Stockholm in 1972, there were at the time of the conference few methods widely available to carry out this type of activity. Subsequently, FAO, UNESCO, and UNEP working in concert have developed methods for assessing the state and rate of soil degradation.

Similarly, methods were developed based on appropriate combinations of satellite and aerial photographic interpretations. Aerial observations and ground surveys were conducted and applied, through FAO's cooperation, to the inventory and monitoring of forest cover in some tropical countries of west Africa. In 1980, FAO and UNEP assessed the state of tropical forest resources at world and regional levels using existing knowledge. This effort employed both published and unpublished sources and was supplemented and verified by analysis of carefully selected satellite images for areas for which little data were known to exist. Data generated by this effort then forms a baseline against which future changes in tropical forest cover can be measured and provides the basis for an ongoing world forest resources data bank.

GEMS's concern with the renewable resources of arid and semiarid ecosystems has involved the monitoring in two closely related fields— rangeland assessment and desertification. Here, the operational basis is in each case the Ecological Monitoring Unit (EMU) approach. This EMU approach has been developed over the last 15 years for use in the arid lands of tropical Africa. Basically, the Ecological Monitoring Unit approach involves repeated multistage samplings from three data acquisition levels—space, air, and ground. Information is obtained from earth resource satellites, low-level (100 m) systematic reconnaissance flights, and representative ground sites by field survey. For each site, data are collected on important resource and habitat variables related to climate, biota, physical features, and human activities. Monitoring is done in the short term, thus providing information on seasonal processes and their inherent variability. These repeated short-term surveys are, however, the basis of long-term monitoring to establish trends and changes through time (e.g., livestock population increases) for the sites. This type of monitoring can then be employed to examine the spatial nature of the changes occurring in an area and to assess the rates and directions of these changes through correlations of a number of the variables collected in the sampling program.

A FAO/UNEP pilot project on pastoral ecosystem monitoring in west Africa has recently drawn to a close. This project served two purposes: 1) to allow monitoring methods developed elsewhere to be tested and adapted to the rangeland conditions peculiar to the Sudano-Sahelian zone and 2) to show the countries of the region the practical benefits related to development that can accrue from the initiation of an EMU. A new phase of this project is currently being implemented with support from several sources.

A number of activities and projects utilizing the EMU approach have been established in various parts of the world, notably in west Africa and Latin America, and advice on how to carry out such programs has been provided to a number of governments and organizations in these and other areas as well. As part of this program, GEMS is also involved in collaborative work with other organizations to ascertain the suitability of

United States National Oceanic and Atmospheric Administration (NOAA) satellite data for monitoring vegetation change. GEMS is also assessing the possibility of perhaps even utilizing these data in the development of an early warning system for detecting the onset of drought conditions. The first of the many necessary calibration exercises that will be required to determine the feasibility of these concepts has begun in Ethiopia and Kenya.

Another resource monitoring activity within GEMS is the determination of the status of threatened species of plants and animals—both domestic and wild—and endangered habitats. In this, UNEP works through the International Union for the Conservation of Nature and Natural Resources and its Conservation Monitoring Centre.

OCEAN MONITORING

Most monitoring activities related to oceans are organized through UNEP's Action Plans for Regional Seas and are the responsibility of the UNEP Program Activity Centre for Regional Seas.

At present, the most developed regional monitoring program is still the Mediterranean Pollution Monitoring and Research Program (MED POL). In this, 83 marine science institutions from 16 Mediterranean countries and the European Economic Community participate in eight pilot projects initiated in 1975 under the technical guidance of FAO, WHO, WMO, the Intergovernmental Oceanographic Commission of UNESCO, and the International Atomic Energy Agency. Similar monitoring programs are being developed for other regional seas. These programs include data gathering on coastal and living marine resources.

This short overview has only highlighted the main activities in which GEMS PAC is directly involved. These include most of the monitoring activities carried out under the aegis of the United Nations system. They have been either initiated or considerably boosted by UNEP's contribution, but it is important to emphasize that UNEP support to any of these activities is not open-ended and will inevitably be reduced or phased out altogether. This will occur when the scope for each of these activities reaches its preassigned limits, when the expansion of the activities becomes largely self-sustaining, or when it is realized that the investment in a certain activity is not commensurate to the returns obtained. The time limits imposed on UNEP support, which obviously vary from activity to activity, are dictated by the fact that UNEP's financial resources do not grow and in real terms are actually decreasing, which places limitations on the funds available for the support of monitoring activities. In a nongrowth budgetary situation, decreased commitments to certain activities imply the release of resources for new activities and therefore the continuation of UNEP's catalytic role.

Each of the global networks within GEMS forms an integral part of it, a data base into which monitored data are placed and from which they are withdrawn for subsequent analysis. The data bases, for budgetary reasons, are not centrally placed nor are they linked. Thus, for example, the data base for GEMS-Water is at the Canada Centre for Inland Waters, Burlington, Ontario; several of the air pollution data bases are operated by the United States Environmental Protection Agency (EPA); and those for threatened species, threatened habitats, and Parks and Reserves are stored at the IUCN Conservation Monitoring Centre in Cambridge, United Kingdom.

Monitored data are only as valuable as the use made of them. Data are thus drawn from the data bases, analyzed, and used to produce assessment statements on a periodic basis—for example, ozone every year, tropical forests every five years, and glaciers every twenty years. These assessment statements not only give an indication of the state and trends of the global environment, but they also reveal the adequacy of the monitoring system used to evaluate them. In this fashion, monitoring system defects can, hopefully, be remedied or new directions taken. Assessments also suggest directions for future management actions. After such actions have been implemented, subsequent monitoring and assessments show the efficiency of the management actions and indicate whether further changes in management practices are required.

As GEMS has grown, it has been realized that the power and usefulness of the various GEMS data sets, held in the data bases of each of the GEMS networks, could be increased many times over if they could be related to each other in a meaningful way. Similarly, the same data sets could be related to particular environmental problems and to the problems of particular areas of the Earth. Such data, especially when joined with other information, can become an extremely powerful management tool when focused upon a specific area or resource. Hardware and software development have now reached the stage where UNEP and GEMS PAC consider that the establishment of such relationships and the focusing of data are feasible. The bridge between resource monitoring and resource management, for example, can now be made in a more meaningful and cost-effective way by utilizing advanced information systems technologies. A key to these actions involves the application of a geographical information system approach and the development of a distributed data system closely coupled by networking within GEMS. Such a system would enable participating countries to have access to the various data holdings of GEMS and to be able to employ this information for their own development and management purposes. Such a system will undoubtedly be of great practical benefit to developing countries. At the same time, a program such as this would help to generate new and better information to put into the GEMS data bases, thus enabling more reliable global assessment statements to be made. In consequence, there has now been created within GEMS a geographical information system—now known as the Global Resource Information Database (GRID).

GLOBAL RESOURCE INFORMATION DATABASE

The Global Resource Information Database is a new project with the objective of providing information to people making decisions that affect the well-being of our planet. As part of the United Nations, GRID will provide data and information to scientists, planners, and decision makers in their job of managing and assessing human impacts on critical environmental issues.

Traditional access to environmental data—in shelves of reports and proceedings—and in fast-aging maps and charts—no longer meet the demands of planners faced with a world where the nature of environmental change is infinitely complex. With the development of computers that can handle large quantities of data, a global data base is now possible.

GRID is designed to make environmental data available to a large user community. Users may be scientists trying to understand the functioning and behavior of the global environment or planners making important management decisions about resources in regions under their jurisdiction. Each decision relative to use of resources and the environment must be made with concern for the future and the generations who will inherit the legacy of today's environmental planning.

It is intended that GRID will enhance the relationship that already exists within the United Nations by effectively giving a wider audience access to data sets that are being built up within the United Nations Environment Programme/Global Environment Monitoring System (UNEP/GEMS), other UN agencies, and other international and national organizations. GRID will be a dispersed system, with facilities linked by telecommunications, eventually sending data to, and receiving data from, nodes throughout the world. It is hoped that this distribution network will help to build a useful picture of the state of the global environment and at the same time enable planners to make better environmental decisions and to manage resources more effectively.

During the pilot phase, three main functions have been identified for GRID: 1) bringing together existing environmental data sets; 2) analyzing existing information in order to pinpoint areas of environmental concern; and 3) training people from both developing and developed countries in the use of the technologies involved in the application of geographically based information systems.

THE GRID DATA

Data for GRID are collected through numerous organizations, both national and international. United Nations Organizations such as UNEP, UNESCO, UN Statistics Office, FAO, WHO, and WMO collect data on a global scale concerning major environmental parameters. These parameters include soils, tropical forest, rainfall, temperature, hydrology, background air pollution, urban air pollution, water quality, worldwide glacier status, human population statistics, agricultural production, economic statistics, and many others. Other international organizations and regional organizations also provide important information. The IUCN maintains records on endangered species, both plants and animals, on a global basis, and International Livestock and Cattle Association (ILCA) maintains statistics on domestic animals in northeast Africa. National organizations collect and maintain national environmental data and some national organizations such as NOAA in the United States collect and analyze satellite data of the entire world.

All environmental assessments need baseline data sets. The first item is a base map of the world, a framework for geo-referencing. To this base map information on elements such as soils, vegetation, plant and animal species, climate, and human settlements can be registered geographically. Initially, it is envisioned that data will flow in from international data sets to which the United Nations Environment Programme already has access. For example, the five program areas of GEMS: 1) climate-related monitoring; 2) long-range monitoring of pollutant transport; 3) health-related monitoring; 4) ocean monitoring; and 5) terrestrial renewable resources monitoring will provide data sets for their respective areas. Using these data sets in the pilot phase, GRID managers will be able to establish priorities for collecting new data to expand the data base. In the future, sufficient data will be available to allow a response to many questions about the environment. Answers will lead to new questions and improve the process of upgrading data sets. The attention of managers will be drawn to gaps in the data base and hopefully action will be taken to fill them. Upgrading is a process that will be part of the day-to-day working of GRID. The goal of this process leads to decisions based on increasingly reliable and comprehensive data.

THE GRID SYSTEM

The GRID system is specifically designed to be a distributed network of data base nodes around the globe. The central nodes have been established within UNEP for the pilot phase. One of the centers is located in the GEMS PAC offices at UNEP headquarters in Nairobi, Kenya. This center, GRID Control, performs the overall policy and priority fixing functions as well as serving as the principle center for Africa. The second center, GRID-Processor, located in Geneva, Switzerland, is the central processing facility and, in addition, will serve the areas of Asia and Latin America until operational nodes are established in these regions. (A NASA center, the Earth Resources Laboratory, located at the National Space Technology Laboratory in Mississippi, United States, is participating in the current pilot phase and provides both technical support and expertise under the NASA Earth Science and Application Program.) Information about the hardware and software to be used is presented in the Appendix.

DATA ANALYSIS

Like most computer-based geographic information systems, GRID has a number of basic features: data capture, storage and retrieval, manipulation and analysis, and output and display.

Data capture involves putting information into the computer and organizing it in memory. Data in GRID are geo-referenced. That is, each feature entered into the system is located with respect to a given projection of the globe as X, Y coordinates. The software recognizes every piece of information in terms of its position on the surface of the Earth. All data entering the system must have this in common.

Data are stored as geo-referenced points. The points, however, are arranged by the software to represent points, lines, or areas. Thus, rain gauges and houses may be stored as points, lines such as roads and rivers are stored and recognized as a series of points, while areas have points defining their perimeters (polygons). GRID software can deal with areas of any shape or size. Data storage also involves registering characteristics. Every point, line, and area is actually described, be it a rain gauge, road, or lake. Any description is possible and any point, line, or area may be described over and over again and given new and revised attributes.

Retrieval is the recovering of ordered data from the computer's memory or from mass storage discs or magnetic tapes so that these data can be analyzed. Retrieval involves asking the computer to search and query both by location and attribute (for example, all the lakes in Gambia, all the man-made lakes or reservoirs, those that are within 10 kilometers of a major road, those that contain a certain species of fish, or those that suffer from high acidity). These are questions that could be asked of the GRID GIS.

The main feature of GRID is its capacity to manipulate and analyze data and then display information. This involves the retrieval of data files or parts of them in any combination and their analysis together or separately to generate tables, graphs, maps, or charts that will facilitate the analyst's ability to examine, manipulate, and understand the problem they are investigating. GRID will be able to handle almost any question couched in geographical terms. Its ability to answer will be more a reflection of data availability than of analytical flexibility. Three basic techniques of data analysis supported by GRID are modeling, overlay, and the use of statistics. For any particular area, any data sets held in the system can be overlaid, with different information from the same area. For example, in a study of soil erosion, the computer will be able to compare the degree of erosion with soil quality, water runoff, temperature, human settlements, or domestic animal distribution by overlaying them one by one (or all together).

The software's capacity to display analyzed data will ultimately determine the way in which they will be used. Perhaps the most common output will be a map, but many other types of output will be possible (e.g., graphs, charts, tables, and displays on color monitors).

APPLICATIONS

Applications of the GRID system will determine its worth. The number of possible applications is only limited by data availability, accuracy and timeliness, and the imagination of GRID users. The system can be used at any scale from the global level (for general assessment statements about worldwide trends in deforestation and desertification) to the local level for analyzing problems within countries (an environmental impact of reservoir construction). Present applications fall into four generic categories:

• mapping (for spatial distribution and relationships)
• inventory (for status)
• monitoring (for change analysis)
• modeling (for cause/effect analysis)

An example of inventory and monitoring can be illustrated by Figures 16–1, 16–2, and 16–3 (see color insert). Figure 16–1 is the extent of tropical forest in Costa Rica in 1940 as determined by areal survey. Figure 16–2 is the same phenomena in 1977 as determined from Landsat satellite data. Figure 16–3 is the amount of change that occurred between the two dates.

An example of modeling results is shown in Figure 16–4, (see color insert) which is the final product of applying the Universal Soil Loss model to a watershed in Iowa, United States. Figure 16–5, (see color insert) a soils map for Africa, is an example of a continental scale data set. This data set, along with data sets of rainfall, number of wet days, wind speed, temperature, land use, vegetation, and climate zones was used in a model to produce the desertification hazard map shown in Figure 16–6 (see color insert).

SUMMARY

The Global Environment Monitoring System is a collective effort of the world community to acquire the data needed for rational management of our global environment. The GEMS Program is a part of the United Nations Environment Programme. In 1975, UNEP began to monitor the environment in a systematic manner. Of the areas of climatic monitoring, tracking long-range transport of pollutants, monitoring food contamination, renewable national resource monitoring, and ocean monitoring, GEMS has initiated the Global Resource Information Database project to assist in these efforts.

The first GRID computer was placed in service on September 27, 1985, in Geneva. The second system in Geneva was placed in service in February 1986, and both Nairobi computer systems were in service in February 1986. Thus, the process has started and its success will depend on many factors. In the beginning, the most important activity is collecting the appropriate data and constructing the geographic data sets. This will require much cooperation between GRID and the international, regional, and national organizations that hold the needed data sets and several years of careful work in their construction. In the end, its success will be judged by the applications and users of the data base.

APPENDIX

The following describes the hardware and software that will be used at the central nodes of the GRID system.

HARDWARE

The Nairobi center is currently equipped with one Perkin Elmer computer system and one Prime computer system. The Perkin Elmer is a model 3220 minicomputer with one megabyte of main memory, 170 megabytes of disk storage, interactive image analysis station, tape drives, printer plotter, color camera, and an *XY* digitizer. The Prime system is a model 2250 with 2 megabytes of main memory, 158 megabytes of disk storage, tape drive, printer, and a graphics system. A third microprocessor system by ERDAS Inc., which utilizes an IBM PC/AT, was added to the Nairobi center in 1986 and will be used as the primary system for training participating scientists and managers.

The Geneva center is also equipped with one Perkin Elmer computer system and one Prime computer system. The Perkin Elmer is a model 3241 with 4 megabytes of main memory (capable of 8 MB), 1200 megabytes of disk storage, tape drives, printer, electrostatic printer plotter, interactive image analysis system, *XY* digitizer, and a color copy camera. A second interactive image memory was added in 1986. The Prime computer system is a model 750 with 6 megabytes of main memory, 1350 megabytes of disk storage, tape drives, printer, and an interactive image display. A graphic system and *XY* digitizer was added in 1986. A microprocessor system by ERDAS Inc. has also been added and will be used for training scientists and resource managers who are participating in GRID or UNEP Programs around the globe.

SOFTWARE

The Perkin Elmer computers in Nairobi and Geneva employ the ELAS software systems. ELAS was designed and developed by NASA at the Earth Resources Laboratory located at National Space Technology Laboratory in Mississippi, U.S. ELAS contains both an image analysis system and a geographic information system. Data input can be accommodated from tapes, disk, terminal, and *XY* digitizer. ELAS is capable of handling all types of digital image data as well as line, point, and polygon data from any geographically referenced source. Data are stored in the data base in the most efficient form, grid (raster) or polygon (vector).

However, when these data are manipulated, the processing is accomplished in the grid format within the GIS modules. ELAS presently contains more than 200 executable modules and overlay programs. The software package is written in Fortran for ease of transfer to various computers.

The Prime computers in both Nairobi and Geneva employ the ARC/INFO software system. ARC/INFO was developed by Environmental Systems Research Institute (ESRI) in Redlands, California, U.S. (INFO was designed and developed by Henco Software, Inc., in Waltham, Massachusetts, U.S.) ESRI is a commercial firm that sells and services the software systems. ARC/INFO brings together geographic analysis and modeling capability with a complete interactive system for entry, management, and display of spatial data. It is a geographic information system and a data base management system. Data input can be accommodated from tape, disk, terminal, XY digitizer, or graphics terminal. ARC/INFO is capable of handling digital data from any geographically referenced source. The relational data base management system is interactive with spatial and attribute query, tabular analysis, and report generation.

EDITORS' INTRODUCTION TO:
L.R. TINNEY AND J.G. LACKEY
The Use of Remote Sensing to Assess Environmental Consequences of Nuclear Facilities

Nuclear production reactors pose a number of potential large-scale environmental problems, including thermal pollution, hazardous wastes, and a variety of forms of toxic substances that can pollute water. In this chapter, a major remote sensing program of the Department of Energy is described and examples of processed data are presented. Messrs. Tinney and Lackey, who conduct the Department of Energy research in multispectral remote sensing for EG&G Energy Measurements, Inc. discuss this program, which is termed Comprehensive Integrated Remote Sensing and involves equipment and personnel of the Remote Sensing Laboratory. They focus on the applications of remote sensing at a more local level than was presented in the previous chapter by Brooner. Data are being collected to support environmental and emergency response functions for large facilities. Integration of the data is being enhanced by implementation of a geographic information system.

17 THE USE OF REMOTE SENSING TO ASSESS ENVIRONMENTAL CONSEQUENCES OF NUCLEAR FACILITIES

L.R. TINNEY AND J.G. LACKEY
EG&G Energy Measurements, Inc.
Las Vegas, Nevada

INTRODUCTION

In response to United States national environmental policies, the Department of Energy must conduct its operations in conformance with environmental and public health laws and regulations. This entails an increasing number of environmental assessments and continual monitoring of activities. Remote sensing is a powerful tool for conducting environmental surveys. Accordingly, several aspects of the Department of Energy's environmental survey programs are now supported by remote sensing and related technologies.

The United States government's Department of Energy conducts both nuclear and non-nuclear programs involving research and production activities. These programs are conducted at more than 30 sites located throughout the United States and pursue a variety of missions. Department of Energy sites include industrial complexes that rank among the world's largest.

Some examples of the environmental impacts of Department of Energy activities include radioactive and nonradioactive hazardous wastes, thermal pollution from nuclear materials production reactors, and major land use and land cover impacts associated with large-scale engineering projects, such as processing facilities, cooling reservoirs, and waste storage sites.

In 1960, the Department of Energy established a Remote Sensing Laboratory in Las Vegas, Nevada. Of particular interest to the laboratory are measurements of the behavior of man-made radionucleides in the environment.[1] The laboratory maintains a small fleet of specially equipped aircraft for remote sensing, including helicopters, light and medium-size aircraft, and a small business jet suitable for high altitude flights (Figure 17–1).

In recent years, the expertise of the laboratory has been broadened to include special technologies developed at the Remote Sensing Laboratory for environmental measurements and analysis: optical and infrared remote sensing; nuclear intensity and spectral mapping; airborne gas and particulate sampling; data processing and analysis; environmental data base development; and sensor system development.

The extensive suite of remote sensor systems available provides a unique opportunity for airborne collection of comprehensive sets of data that are well suited for Department of Energy environmental

FIGURE 17–1   Primary aircraft of the Aerial Measurement Systems (AMS) program. A small fleet of both rotary and fixed-wing aircraft are maintained by the AMS program as remote sensor system platforms.

applications that range from emergency response and disaster control planning to environmental monitoring and protection.

A trend in the analysis of these data for environmental applications is the increasing integration of the remotely sensed data with other data. The multidisciplinary nature of remote sensing data is becoming evident, and the ability to extract useful information from these integrated data sets is proving valuable for many types of analyses.

COMPREHENSIVE INTEGRATED REMOTE SENSING

In 1982, the Remote Sensing Laboratory established a program called "Comprehensive Integrated Remote Sensing," whose overall objective was to provide remotely sensed data for all large Department of Energy sites. As implied by the title, the program involves data acquired from a number of different sensor systems[2] including: large format aerial photography; color video imagery from aerial platforms; thermal video imagery; multispectral scanner imagery; airborne nuclear radiometric surveys; Landsat and Spot satellite imagery. A description of the comprehensive Integrated Remote Sensing data base is included in the Appendix.

ENVIRONMENTAL PROTECTION STUDIES

Thermal effluents at Savannah River Plant in Savannah, Georgia, are significant and have had major impacts on wetland vegetation. When released directly into the natural environment, thermal discharges have damaged and killed vegetation along three stream channels and in portions of the Savannah River swamp system. Historical photography has been interpreted and processed for Steel Creek. The historical coverage, in conjunction with interpretations from more recent photography and multispectral imagery, documents the removal and regrowth of vegetation in response to reactor effluent temperature and flow rates.[3]

As part of the recent renovation of a nuclear reactor at the Savannah River Plant, a 400-hectare cooling reservoir was constructed in the upper basin of Steel Creek. As part of the Comprehensive Integrated Remote Sensing program, a land cover map of the Steel Creek corridor region of Savannah River Plant was prepared using vertical photography (taken prior to dam construction) and conventional photo interpretation techniques. This map was digitized for analysis purposes and entered into a geographic information system data base. Statistical summaries could then be generated by specified classes and areas. A color-coded version of the map is presented as Figure 17–2 (see color insert).[4]

Land cover data from the map has been used to evaluate wildlife habitat impacts of the cooling reservoir. Land cover data is an important input to habitat evaluation studies undertaken at the site. An update of the land cover map was also prepared using relatively simple and quick visual change detection techniques and photography acquired after the dam site was cleared. This detailed land cover map was recently expanded to include approximately one-third of the 77,700 hectare Savannah River Plant site.

Initial acquisitions for the Comprehensive Integrated Remote Sensing program have focused upon the Savannah River Plant. Primary applications of the data base to date include site development planning, environmental protection studies, emergency response and disaster control planning, and waste disposal planning and monitoring.

Periodic surveys have been undertaken to monitor thermal effluents released at the Savannah River Plant.[5] A recent water quality demonstration study has made use of Comprehensive Integrated Remote Sensing thermal infrared data to assess possible thermal effluent impacts on aquatic macrophyte vegetation in the Par Pond cooling reservoir. Par Pond is approximately 1,000 hectares in size and receives effluents from a precooler pond. Effluents entering Par Pond can be in excess of 40°C. A map of aquatic macrophytes, grouped into floating-leaved and emergent macrophyte classes, was prepared using aerial photography and multispectral scanner imagery. The map was registered to thermal imagery (Figure 17–3, see color insert) and the abundance of vegetation for three selected thermal regimes was evaluated. No major differences in the areal extent of aquatic macrophytes were noted in the three thermal regimes.[6]

Comprehensive Integrated Remote Sensing data is useful for several aspects of emergency response and planning. Much of the data is provided directly to appropriate site users for such purposes. Under emergency situations, many of the same resources would be used to review prior conditions or rapidly provide additional surveys or updates of dynamic conditions.

Some Comprehensive Integrated Remote Sensing data is also used as input to a companion Remote Sensing Laboratory project called the Graphic Overview System,[7] which compiles information concerning environmental programs and related information into sets of graphic prints and overlays. This provides a tool with which Department of Energy management can place key aspects of emergency situations into proper perspective to evaluate potential public and environmental impacts.

The disposition of man-made radionucleides is of great concern at Department of Energy nuclear facilities. The Comprehensive Integrated Remote Sensing program has conducted several surveys of the Savannah River Plant site and surrounding areas. In 1983, the entire Savannah River floodplain from Augusta to Savannah, Georgia, was surveyed[8] by airborne measurements of both natural and man-made gamma radiation emanating from the surface. Results of this survey have been graphically portrayed as isopleths of isotopic concentrations in the floodplain area superimposed on maps and aerial photographs of the area.

A sample of the results are shown for the Steel Creek delta region (see Figure 17–4). This area was of special concern because of potential resuspension and deposition of isotopic materials when effluents from the renovated reactor were resumed. The region was resurveyed at a later data to monitor the movement of isotopic materials.

SUMMARY

The U.S. Department of Energy's Remote Sensing Laboratory operates a unique assemblage of remote sensing equipment and aircraft. The Comprehensive Integrated Remote Sensing program involves the application of several sensor systems to meet the environmental and emergency response needs of large Department of Energy facilities. The multidisciplinary nature of these applications has increased the need for integrated data analyses.

Image processing techniques are used for merging the digital imagery that includes both digitized photography and multispectral scanner imagery.

The material presented here provides only brief examples of the potential application of remote sensing data and associated information system technologies to the study of environmental and emergency response problems. As efforts continue to use our environmental resources more wisely, programs such as Comprehensive Integrated Remote Sensing should prove beneficial for better assessment and monitoring of our environment.

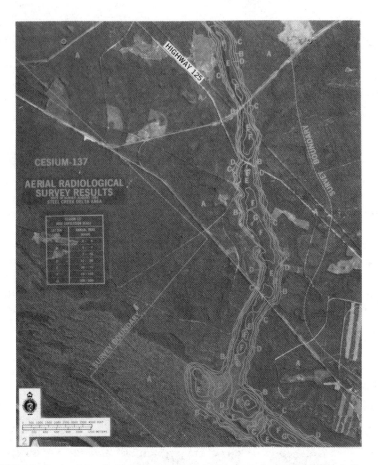

FIGURE 17–4 Aerial radiological survey results for a portion of Steel Creek, Savannah River Plant. A photographic base has been used for the Cesium-137 contour results. The lower region is where the stream flows into the Savannah River floodplain.

APPENDIX: TECHNICAL CONSIDERATIONS

The successful development of a Comprehensive Integrated Remote Sensing (CIRS) data base for a given Department of Energy site requires joint planning and coordination between the Remote Sensing Laboratory and environmental scientists at the site. Careful and early attention to the definition of site user needs is fundamental to the overall utility of the data base. Although data set standardization is strongly encouraged, Comprehensive Integrated Remote Sensing researchers understand that each site has different environmental conditions. In addition, these sites often have specific problems that may require unique data collection and analysis efforts.

Processing and analysis of the data also require close cooperation, involving many scientific disciplines and personnel from both the Remote Sensing Laboratory and the site. Some Comprehensive Integrated Remote Sensing data are transferred directly to site personnel after only minimal or routine preparations. As an example, video imagery from aerial platforms may only require duplication prior to distribution; all original imagery is maintained by the Remote Sensing Laboratory for archival purposes. Similarily, aerial photography often requires only development and duplication prior to distribution. In both cases, however, more extensive Remote Sensing Laboratory processing is available.

Systematic preprocessing steps for the multispectral scanner data include: conversion from the scanner's high-density digital tape (HDDT) to standard computer-compatible tape (CCT) format; line-by-line radiometric calibration of the thermal data based on the scanner's two internal blackbodies; and systematic geometric corrections to remove panoramic (S-bend) and aspect (velocity/height-related) effects. Additional processing of the data on site requires digital image processing equipment. Formerly requiring mainframe or minicomputer systems, image processing hardware and software is now available for relatively low cost microcomputer-based systems.

The nuclear radiation data require somewhat more extensive preparation prior to transmittal. Specialized equipment has been developed by Remote Sensing Laboratory for simultaneously collecting locational and gamma spectral radiation data. This data is routinely calibrated to represent surface level intensity levels as well as exposure rates for specific radionucleides.

Much of the Comprehensive Integrated Remote Sensing data does undergo some level of specialized preparation prior to transmittal. These data are processed by Remote Sensing Laboratory scientists familiar with the specific data types acquired. In some cases, full analyses of the data are prepared by Remote Sensing Laboratory and only summary reports are issued. Department of Energy sites vary in terms of how much assistance is required for their use of Comprehensive Integrated Remote Sensing data.

Management of a Comprehensive Integrated Remote Sensing data base entails various levels of integration for the multiple data types. The data types available include analog videotape and film or print format photography, digital point data (nuclear radiation data) and contour maps generated from them, and digital raster imagery from the multispectral scanner or digitized photographs. In addition, numerous acquisition parameters and related field data need to be handled. For complex analysis purposes, much of these data may need to be integrated with existing collateral data such as soils and land cover conditions.

Functions necessary to manage the Comprehensive Integrated Remote Sensing data base include indexing, searching, scaling, registration, rectification, and data set merging.

Several different organizational departments are involved in the collection and processing of Comprehensive Integrated Remote Sensing data. As a result, multiple systems are currently in use to handle the data and full integration is limited to specific analysis tasks where integration is considered most critical.

Efforts are underway to simplify and thus make data integration more routine for digital data. Two primary technologies are being used as a basis for accomplishing this goal: 1) digital image processing and 2) automated Geographic Information Systems (GIS). Conversions between vector and raster formats allow data to be transferred between the GIS and image processing systems. Remote Sensing Laboratory supports several different image processing systems, with a set of ESL VAX-based IDIMS workstations responsible for most Comprehensive Integrated Remote Sensing work. The VAX version of the ARC/INFO package developed by ESRI has been implemented for GIS support.

The integral data base management system and spatial data handling capabilities of the GIS can readily handle much of the integration needs of the Comprehensive Integrated Remote Sensing program, once the remote sensing data have been geographically referenced. As an example, maintaining a centralized indexing of Remote Sensing Laboratory remote sensing data is a relatively simple task for the GIS. In this case, however, only a reference to the data need be spatially defined in the data base.

The point sampled nuclear radiation data is collected in conjunction with suitable locational data to geographically reference it and merge it with other Comprehensive Integrated Remote Sensing data. Integrated analyses of Comprehensive Integrated Remote Sensing photography and multispectral data are hampered by the geometric distortions present in both aerial photography (tilt and relief distortions) and, especially, multispectral scanner imagery (pitch, roll, yaw, velocity, and altitude variations). The increased use of orthophotography techniques are being explored for the former, while efforts are being directed towards improved digital image registration capabilities for processing aircraft multispectral scanner imagery. Effective, albeit labor-intensive, registration algorithms from the VICAR image processing system have been implemented, but automated geometric corrections based upon an aircraft attitude model will be necessary for more routine preparation of rectified scanner data sets.

NOTES

[1] The laboratory is operated under contract by the Aerial Measurements Operations group of EG&G Energy Measurements, Inc., with program direction provided by the Nuclear Systems Division of the Nevada Operations Office of the Department of Energy; J.E. Jobst, 1979, "The Aerial Measuring Systems Program," *Nuclear Safety*, Vol. 20, No. 2, March–April, pp. 136–147; J.E. Jobst, 1985, "USDOE Aerial Radiation Monitoring Programs," *Proceedings Institute of Environmental Sciences*.

[2] J.G. Lackey and Z.G. Burson, 1984, "Comprehensive Integrated Remote Sensing at DOE Sites," in *Proceedings of the Fifth DOE Environmental Protection Information Meeting*, Albuquerque, New Mexico, November.

[3] L.R. Tinney, C.E. Ezra, and H.E. Mackey, 1985, "Stream Corridor and Delta Wetland Assessments—Savannah River Plant, Aiken, South Carolina," EG&G Energy Measurements, Inc., Las Vegas, Nevada, Department of Energy/(ONS-SRL)-8513, 49 pp.; E.J. Christensen, M.E. Hodgson, J.R. Jensen, H.E. Mackey, and R.R. Sharitz, 1984, "An Evaluation of Steel Creek Delta Growth and Recovery Using Photogrammetric and Geographic Information System Techniques, Savannah River Laboratory," DPST-83-1027, January, 22 pp.; E.J. Christensen, M.E. Hodgson, J.R. Jensen, H.E. Mackey, and R.R. Sharitz, 1984, "Pen Branch Delta Expansion, Savannah River Laboratory," DPST-83-1087, February, 19 pp.

[4] C.E. Ezra and L.R. Tinney, 1985, "Steel Creek Land Cover Data Base—Savannah River Plant, Aiken, South Carolina," EG&G Energy Measurements, Inc., DOE/ONS-8510, July, 34 pp.; the original GIS map was coordinate-digitized; the presentation color-coded map was in raster format.

[5] D.S. Negri, J.E. Shines, and L.R. Tinney, 1985, "A Thermal Infrared Survey of the Savannah River Plant—Aiken, South Carolina," EG&G Energy Measurements, Inc., Las Vegas, Nevada, DOE/ONS-8505, February, 39 pp.

[6] C.E. Ezra and L.R. Tinney, 1986, "Par Pond Macrophyte Study—Savannah River Plant, Aiken, South Carolina," EG&G Energy Measurements, Inc., Las Vegas, Nevada, Department of Energy (ONS-SRL)-8604, 5 pp.

[7] Z.G. Burson and D.R. Elle, 1980, "Graphic Overview System for DOE's Effluent and Environmental Monitoring Programs," EG&G Energy Measurements, Inc., Las Vegas, Nevada, EGG-1183-1760, March, 33 pp.

[8] P.K. Boyns, 1984, "An Aerial Radiological Survey of the Savannah River Floodplain," EG&G Energy Measurements, Inc., Las Vegas, Nevada, EGG-10282-1049, December, 39 pp.

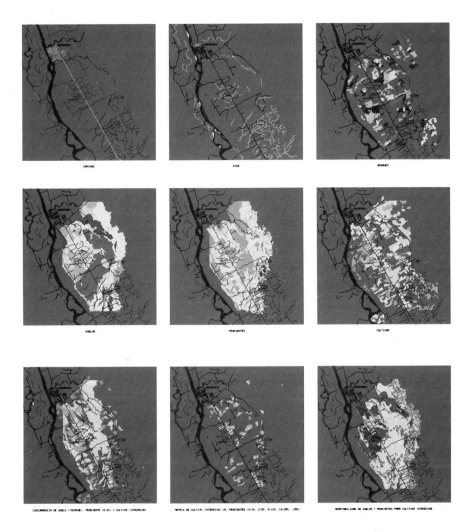

FIGURE 13–1 Information Derived for the Town of Aucayacu in the Huallaga Valley in East Central Peru.

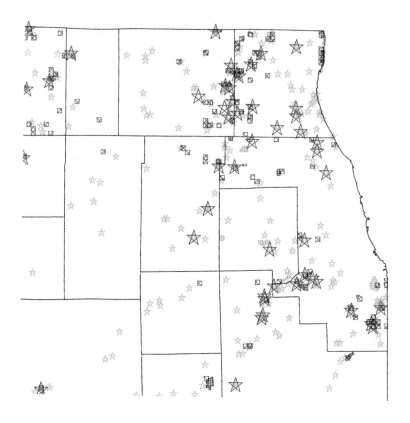

☒ T&E SPECIES LOCATIONS

▓ NATURE PRESERVES (polygon)

▦ NATURAL AREAS (polygon)

▨ NATURE PRESERVES (point)

▨ NATURAL AREAS (point)

FIGURE 15–1   Spatial relationship between locations of northeastern Illinois threatened and endangered plant species and Nature Preserves or Natural Areas. Polygons are areas greater than 300 acres in size and points are smaller areas.

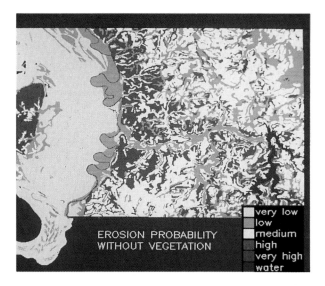

FIGURE 15–2   Probability for excessive erosion in southwest Jackson County, Illinois, if the region were devegetated.

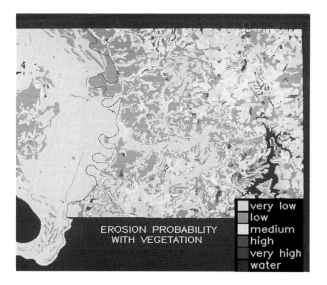

FIGURE 15–3   Probability for excessive erosion in southwest Jackson County, Illinois, with current land use/vegetation patterns.

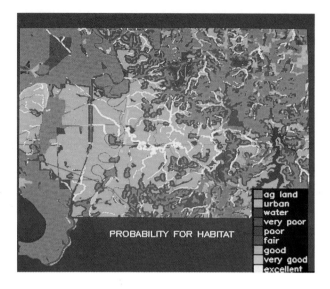

FIGURE 15–4 Probability for good quality endangered plant species habitat in southwest Jackson County, Illinois.

FIGURE 15–5 Estimates of mean annual increment (MAI) in Pope County, Illinois, based on Landsat TM and soils data.

FIGURE 16–1 Extent of tropical forest in Costa Rica in 1940.

FIGURE 16–2 Extent of tropical forest in Costa Rica in 1977.

FIGURE 16–3 Change in the extent of tropical forest in Costa Rica between 1940 and 1977.

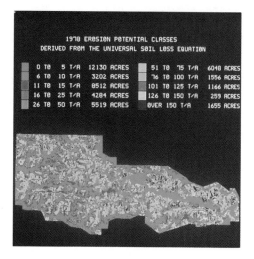

FIGURE 16–4   Application of the Universal Soil
Loss model to a watershed in Iowa.

FIGURE 16–5   Soils map for Africa.

Changing the Global Environment:
Perspectives on Human Involvement

Edited by D. Botkin, M. Caswell, J. Estes, and A. Orio

Erratum

The following reference was inadvertently omitted from the Landsat images that appear on page 225:

Woodwell, G.M., R.A. Houghton, T.A. Stone, R.F. Nelson, and W. Kovalick, 1987, "Deforestation in the Tropics: New Measurements in the Amazon Basin Using Landsat and NOAA Advanced Very High Resolution Radiometer Imagery," Journal of Geophysical Research, 92: 2157-2163, and related studies by the same authors. (Their images were obtained from the Brazilian Space Agency (MCT/CNPQ-INPE) under the general agreement covering the operation of Landsat, and their research was supported by the U.S. Department of Energy.)

FIGURE 16–6 Desertification hazard map for Africa.

STEEL CREEK
CORRIDOR

15 MAY 1989
△ EG&G/DOE RSL 5095.01

───────────────

LEGEND

■ WATER
■ NON-PERSISTENT EMERGENT
■ PERSISTENT EMERGENT
■ SCRUB-SHRUB
  BOTTOMLAND HARDWOOD
  SWAMP FOREST
■ DECIDUOUS FOREST
■ EVERGREEN FOREST
  CLEAR CUT
■ MIXED FOREST
  TRANSITIONAL
  INDUSTRIAL
■ UTILITIES
■ RAILROADS
■ ROADS

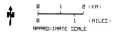
APPROXIMATE SCALE

FIGURE 17–2 Land cover map of the Steel Creek corridor region of the Savannah River Plant, located near Aiken, South Carolina. This map was photo interpreted from multiple date photography and entered into a geographic information system for analysis.

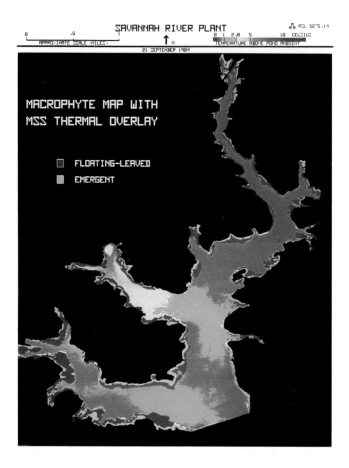

FIGURE 17–3 Macrophyte vegetation map of Savannah River
Plant's Par Pond cooling reservoir merged with color-coded aircraft
thermal infrared imagery.

Vegetation classes were photo interpreted using both aerial photography and multispectral scanner imagery. Five flight lines of
thermal infrared imagery were mosaiced and rectified to the vegetation map.

EDITORS' INTRODUCTION TO:
M.I. DYER AND D.A. CROSSLEY, JR.
Linking Ecological Networks and Models to Remote Sensing Programs

Previous chapters in this section suggest that technology can play a role in facilitating our understanding of the global environment. Having discussed what is, it is useful to look to the future and discuss what might be, what experts hope will be, and what is planned for the next decades. The next two chapters discuss future applications of technology. In the first, Drs. Dyer and Crossley propose an explicit research program whose goal is to improve the utility of remote sensing for the analysis of global environmental issues. They suggest that the technological capabilities of remote sensing, at present, greatly exceed the ability of scientists to put the method to work. They also suggest that a lack of communication between the engineers who have been the developers of remote sensing and the ecological scientists who could apply remote sensing to environmental issues has thwarted the advance of knowledge of our global environment, a concern also addressed in the chapter by Dr. Mar. The project they propose makes use of an existing international activity, the Man and the Biosphere Program, a program of international ecological research reserves. In this way, their chapter suggests not only a technological development, but also a social and political development, which follows from the programs discussed by Gywnne and Mooneyhan in the preceding chapter.

Drs. Dyer and Crossley are ecologists who have studied forest ecosystems of the southeastern United States. Both are associated with the Institute of Ecology, University of Georgia in Athens, Georgia.

18 LINKING ECOLOGICAL NETWORKS AND MODELS TO REMOTE SENSING PROGRAMS

M.I. DYER*
Biosphere Research, Inc.
Lenoir City, Tennessee

D.A. CROSSLEY, JR.
University of Georgia
Athens, Georgia

INTRODUCTION

Engineers and scientists working with high-altitude and space-oriented hardware and software have surpassed the abilities of environmental scientists and ecologists to utilize remote sensing information, while at the same time not really understanding fundamental problems on the ground. Developments in technology of remote sensing are now at a stage where these methods can be applied to large-scale environmental issues, but the ability to attribute the images to specific features on the Earth's surface and, even more importantly, to general conditions at other locations, had lagged behind.[1]

Ecological dynamics and global-to-local-scale variables associated with environmental change must be identified if goals of the newly formed International Geosphere-Biosphere Programme (IGBP) are to be met.[2] A focus must be placed on identification and analysis of processes, rather than a continued reliance on monitoring of state variables, one of the primary goals of many large-scale programs studying global, regional, or even local environmental dynamics. New technologies hold some prom-

* We thank the U.S. MAB Consortium for support for the workshop held to discuss the subjects reviewed here. We also thank the many workshop conferees for contributing to the discussions that we abstracted in this chapter.

ise for increasing our abilities to acquire information about vegetation dynamics,[3] detection of a variety of stresses in plant communities,[4] the identification of physical features on the Earth's surface that can be mapped to specific locations,[5] and the carbon and nutrient state of plants and inferred soil responses.[6] However, in addition to this technological development, we also need to develop a protocol that can integrate information about terrestrial and aquatic communities, geology, and soils in order to predict with reasonable accuracy changes in space and time. Thus, for studies of ecological phenomena with complex spatial patterns and changes across large areas, we need accurate determination of variables in space and the definition of processes by which change on the landscape is induced in time. In order to use remote sensing for developing this information, new process and modeling studies will be necessary to complement studies that can be carried out now.

THE BASIC PROBLEM

In developing ecological theory for landscapes and the technology of remote sensing, unwittingly a situation has been created where scientists find it difficult to couple the two subjects. On one hand, we have the extraordinary ability to engineer extremely capable satellite and remote sensing systems, while on the other we have gained enormous insight into how ecological associations come together and function. Olson[7] compare the situation to digging from opposite sides of a mountain to construct a tunnel. This task is possible, but only when the two teams have received and understood the same set of instructions. For the design of the study of large-scale environmental problems where both remote sensing and ecological theory play a role, seldom have the experts in each field gotten together to plan for common goals.

The need for this synthesis may seem strange at this stage of development of remote sensing of the environment, but it is still not apparent that scientists and engineers with the ideas and capabilities of accomplishing useful monitoring of global environmental change have developed a rational program. There is considerable progress in some areas,[8] but it is not sufficiently broad in scope. The extent of the problem is perhaps best understood by contrasting two well-known systems that have received wide attention in the past few years, coverage of the Earth's surface by Landsat and NOAA satellites and their on-board instruments.[9] In one [Landsat with Thematic Mapper (TM)] there is high spatial resolution (30 m), which allows for detailed examination of surface features, but the temporal resolution is very low (at best a 16-day repeat potential). In the other [NOAA series with Advanced Very High Resolution Radiometer (AVHRR) capabilities], there is excellent temporal resolution for a large portion of the Earth because images can be collected daily, but the spatial resolution is very low (1 km). To describe ecological systems, which of these two remote sensing systems is best? The question can be answered only when the objective is clearly defined. In actual practice the two systems approaches are complementary and should not be regarded as alternatives in the study of ecological problems, particularly when examining problems with several levels of spatial and temporal scale. But, because the two systems were designed without thought being given to the ecological scale of the problems, their combined use can be achieved only with a great degree of effort after costly research and development.[10]

For the future, programs designed around the strongest attributes of remote sensing and ecosystem analysis are needed. In such a program, scientists from a variety of disciplines should be invited to assemble a proposal for coupling these two different approaches. Experts on both Landsat TM (along with France's SPOT technology) and AVHRR should meet with experts on physical and biotic function of specific ecosystems. The ecological focus should be in ecosystems where there already is a great deal of available information about structure and function, especially where ecosystems have been set aside as research reserves. The best candidates for such new research and development are the Man and the Biosphere Reserves[11] and the United States National Science Foundation's Long-term Ecological Research Sites (LTER).[12] (There is considerable overlap in the United States; eight of fifteen LTER sites are Biosphere Reserves.)

As a measure of concern about how people contribute to change in global environment, it is noteworthy that 30 years ago none of the major facilities we now have to measure and describe the degree of global environmental change existed. Since then, the International Geophysical Year, the International Biological Program, and the International Hydrological Decade, to name only a few prominent programs, have come and gone. Now we have the Man and the Biosphere Program with its Biosphere Reserves, many national projects, such as the National Science Foundation Long-term Ecological Research project, and recently the advent of the International Geosphere-Biosphere Programme. All these programs have required research sites and have sought to link their findings through a network. With the growing desire to develop more robust links between remote sensing and ecological principles, it is logical to turn to the sites that have accumulated the most information over the past several decades. To start discussion on this design, experts from four U.S. Biosphere Reserves (three of them now LTER sites) and researchers and agency staff from NASA attended a workshop in Athens, Georgia, in May 1985. Some of what we report here emerged from that workshop.[13]

BACKGROUND AND APPROACH

The four research sites discussed here, particularly the three U.S.D.A. Forest Service research watersheds, are widely scattered, but they all collect the same types of ecological information. Because of this spatial displacement, there are certain similarities and differences that must be considered, both for ecological and remote sensing problems. All are forested systems, even though the community type and species composition vary greatly among sites. All have steep elevation gradients and highly variable aspects, with streams both controlling and integrating several landscape and ecological features. Nurients are highly variable, both from site to site and on yearly and seasonal bases within sites. Green-leaf surface area is high because of dense, thick forest canopies and is also highly variable. None of these four systems is homogeneous. Accordingly, they display a large site-specific variability in their ecological parameters. Indeed, there is likely as much variation within each of the sites as there is among them. Thus, the overall problem reduces to asking questions about what dominant land and ecological features exist at each site, how these vary spatially and temporally within a site, and, lastly, how they vary from site to site. Even though the task for defining this variance is not easy, it is one that is amenable to hierarchical ordering, a subject examined later in the chapter.

SETTING

To assess the problem, four sites were chosen from forested U.S. Biosphere Reserves: Coweeta Hydrological Laboratory, North Carolina; Great Smokey Mountains National Park, Tennessee; H.J. Andrews Experimental Forest, Oregon; and Hubbard Brook Experimental Forest, New Hampshire. Great Smokey Mountains National Park is administered by the U.S. Department of Interior, National Park Service; the other three are administered by the U.S. Department of Agriculture, Forest Service. The three U.S. Forest Service sites are widely known for their long-term research base on mountain watersheds; a great amount of work, funded by the National Science Foundation, has been conducted on basic ecosystem research projects at each site. The Great Smokey Mountains National Park is known for its diverse biotic communities and, even though there are no experimental watersheds, it has a long research tradition. Short synopses are presented to acquaint readers with the characteristics of the four chosen sites.

COWEETA HYDROLOGIC LABORATORY

This 2185-hectare site is located in a basin of the Nantahala Mountain Range of western North Carolina within the Blue Ridge Mountain chain, a part of the Eastern Deciduous Forest ecoregion.[14] The area has a moderate climate with cool summers and mild winters, and abundant rainfall is present in all seasons, averaging 1,780 mm at lower elevations and 2,500 mm at higher elevations. Its terrain is steep, highly dissected with elevations ranging from 686 to 1,600 m. There are 69 km of first- to third-order streams in the basin, many with active gauged weirs operating full time. Soils are deep and occur in two orders, fully developed Ultisols and young Inceptisols. Vegetation belongs to the Eastern Deciduous Forest province and is Appalachian oak forest[15] with abundant oak and hickory. The forests are diverse, with their distributions highly associated with moisture gradients, elevation, and aspect.[16]

GREAT SMOKEY MOUNTAINS NATIONAL PARK

The Great Smokey Mountains National Park is a 208,000-hectare preserve in the southern Blue Ridge Mountain chain, just slightly north of Coweeta Hydrologic Laboratory. Elevations range from 260 to 2,021 m. Climate varies from mesothermal-humid at low elevations to microthermal-perhumid at high elevations.[17] The terrain is steep and highly dissected with many streams throughout the preserve, ranging from first to approximately fifth or sixth order. Vegetation is complex. The reserve lies in the Appalachian oak forest section of the Eastern Deciduous Forest[18] but contains mixtures of evergreen; needle-leaved forests; and deciduous, broad-leaved forests. Two nonforest types, heath and grassy balds, are also present. Its complexity is further increased by the fact that much of the lower elevations of the preserve has been highly disturbed in the past, being either burned over, logged, grazed, or cleared for small-scale agriculture and subsequently allowed to revegetate. As was the case in the southeastern United States, the once dominant chestnut (*Castanea dentata*) was all but extirpated by a blight 50 to 60 years ago and has been replaced by other broad-leaved species. Elevation and site moisture class dominate the environmental gradients, as noted for Coweeta. As a result, biotic communities in the preserve are quite patchy.[19]

H.J. ANDREWS EXPERIMENTAL FOREST

This experimental forest is a 6,050-hectare watershed located in the central-western Cascade Mountains of Oregon. The climate is maritime. Winters are mild and wet, with warm, dry summers. Precipitation is high, on the order of 2,300 mm at lower elevations to over 2,500 mm on higher ridges. Its terrain is also steep and highly dissected with elevations ranging from 500 to 1,600 m. Several first- to third-order streams, many fitted with gauging stations to measure stream flow, are being studied intensively. Bedrock and soils are complex throughout the area, owing to a large number of volcanic flows in the Pliocene and Pleistocene. The vegetation is in the silver fir/Douglas fir section of the Pacific Forest,[20] and the H.J. Andrews site is dominated by Douglas fir (*Pseudotsuga menziesii*) in old-growth forest areas. Second growth, following fire or logging, is dominated by Douglas fir or noble fir (*Abies procera*).[21]

HUBBARD BROOK EXPERIMENTAL FOREST

This experimental forest is an approximately 3,000-hectare preserve situated in the White Mountains of New Hampshire. Its climate is continental, with an average rainfall of 1,300 mm per year, one-quarter to one-third of which is snowfall. Its terrain is also steep, dissected by first- to third-order streams, several of which have gauged weirs. Bedrock for the site is coarse-grained gneiss covered by a shallow layer of glacial till. Soils are relatively thin and well-drained, of a type called "Spodosols," with a sandy loam texture. The forest floor is also thin, and surface topography is rough because of pits and mounds caused by tree falls and surface boulders. Vegetation is complex, belonging to the Northern Hardwoods and the Spruce-Fir Forests.[22] The forest composition is correlated with elevational gradients, ranging from deciduous northern hardwoods at lower elevations to spruce and fir typical of the boreal forest at higher elevations.[23]

The three U.S. Forest Service sites have emphasized watershed research programs for decades; thus, an unequaled data base of terrestrial, aquatic, and geochemical processes exists for them. One of the main approaches in the three research programs has been to study effects of perturbations to entire watersheds. The most prominent perturbation has been removal of the forest trees. Clear-cuts and a variety of partial tree cuts have been employed over several decades. Each site, thus, has a strong published history of effects of forest community perturbations on successional patterns, ecosystem productivity, hydrology, water chemistry, and general models of the biogeochemistry of the area. This information is available for synthesis and new study efforts, which are needed to extrapolate the site-specific findings to other regions.

STUDY COMPONENTS

As we develop programs to examine ecosystem function through remote sensing technologies, it is apparent that we need to know a great deal about both topics. Not only that, we have to know what surrogates of ecosystem function we can measure from a remote sensing standpoint, since ecosystems have no metric per se. Measurement of green-leaf biomass has been one of the most useful methods developed to date.[24] Also, soil spectra have been measured to help resolve green-leaf biomass[25] but might be useful for other ecosystem parameters as well.[26] More recently, methods utilizing imaging spectrometer have been developed that can identify a wide variety of surface materials, including vegetation, soils, and minerals,[27] and perhaps total nitrogen and how it is partitioned in living plants.[28] However, to date, other than green-leaf biomass changes over time in an area, no ecosystem function properties have been examined in depth.

From changes in green-leaf biomass, it is possible to construct a view of ecosystem functioning using various ratios of spectral bands currently contained by Landsat satellite data. Botkin et al.[29] expanded this potential into a spatially hierarchical problem by calibrating green-leaf reflectance at a single site so that an area with low vegetation heterogeneity could be examined from a variety of heights with remote sensing equipment. Tucker and his coworkers[30] have followed seasonal changes of green-leaf biomass on the entire continent of Africa for a 19-month period. They have also viewed changes in large portions of the rest of the globe[31] using NOAA AVHRR satellite information. Also, crop coverage of the wheat and corn belt of the United States and other areas of the world has been reported for several years.[32] With the exception of inferences derived from some of the AVHRR work conducted by Tucker and his coworkers,[33] few of these reports really address the problems of ecosystem function that is needed for future monitoring and predictive purposes when we turn our attention to global change programs.

In order to build a program where ecosystem function can be monitored by remote sensing, it is necessary to consider the fundamental requirements. One of the first is recognition of the fact the problem is hierarchical.[34] Without the definition of the problem components, it is not possible to address proper ecological questions. This approach has been described in some detail for forested systems, such as represented at the four Biosphere Reserves discussed here. As Olson[35] indicated, we must work at more than one level in such a problem, but the question is how to integrate these levels. Highly detailed information being collected at the ground level for "bottom-up" research purposes must be linked to lower resolution information from Landsat or AVHRR satellite images for "top-down" assessment. Both approaches must deal with patchiness of information. The main problem is to be able to assign the linkages to either of the two approaches.[36]

In order to create the linkages, models must couple the two information sources.[37] To design this approach, it is necessary to define the state variables of the system being studied and then to structure the flows (processes) so they can be linked. If there is a hierarchical structure, then each of the levels in the hierarchy must in turn be linked. In practice for remote sensing purposes, it will be necessary to design and carry out ground-level research to make certain that either the changing state of key state variables can be measured or that some measure of the key processes themselves can be measured. Only in this way will it become possible to ascribe any degree of dynamics to a living system.

Once we can measure these systems adequately and learn to link them together with simulation models, we must consider whether the phenomena being measured can be made to represent similar conditions elsewhere. Perhaps one of the biggest problems to be encountered in this entire array of programmatic developments is the ability to extrapolate point-source information to larger areas. This step must be guided by model representation as well because there simply are not enough resources to repeat these intensive site examinations over and over to fill in gaps with statistical interpolations. Several approaches make this important step possible, although all are in early development. One is the use of fractal geometry to investigate whether patterns of state variables or processes are repeated in nature.[38] Building new regional models from those model systems developed at point locations will be daunting, but it is still possible.

PROGRAM DESIGN

In this chapter, we have promoted codevelopment of technological, environmental, and ecological projects. We urge new work at sites where ecosystem-oriented research is being conducted or at other sites with key system-level experiments called for by Waring et al.[39] At least two biome types should be considered, grasslands and temperate forests. They contain the best and longest-term data sets available, and there is in place a network from which such new work can be structured. Since we deal with temperate forests in this chapter, we will concentrate on them. However, programs exist for other areas. For example, in 1987, NASA, in association with the International Satellite Land Surface Climatology Project (ISLSCP), launched a three-year program in grasslands at the Konza Prairie Conservation Area and Biosphere Reserve in Kansas.

The foregoing sections concerning the three U.S. Forest Service and Great Smokey Mountains National Park give a synoptic review of their strengths for developing whole-system projects to be coupled with remote sensing. The work of Botkin et al.[40] suggests that correlations of state variable information with ground-level to high-altitude sensors should be an early part of this work, although it should not necessarily be the prime goal, since such correlations are lacking in predictive capabilities, one of the main aims of such work. Early in any new projects, it will be necessary to formulate conceptual models that can represent the essential hierarchical aspects of the entire problem. A simulation model structure should then be sought to represent the conceptualizations. One of the best potentials for this simulation approach is the development of what is currently being considered as the "telescoping model," one that can describe a series of hierarchical levels in an ecosystem, ranging from individual tree and small-plot performance to entire watersheds, all without having to change the model structure.[41] There currently is sufficient information about the state variables and processes at the U.S. Forest Service sites to commence this model development.

Remote sensing coverage for each site should progress at the same time. Thematic Mapper (TM) data exist for the H.J. Andrews Experimental Forest and should be developed more broadly in the other sites. Aircraft flights equipped to provide Multispectral Scanner (MSS) information exist for Great Smokey Mountains National Park.[42] There is a large library of information for AVHRR coverage for all sites. Along with this satellite coverage must be considered other types of remote sensing work, but with specific experiments at ecosystem-level research projects in mind. Finally, investigations into new modes of sensing ecosystem parameters remotely must be considered. A large variety of potentials exists, and each will have to be evaluated in the context of the whole-system properties and the ability to develop the technology to measure them.

Once these site-specific and cross-site programs have been initiated, it will be necessary to develop methods for broadening the scale and extrapolating the findings outward to surrounding areas to use this information for regional models.[43] It is also necessary to expand the measurements into other Biosphere Reserves so that a network of information can be built up. This expansion will constitute the Geosphere-Biosphere Observatory network being considered for the International Geosphere-Biosphere Programme (IGBP). These will be crucial steps, because it is only at this point that we can begin to utilize the full potential of satellite-based remote sensing of long-term change in the biosphere and geosphere.

NOTES

[1] L.B. Slobodkin, 1984, *Bioscience*, Vol. 34, pp. 484–485; T.F. Malone and J.G. Roederer, 1985, *Global Change*, Cambridge University Press; International Council of Scientific Unions, 1986, *The International Geosphere Biosphere Programme: A Study of Global Change*, 21 pp.; NASA, 1984, "Earth Observing System," Tech. Memo. 86129, Goddard Space Flight Center, Greenbelt, Maryland, Vol. I, Part 1, 51 pp.

[2] International Council of Scientific Unions, 1986, *op. cit.*

[3] J.S. Olson, J.A. Watts, and L.J. Allison, 1983, "Carbon in Live Vegetation of Major World Ecosystems," ORNL-5862, Oak Ridge National Laboratory, Tennessee, 161 pp.

[4] B.N. Rock, J.E. Vogelmann, D.L. Williams, A.F. Vogelmann, and T. Hoshizaki, 1986, *Bioscience*, Vol. 36, pp. 439–445.

[5] A.F.H. Goetz, G. Vane, J.E. Solomon, and B.N. Rock, 1985, *Science*, Vol. 228, pp. 1147–1153.

[6] R.H. Waring, J.D. Aber, J.M. Mellillo, and B. Moore III, 1986, *Bioscience*, Vol. 36, pp. 433–438.

[7] J.S. Olson, 1986, in *Coupling of Ecological Studies with Remote Sensing*, M.I. Dyer and D.A. Crossley, Jr. (eds.), Dept. of State Publ. 9504, 143 pp.

[8] H.J. Bolle and S.I. Rasool, 1986, ISLSCP. "Parameterisation of Land-Surface Characteristics; Use of Satellite Data in Climate Studies; First Results of ISLSCP," European Space Agency, 584 pp.

[9] N.E.G. Roller and J.E. Colwell, 1986, *Bioscience*, Vol. 36, pp. 468–475.

[10] D.B. Botkin, J.E. Estes, R.M. MacDonald, and M.V. Wilson, 1984, *Bioscience*, Vol. 34, pp. 508–514.

[11] M. Batisse, 1986, *Nature and Resources*, Vol. XXII, pp. 1–10.

[12] J.T. Callahan, 1984, *Bioscience*, Vol. 34, pp. 363–367.

[13] M.I. Dyer and D.A. Crossley, Jr., 1986, "Coupling of Ecological Studies with Remote Sensing," Dept. of State Publ. 9504, 143 pp.

[14] R.G. Bailey, 1978, "Descriptions of the Ecoregions of the United States," U.S.D.A. Forest Service, Intermtn. Reg., Ogden, Utah, 77 pp.

[15] Bailey, 1978, *op. cit.*

[16] Aboveground biomass is approximately 140 metric tons per hectare (t/ha) with a leaf area index (LAI) of 6.0 m2/m2; however, this reduces at higher elevations to approximately 3.5 m2/m2. Annual Net Primary Production (NPP) ranges from 4.1 t/ha in young hardwoods to 15 t/ha in older even-aged black locust stands. See W.T. Swank and D.A. Crossley, Jr., 1988, Forest Hydrology and Ecology at Coweeta, Springer-Verlag.

[17] P.W. White and M.D. MacKenzie, 1986, in "Coupling of Ecological Studies with Remote Sensing," M.I. Dyer and D.A. Crossley, Jr. (eds.), Dept. of State Publ. 9504, 143 pp.

[18] Bailey, 1978, *op. cit.*

[19] White and MacKenzie, 1986, *op. cit.*

[20] Bailey, 1978, *op. cit.*

[21] Standing crop biomass values are large, averaging approximately 868 t/ha in old growth forests. LAI values are among the highest in the world, reaching maximum levels of 15 m2/m2 or greater. The total organic material in this system is substantial because of the large trees and great amount stored in standing snags and rotting logs on the forest floor. See J.F. Franklin and R.H. Waring, 1979, in *Forests: Fresh Perspective from Ecosystem Analysis*, R.H. Waring (ed.), Oregon State University Press, Corvallis, pp. 59–86.

[22] Bailey, 1978, *op. cit.*

[23] In 50- to 60-year-old forests, standing crop biomass is 325 to 390 t/ha, with annual NPP approximately 3.6 t/ha. Almost all of the forests on the watershed have been cut at least once. A large data base exists for biogeochemical cycling research, one of the main research emphases at the site in the past two decades. See F.H. Bormann and G.E. Likens, 1979, *Pattern and Process in a Forested Ecosystem*, Springer-Verlag, New York, 253 pp.

[24] C.J. Tucker, 1977, *Remote Sensing of the Environment*, Vol. 6, pp. 11–26.

[25] C.J. Tucker and L.D. Miller, 1977, *Photogram. Eng. and Remote Sensing*, Vol. 43, pp. 721–726.

[26] See White and MacKenzie, 1986, *op. cit.* for survey of potentials.

[27] Goetz et al., 1986, *op. cit.*

[28] Waring et al., 1986, *op. cit.*

[29] Botkin et al., 1984, *op. cit.*

[30] C.J. Tucker, W.H. Jones, W.A. Kley, and J.J. Sundstrom, 1981, *Science*, Vol. 211 pp. 281–283; C.J. Tucker, J.R.G. Townsend, and T.E. Goff, 1985, *Science*, Vol. 227, pp. 369–375; C.O. Justice, (ed.), 1986, *International Journal of Remote Sensing*, Vol. 7, pp. 1383–1622.

[31] C.O. Justice, J.R.G. Townsend, B.N. Holben, and C.J. Tucker, 1985, *International Journal of Remote Sensing*, Vol. 6, pp. 1271–1318; C.J. Tucker, I.Y. Fung, C.D. Keeling, and R.H. Gammon, 1986, *Nature*, Vol. 319, pp. 195–199; Justice (ed.), 1986, op. cit.

[32] AGRISTARS Research Report, Fiscal Year 1983, 1984, NASA/JSC 18920, Johnson Space Center, Houston, Texas, 74 pp. plus appendices.

[33] Tucker, Townsend, and Goff, 1985, *op. cit.;* C.O. Justice, (ed.), 1986, op. cit.; Tucker et al., 1986, *op. cit.*

[34] R.V. O'Neill, D.L. DeAngelis, J.B. Waide, and T.F.H. Allen, 1986, *A Hierarchical Concept of the Ecosystem*, Princeton University Press.

[35] Olson, 1986, *op. cit.*

[36] D.L. Urban, R.V. O'Neill, and H.H. Shugart, Jr., 1986, in Dyer and Crossley, 1986, *op.*

cit.; D.L. Urban, R.V. O'Neill, and H.H. Shugart, Jr., 1987, *Bioscience,* Vol. 37, pp. 119–127.

[37] H.H. Shugart and T.M. Smith, 1986, in Dyer and Crossley, 1986, *op. cit.*

[38] J. Pastor and M. Huston, 1986, in Dyer and Crossley, 1986, *op. cit.;* J.R. Krummel, 1986, in Dyer and Crossley, 1986, *op. cit.;* S.N. Goward, C.J. Tucker, and D.G. Dye, 1986, in Dyer and Crossley, 1986, *op. cit.*

[39] Waring et al., 1986, *op. cit.*

[40] Botkin et al., 1984, *op. cit.*

[41] Shugart and Smith, 1986, *op. cit.*

[42] White and MacKensie, 1986, *op. cit.*

[43] Krummel, 1986, *op. cit.*

EDITORS' INTRODUCTION TO:
A. J. TUYAHOV, J. L. STAR, AND J. E. ESTES
Observing the Earth in the Next Decades

Scientists have been limited in their ability to understand our effects on the global environment because they have lacked data of appropriate spatial, spectral, and temporal scales of resolution. The ability to collect such data and to modify existing data in geographic information systems (as discussed previously by Gwynne and Mooneyhan) or in other more conventional forms is critical if we are to advance our understanding of the Earth. This chapter discusses the tendency of science to move toward increasing specialization and the present need for individuals capable of working across disciplines to address such complex problems as deforestation, desertification, or transnational pollution. The authors discuss advanced techniques that can help in this process, including the Earth Observing System, a complex of sensor systems on platforms associated with the United States Space Station project. Also discussed are the types of analyses that could be applied to remotely sensed and ancillary data as well as mapping, monitoring, and modeling of features at scales from local to global. The authors conclude that the technology exists to accomplish these goals; the challenge is for an international science community to develop both the cooperation and expertise to allow effective use to be made of the information provided by technology.

Mr. Tuyahov is the program manager of the Earth Observing System for the National Aeronautics and Space Administration. Dr. Star is a senior developmental engineer in the Geography Remote Sensing Unit and an expert in image processing and analysis systems. Dr. Estes is the director of the Geography Remote Sensing Unit at the University of California at Santa Barbara and an expert in the analysis and application of remotely sensed data.

19 OBSERVING THE EARTH IN THE NEXT DECADES

ALEXANDER J. TUYAHOV*
National Aeronautic and Space Administration, Washington, DC

JEFFREY L. STAR and JOHN E. ESTES
University of California Santa Barbara, California

INTRODUCTION

An understanding of Earth as a system, as discussed throughout this work, requires that research be conducted from an interdisciplinary focus. Such research must be directed toward the processes operating across the boundaries between the atmosphere, oceans, and land surfaces, as well as on man's impact on these processes. Examples of such cross-boundary problems include the increase in atmospheric carbon dioxide, depletion of the ozone layer, and deposition of acid rain. Study of these Earth processes must be approached in quantitative terms, on a unified, global basis, and over several decades. The purpose of this chapter is to review the historical trends leading to the current interest in interdisciplinary Earth science and the development of a plan for observing the Earth in the next decades. In the United States, work is proceeding to launch an Earth Observing System (EOS), a program related to a planned National Aeronautics and Space Administration

*Work on this paper was supported in part by the U.S. National Aeronautics and Space Administration through NASA Grant NASW–455. The germination of some of the ideas in this chapter came from discussions at the AIAA/NASA Earth Observing Systems Conference, Virginia Beach, Virginia, October 1985 and the conference in Venice, *Man's Role in Changing the Global Environment*.

(NASA) Space Station program for the 1990 and 2000 decades. EOS will provide the observational and information system capabilities needed for an understanding of Earth processes, with an emphasis on those global processes that operate at or near the Earth's surface.

The newest and most important initiatives in the U. S. civilian space program currently revolve around the space station complex. The space station complex includes the station and its associated co-orbiting and polar satellite platforms. This proposed suite of platforms and support systems offers an unique potential for facilitating long-term, multidisciplinary scientific investigations on a truly global scale.

The human-tended systems that are proposed for these future platforms will have new and unique capabilities to provide a wide range of data from both operational and research sensors. The large volumes of multispectral, multitemporal data expected from these systems, if supported by efficient and effective data systems, may provide a continuity of data that has been substantially lacking from previous satellite remote sensing systems, which operated on independent free-flying platforms. The challenge to the remote sensing community is twofold. The first challenge is to prepare for the large volumes of data that will become available in the late 1990s. The second challenge is to bring the tools we are developing to a broader constituency, in the service of what we call *global science* or as discussed by Botkin et al. in "The Science of the Biosphere,"[1] which we define as the science of the large-scale planetary system that includes and sustains life.

For those scientists concerned with the Earth's surface, the most important component of the proposed space station complex is the Earth Observing System (EOS).[2] EOS, based on the current design concept, has both active and passive earth surface sensor systems, as well as atmospheric sounding systems (Table 19–1). EOS is an evolutionary step in our efforts to remotely sense the Earth; it can provide a large scientific community with data in support of multidisciplinary research on an unprecedented scale. Unlike the previous generation of satellites, designed for relatively limited constituencies (such as the Landsat series for the land scientist and Seasat for oceanographers), EOS has the potential to provide an integrated source of information for investigating the dynamic interactions among the oceans, land surface, and atmosphere.

In the same way that EOS represents an evolution in earthward-looking satellite technology, we believe that EOS may help begin an evolutionary improvement in our understanding of our planet. Traditional branches of the earth sciences have been limited in scope to

TABLE 19-1   CANDIDATE EOS INSTRUMENTS

Moderate Resolution Imaging Spectrometer (MODIS)
   MODIS-N   Nadir Viewing
   MODIS-T   Off-Nadir Viewing
High Resolution Imaging Spectrometer (HIRIS)
Thermal Infrared Imaging Spectrometer (TIMS)
High Resolution Multifrequency Microwave Radiometer (HMMR)
Special Sensor Microwave Imager (SSMI)
Synthetic Aperture Radar (SAR)
Radar Altimeter (ALT)
Scatterometer (SCAT)
LIDAR Atmospheric Sounder and Altimeter (LASA)
Correlation Radiometer (CR)
Nadir Climate Interferometer Spectrometer (NCIS)
IR Radiometer (IR-RAD)
Microwave Limb Sounder (MLS)
UV/Visible Spectrometer (VIS/UV)
High Resolution Doppler Imager (HRDI)
Magnetosphere Particles Detector (MPD)
Magnetosphere Currents/Fields (MAG)
Earth Radiation Budget Instrument (ERBI)
Doppler LIDAR (DOPLID)

---

modest areas and to relatively narrow ranges of biophysical, geochemical, and socioeconomic processes by the extent to which technology allowed the measurement to map, monitor, and model those processes. It is our hope that EOS and other future satellite sensor systems that will evolve from this effort will foster and expand collaboration among scientific disciplines, continuing recent trends toward interdisciplinary science on an international scale.

HISTORICAL PERSPECTIVE

The history of science and technology shows a general trend towards specialization, with individuals developing greater expertise in increasingly narrow fields. A portion of this specialization has been enhanced by technological developments. The microscope expanded our perspectives inward: early optical microscopes evolved into today's computer-controlled electron microscopes and microprobes. The telescope expanded our horizons outward: technology was brought us to a time of electronically controlled active mirror telescopes and radio telescopes to probe the distant reaches of the universe.

Over the last decade, however, we have become more aware of problems that are fundamentally interdisciplinary, such as the greenhouse effect, regional deforestation, and groundwater pollution (discussed in the first section of this book). An understanding of the greenhouse effect, for example, requires not only knowledge of the effect of the atmosphere's composition on radiative heat balance, but also details of the circulation of the atmosphere, land/atmosphere interactions, ocean/atmosphere interactions, and biogeochemical cycles on land, in the air, and in the ocean. The EOS program as presently constituted represents a means to provide the data needed for such complex, large-area problems and an attempt to develop the infrastructure needed to address these problems.

The history of remote sensing mirrors those trends that have occurred in science and technology at large. The tethered balloons of the 1850s evolved to the aircraft of the early 1900s and then to the first satellite platforms available in the middle of this century. The space station currently being planned for the 1990s is based on a permanent human presence in space. This station complex, with its human-occupied core, and co-orbiting and polar platforms, will represent a major step in our observational potential.

Today's remote sensing practice uses virtually every technique developed in the past 100 years. Balloons, aircraft, and satellites all carry sensors ranging from cameras to electronic scanners, sounders, and synthetic aperture radars, using virtually all the electromagnetic spectrum. Resulting data are analyzed by people and machines, using both analog and digital techniques. In a modern remote sensing laboratory, the light table and stereo viewer are found next to the computer terminal—and the modern student of remote sensing science recognizes the potential of each.

The field of statistics developed in the seventeenth and eighteenth centuries provided science with a vital tool for understanding natural processes. In the 1920s and 1930s, the development of sampling theory furthered applications of statistics. These developments, along with computer technology in the 1950s and 1960s, provided the remote sensing experts with necessary tools for testing hypotheses and for the design of field work. Further, statistics provide a theoretical background as well as a suite of tools to move from simple identification of single source data to complex problem solving using multiple data sources.

Within the context of the science of the biosphere, vigorous application of sampling theory and statistical accuracy verification are required for at least two reasons. First, we are beginning to demonstrate that existing maps are woefully inadequate to the task of providing baseline information for monitoring and modeling those dynamic processes that help to sustain life on Earth.[3] Second, the multidisciplinary work we anticipate in the future must be rigorously based on ground truth and accuracy verification.

Applications of multisource data are most important in modern remote sensing, and we often use the phrase "information system" to describe our working concept.[4] An information system encompasses the entire flow of data—from sensor systems through data processing through dissemination of derived information—to an end user and a decision process. An important element of a new direction in remote sensing research is found in the recommendations of the EOS Science and Mission Requirements Working Group: "The Earth Observing System should be established as an information system. . . ."[5] This statement recognizes that if EOS is viewed simply as a sensor platform without considering the processing and distribution of resulting data and information to a user community, the potential of EOS will never be realized.

CURRENT TRENDS

The popular book *Megatrends*[6] reports on the new directions that are transforming modern human beings and our planet. Several of these megatrends are directly relevant to the challenges to be met by the remote sensing scientists as we move to take full advantage of evolving remote sensing and related information science technology.

The first of these megatrends is a global move from an industrial to an information society. In Naisbitt's words, "None [of these megatrends] is more subtle, yet more explosive than the megashift from an industrial to an information society." This information society, says Naisbitt, had its beginning in 1956 and 1957, which was the time of the launch of Sputnik. About that time, we began to move from using the term *aerial photographic interpretation* to the term *remote sensing*.

Remote sensing is an information-generating technology. One only has to examine the Applications volume of the recent *Manual of Remote Sensing*[7] to see the tremendous variety of information being generated from this technology. However, many professionals in this field feel frustrated. We feel that if we could find our data more efficiently, manage it better, process it more efficiently, and use it in a better fashion, we could do so much more. Better information systems are needed that link scientists at institutions around the globe.

In remote sensing we are also moving, albeit most slowly, from forced technology to what is currently referred to as "high tech/high touch." In the development of this technology, the users have not always been well served. Often scientists have been presented with systems by the engineers and asked "What can you do with this?" This is the ultimate in forced technology. Such efforts hamper the adoption of new technologies in a variety of ways, some of which are discussed by Caswell in the section that follows. While this has changed somewhat in recent years, scientists and other users must be brought into the planning process at the earliest possible moment. There is still a nagging suspicion on the part of many remote sensing scientists that our voices are not always heard.

Other unresolved issues in the United States are the impact of commercialization on long-term science access to satellite data; the continued funding of hardware systems—space stations and the associated systems, such as EOS; and whether this funding will be at the expense of other important projects.

Within this "high tech/high touch" trend, we see an increase in the use of artificial intelligence in the effective use of remotely sensed data. Particularly, work in the areas of expert systems and natural languages is showing potential to make the use of remotely sensed data easier and more understandable. These techniques, if properly applied, may allow less technically trained individuals to take better advantage of remote sensing information.

Analogous to Naisbitt's short-term/long-term megatrend are the shifts we have seen from applied to basic research within NASA since the launch of Landsat 1. Prior to 1972, many researchers in remote sensing did fundamental work on the digital processing of aircraft multispectral scanner data. Overnight, Landsat 1 provided a large volume of satellite data. Instead of building a solid research foundation, the scientists were forced to move directly towards applied topics using a new sensor, which had an inadequate information system and about which there had been insufficient basic research to understand the sensor's true capabilities.

In recent years there has been a shift within NASA to a more basic research emphasis requiring long-range research. The recent Global Biology/Global Habitability Programs illuminate this trend.[8] However, more work needs to be done in the information sciences.

Many researchers now employ a wide variety of spatially referenced data in remote sensing research. The synergism between geographic information system technology and remote sensing truly enhances the potential of each. To be most useful, remote sensing data must be combined with other types of data. In contrast, the quality of geographic information systems depends on the currency of the data they contain. Remote sensing can update data contained within a geographic information system, while geographic information systems can provide for the efficient use of the ancillary data required by remote sensing.[9]

Finally, in the use of remote sensing, we are moving toward addressing issues that are truly global in nature. We now have the potential to collect consistent global-scale data sets from which information may be derived and whose accuracy is verifiable. Past estimates of important global parameters (such as vegetation types, primary productivity, and biomass) have been difficult to develop and virtually impossible to verify. EOS can be one of the keys to unlocking global science. Information systems will allow us to turn this key in the lock. Improved information systems will facilitate our ability to conduct global research in an effective manner.

ANALYTIC FORMS AND OBJECTIVES

Examples of the kinds of analyses that will be performed on EOS data cover a wide range. Such analyses will generally take one of four explanatory forms and be oriented toward four objectives. The explanatory forms are discussed below and include morphometric analysis, cause-and-effect analysis, temporal analysis, and functional ecological systems analysis.[10] Objectives include inventory, mapping, monitoring, and modeling[11]; we discuss some of these below.

MORPHOMETRIC ANALYSIS

Scientific studies typically require measurement to determine the morphology of phenomena (i.e., their form and structure). Measured properties of phenomena may be generally classified as physical, spatial (geographical), or temporal. It is important to obtain quantitative information concerning these parameters in addition to descriptive evaluation. Field investigations are typically costly and site specific, providing only point observations that must be interpolated to yield a geographical surface. Remote sensing, however, can provide information about both point (per picture element) and areal properties. Remote sensing can play an important role in providing information on a number of biophysical properties, such as geometry (size, shape, arrangement, etc.), color or visual appearance, temperature, dielectric nature, moisture content, and organic and inorganic composition.[12]

CAUSE-AND-EFFECT ANALYSIS

Human beings have always examined the processes acting on their surroundings and attempted rational explanations of the causes. The synoptic view has important implications for regional studies that attempt to identify cause-and-effect relationships. Establishing such relationships is important to researchers in all branches of science. Increasing our ability to perceive effects that may be beyond direct visual experience can provide insights that may lead to improved understanding of environmental phenomena and processes.

EOS (and remote sensing in general) offers capabilities to detect and characterize effects that were previously beyond the limits of our perception and effective measurement. For example, thermal infrared scanners can record temperature differences in a river to pinpoint the location and provide a spatial perspective on a thermal plume undetectable by the unaided eye.[13] Similarly, the reflective near infrared has been employed to detect biophysical stress (i. e., effect) before the cause (e. g., loss of moisture from pathogens) is detectable in the visible spectrum.

TEMPORAL ANALYSIS

A concern with time in science stems from two principal considerations: 1) the explanation of observed phenomena typically involve an analysis of processes and sequences that occur through time and 2) the rates of change for a given phenomenon constitute an important characteristic.

Change in many scientific studies is synonymous with process and sequence. To be able to identify and monitor change accurately and consistently within a spatial framework is important. The ability to view objects or phenomena in their spatial context through time in a consistent manner is an important contribution of remote sensing to global science. Inconsistent data plague temporal studies. Planned EOS sensor systems show a variety of temporal resolutions (see Table 19–1). EOS data will be our internally consistent, longitudinal (i.e., temporal) data set.

FUNCTIONAL AND ECOLOGICAL SYSTEMS ANALYSIS

While researchers often require spatially accurate data about phenomena at many different scales, efficient and accurate methods commonly do not exist for collecting these data. Remote sensing systems offer the means to acquire such data in a number of disciplines and are beginning to be applied to systems analysis at both ends of the spatial continuum.

As an example, researchers have been examining the potential of using the advanced very high resolution radiometer (AVHRR) in conjunction with Landsat imagery to map within known accuracy limits the areal extent and spatial distributions of major forests types in the North American Boreal Forest. The combination of these research projects is directed at improving our scientific understanding of the cycling of carbon and other elemental materials.[14] In addition, scientists with remote sensing backgrounds are examining the information gained by the application of models to a number of physical processes and cultural phenomena (such as crop inventories and yield projection, monitoring snowmelt runoff, developing models for monitoring urban expansion, and energy consumption). EOS will greatly facilitate these types of studies.

The use of remotely sensed data as input to numerical models is complex to implement, but attractive in several ways. First, remote sensing data are inherently distributed (i.e., spatially disaggregated). As such, they are incompatible with many conventional models of environmental processes wherein values for a given area are "lumped" in some fashion or assigned to a specific node. Typically, these models do not readily accommodate remote sensing inputs.

Second, distributed models (both because of their greater spatial specificity and because they often are more of the deterministic than of the nodal or index type) may offer the potential of greater forecasting power under extreme conditions. Finally, the combination of remote sensing and modeling within a geographic information system framework (where data is fundamentally organized based on spatial location) has special appeal, as discussed earlier. Thus remote sensing may play an integral part in functional and ecological systems analyses wherein it may act as a key to the interfacing of biophysical, geochemical, social, and economic data for effective modeling purposes.

While the modes of explanation discussed earlier are examples of the scientific analyses that will be conducted employing the EOS system, the objective of these studies will be to achieve an improved knowledge of those biochemical, geophysical, and socioeconomic processes that affect life on this planet. EOS and the scientific programs it will foster can provide significant help in this area. EOS will improve our ability to inventory and map critical resources, facilitate monitoring of critical resources and processes occurring over large and small areas of the globe, and improve the accuracy of our models of the complex processes that impact life on this planet.

MAPPING

Most users involved in any form of spatial analysis, particularly those involved in resource management activities, want to see a map of information relevant to their application. It is in the area of thematic mapping (e.g., land cover, hydrology, soils, etc.) that considerable research is occuring on the use of remotely sensed data. Thematic mapping is an important component of any land resources investigation.[15]

Producing thematic maps from remotely sensed data requires us to develop means to extract specific information of interest from the general-purpose tools of the remote sensing discipline via manual and machine-assisted processing techniques. It is important to note that most maps produced for operational applications of a geographic nature are derived from visual image analysis techniques. Researchers in many disciplines are working to improve machine-assisted classification accuracies.[16] This task, however, is formidable, and there has been a general overselling of remote sensing's ability to provide accurate thematic data in a rapid fashion. For many scientists, there is still a great need for fundamental research in this area. Unfortunately, these topics are not always in vogue with the funding agencies.

MONITORING

Detecting change in land cover (for example, in terms of pattern or biophysical characteristics) is central to our ability to use remotely sensed data for planning and management purposes.[17] Monitoring of agricultural crops during a growing season can lead to the prediction of regional production. Identifying the characteristics of vegetation on a site can lead to better assessment of fire potential as well as better management of resources during a wildland fire.

Rates of change of environmental parameters are highly variable by category and location. As examples, the encroachment of urban land use onto prime agricultural land at the rural-urban fringe occurs at a rapid rate in southern California, while regeneration of clear-cut land to forest in the Boreal Forests occurs slowly. Thus variation in rates of change must be carefully considered from both functional and spatial perspectives in any environmental sampling design.

Interest in the potential of remote sensing for monitoring environmental phenomena has increased in recent years. Recent NASA programmatic interest in Global Biology and Global Habitability and the

National Academy of Science's proposed International Geosphere Bio-sphere Program (IGBP) are largely predicated on the ability of remote sensing to monitor selected environmental conditions on a global scale.[18] From research on the pattern and characteristics of desertification and deforestation, to estimates of global elemental cycling, to determination of those factors affecting climate, these programs call for monitoring and modeling research on an unprecendented scale. It is encouraging to note that these programs recognize the need for long-term research. However, there is an underlying assumption in these and similar documents that the image analysis techniques and processing, storage, and retrieval systems required to support these efforts are in place and only need to be applied. This is unfortunately not the case.

Research using Landsat data for the detection and mapping of changes in land cover have demonstrated significant potential, but much more needs to be done. To date, change detection studies employing machine-assisted processing techniques have been demonstrated for detecting and identifying areas of certain types of environmental change.[19] They have not, however, demonstrated the capability to detect changes consistently and with field verified absolute accuracies in the 80 to 90 percent range in a variety of geographic environments.[20]

MODELING

An important aspect of remote sensing has been to develop models that can be driven by inputs derived from remotely sensed data. Models that directly employ processed remotely sensed data to address specific geographic applications are still largely in the development stage. Considerable research emphasis must take place if we are to extend our understanding from the realm of systems structure—a difficult but primarily descriptive problem—into systems processes and the underly-ing dynamics.

The ability to predict consequences of trends in environmental conditions and assess the impacts of management decisions through simulations is an important step towards understanding the state and dynamics of a variety of geographic phenomena.

Remote sensing techniques have been applied to provide inputs to land capability and suitability models. Most operational usage, however, is based on manual interpretation of aerial photographs. In many instances, acquiring and processing aerial survey data and the sub-sequent manual interpretation prevents the timely and effective opera-tion of both land capability and suitability models. Land use updates typically cost more than half of the original survey costs, which severely restricts their application.[21] Many researchers consider the potential for semiautomated digital updates of land use surveys as the major unfulfilled promise and potential advantage of satellite remote sensing.

All land resources have inherent temporal and spatial components. It is necessary to predict both the quantity of aggregate change which is likely to occur in the future (i.e., the amount of land area likely to leave or enter a particular land cover category) and the most probable geographic location of change. The existing literature on the application of remote sensing to land cover spatial predictive modeling is limited.[22] So, too, is the literature of all modeling using remote sensing that documents the potential of remote sensing inputs to models on a quantitative basis.[23] Research in this area must occur if the application of remotely sensed data to research on the biosphere is to achieve its true potential.

BACKGROUND FOR NEW SENSING SYSTEM FOR 1990s

The need and concept for the proposed system, termed EOS for the Greek goddess of dawn, were defined by a Science Mission Requirements Working Group (SMRWG). This group was chartered by NASA to consider the broad Earth science objectives that could be addressed from a low-Earth orbital perspective in the 1990 and 2000 decades. These Earth science needs and system concepts are described in detail in the EOS Science and Mission Requirements Working Group Report. The SMRWG found that the concept of a permanent space station and accompanying polar orbiting platforms supporting a set of highly capable remote sensing instruments had great promise for addressing a number of problem areas in Earth science. The SMRWG made recommendations for synergistic groupings of instruments to be used to study the various components of the Earth system and the processes that link them together. One strong conclusion of this study was that early and proper emphasis on an information system to handle EOS data would ultimately be the key to the success of the system.

Following the completion of the SMRWG report in summer 1984, an EOS Science Steering Committee was formed to develop the mission implementation strategy. Representation from the U.S. National Oceanic and Atmospheric Administration (NOAA) was added to the steering committee as a first step in studying possible collaboration between NASA and NOAA in the use of the Polar Platform for joint scientific as well as operational remote sensing of the Earth. Under the auspices of the steering committee, five EOS instrument panels and an EOS data panel were formed to refine the respective concepts established by the SMRWG. The preliminary work of these groups was completed in December 1985, and described in separate reports.

An EOS Program Office was established at NASA headquarters in 1983 with a Technical Project Office at the Goddard Space Flight Center (GSFC) in Greenbelt, Maryland, and supporting project activities at the Jet Propulsion Laboratory (JPL) in Pasadena, California. GSFC and JPL are presently conducting a broad range of system studies including preliminary instrument design, instrument deployment scenarios, platform configurations, orbit altitude selection, and the possible joint use of the polar platform for operational weather purposes with NOAA. Planning has also been initiated for the eventual release of an Announcement of Opportunity for instrument development in preparation for a planned launch in late 1993. A vigorous program of EOS data analysis, including acquisition of ground and other in-situ observations and modeling to assist the interpretation of spaceborne data will be an integral part of the program.

THE EOS CONCEPT

Conceptually, the EOS consists of a permanent, highly adaptable, and evolving space facility. EOS will be supported by a ground data base and information management system, which will interact with the broad science community as well as existing national and international science information data base systems. The cornerstone of EOS will reside in the capability of the data and information system to network the scientific research community and, in so doing, provide them with the ability to access, process, and retain diverse Earth science data sets over a long period of time. EOS will be characterized by a coordinated interdisciplinary approach involving the synergistic and interchangeable use of a broad array of sensors operating across the electromagnetic spectrum. As currently planned, EOS will reside on the Polar Platforms of the Space Station in a near polar orbit and will be designed for Shuttle launch and on-orbit servicing, instrument augmentation, and replacement.

The current EOS planning scenario, based on a meeting of an international working group in Ottawa in May 1987, involves a series of polar orbiting platforms. Platforms will be in similar orbits, with complementary suites of instruments on each. Active participants in the program include NASA, NOAA, the European Space Agency, and

Japan. Principal components of EOS, according to current plans, include several imaging spectrometers [Moderate Resolution Imaging Spectrometer (MODIS) and High Resolution Imaging Spectrometer (HIRIS)], Synthetic Aperture Radar (SAR), a High Resolution Multifrequency Microwave Radiometer (HMMR), and atmospheric sounding systems [Lidar Atmospheric Sounder and Altimeter (LASA) and Laser Atmospheric Wind Sounder (LAWS)]. Some of these instruments use existing technology while others are still in the engineering design phase, but are expected to be flight-ready in the 1990s. Groupings of specific instruments on a given polar platform reflect balances between scientific use and platform resource availability (such as power and telemetry).

The imaging spectrometers in the proposed instrument configuration are a full generation more sophisticated than those in current operation. These systems have dramatically higher spectral resolution than current systems such as Landsat Thematic Mapper and SPOT. Both nadir and off-nadir viewing modes are required for science use.

A number of the EOS instruments make measurements in the active microwave mode. This approach will be significant for advances in geology, physical oceanography, sea and land ice sciences, forestry, agriculture, and hydrology. The instruments in this package include a Synthetic Aperture Radar (SAR), a Radar Altimeter (ALT), and a Radar Scatterometer (SCAT). Each instrument in this package is in a highly developed stage except for the multiple-look, multiple-frequency capability of imaging radar, which has yet to be implemented.

The SAR is the most demanding instrument in this package in terms of power, weight, cost, and data rate. The identification and location of ice ridges, floes, and leads with a SAR will provide information needed to determine ice motion and deformation. Geologic mapping and stereo imaging are the key uses of SAR in geology, while the temporal and spatial characterization of the extent and condition of vegetation are the key uses in vegetative land cover studies. In oceanography, short wavelength phenomena such as internal waves, surface currents, swells, surface wind speed, rings, and ocean fronts will be mapped by SAR. Quantitative measures of ocean phenomena mapped by SAR will be possible with the ALT and the SCAT.

Another suite of instruments will focus on the radiation, chemistry, and dynamics of the atmosphere and the linkage of these processes on a global scale. Furthermore, an improved knowledge of the interaction of the atmosphere with the land and ocean surfaces will be possible when the measurements from this package of instruments are combined with others. The instruments include tropospheric composition monitors, a Doppler lidar instrument for sensing tropospheric winds, Fabry-Perot and Michelson interferometers for measuring upper atmospheric winds and temperatures, and upper atmospheric composition instruments that include an infrared radiometer, a microwave limb sounder, and a visible and ultraviolet spectrometer. These instruments will make it possible to improve our predictive capabilities relating to the atmosphere by increasing our knowledge of global wind patterns, the coupling of the stratosphere and the troposphere, the distribution and fate of minor chemically active constituents in the atmosphere, and the response of the atmosphere to changes in its composition.

The instrument payload (see Table 19–1) was not intended to be fixed. It was assumed that a capability to change instruments on-orbit would exist as new scientific knowledge is attained and new observational capabilities emerge. The science working group also emphasized that in-situ and ground-based measurements are needed to complement the three remote sensing instrument packages in order to address successfully the identified science issues. Thus, Table 19–1 specifies a requirement for an Advanced Data Collection and Location System.

SPACE ELEMENT

The deployment strategy presently adopted for EOS is to use the polar platforms of the space station program. The polar platform will be a permanent facility in space with the following characteristics:

- design driven by the science requirements
- evolutionary development for growth
- modular systems and equipment design
- utilization of the space shuttle
- utilization of and commonality with space station developments, systems, hardware, and components
- on-orbit instrument servicing substitution, replacement, or augmentation
- continuous and essentially permanent on-orbit operations over decades

It is anticipated that four platforms will be used to deploy the total EOS payload. The platforms will be launched by the space shuttle in approximately one- to two-year intervals. The platforms will also have intrinsic propulsion sufficient to support the boost of the platform and payload to operating altitudes, deboost for rendezvous with the shuttle, and station keep at operational altitudes for a period of up to three years.

Servicing of the platform has been an area of priority and major study. As noted, some of the Shuttle launches will be used for servicing the platforms and instruments, including repair and replacement of instruments as well as the addition of new instruments. Initially, the polar platform will be serviced at the shuttle altitude via both extravehicular activity and internal vehicular activity using the remote manipulator on board the shuttle. The capability to service will make it possible for the mission life to be greater than ten years, a time interval required for studying some of the long-term processes.

An altitude of approximately 800 km is considered a baseline because it will provide good two-day coverage from the sensors under consideration. In addition, it provides for some continuity with existing long-term programs such as Landsat.

The initial crossing time of the first polar platform, approximately 1:30 PM, was chosen to accommodate the oceanographic as well as the land and biological science communities. It is close enough to noon to avoid or, at least, lessen the sun glint problem over water and at the same time permit good illumination in the visible and near-infrared for passive observations. It is planned that later Polar Platforms will be launched at other times to optimize data acquisition.

DATA AND INFORMATION SYSTEM

In addition to being an observing system, EOS is first and foremost an information system that includes spaceborne and ground observations, a system to integrate space, ground, and ancillary data, and a program of scientific analysis. Meeting the program's scientific objectives will require a data system in a geographically distributed environment. The system shall also facilitate access over the long term to other NASA data bases as well as data from other international organizations and U.S. government agencies.

The most fundamental purpose of EOS is to address the science questions posed by the mission requirements working group. Thus the most fundamental objective in implementing EOS is to provide to the science community the data that is necessary to answer those questions in the format and time scale that is required. In this regard, the following cornerstone recommendations have been made to ensure that EOS will be a capable information system:

1. A program must be initiated to ensure that present time series of Earth science data is maintained and continued. Collection of new data sets should be initiated.
2. A data system that provides easy, integrated, and complete access to past, present, and future data must be developed as soon as possible.
3. A long-term research effort must be sustained to study and understand these time series of Earth observations.
4. The Earth Observing System should be established as an information system to carry out those aspects of the preceeding recommendations that go beyond existing and currently planned activities.
5. The scientific direction of the Earth Observing System should be established and continued through an international scientific steering committee.

In the final analysis, it must be remembered that the EOS system is a truly international system, with active participation by the United States, Canada, Japan, and the countries of Europe. It is precisely this cooperation that will provide support for the International Council of Scientific Unions' program on Global Change. Yet, even beyond this, one needs to understand that the Soviet Union, India, Indonesia, France, Canada, China, and a number of other nations are actively pursuing programs to put sensors in space pointed towards the Earth. These future systems are designed to examine our current resource base, explore new resources, and monitor changes in critical environmental processes. Scientists of tomorrow, who are students in universities today, will have access to imaging spectroradiometers (with more than 100 channels in the visible and near-infrared to search for minerals and to look for signs of increased or decreased vigor in vegetation as well as tracking pollution); multifrequency, multilook-angle, active microwave systems that can view the Earth day or night, rain or shine; and laser altimeters capable of precise measurements.

All this derives from satellite remote sensing, a technology that did not exist until the 1960s, after the publication of *Man's Role in Changing the Face of the Earth*.[24] Remote sensing from geostationary and sun synchronous orbits, combined with aerial imagery and field sample data, is revolutionizing our understanding of the dynamics of our planet and will continue to do so. For the first time, this technology has given us the capability to acquire those globally consistent data sets whose accuracy is verifiable and will truly permit the development of a science of the biosphere.

SUMMARY

In conclusion, Earth science and technology development have progressed to a point where the conduct of a global science appears feasible. As discussed in other chapters in this book, Earth sciences are already faced with problems that are truly global in extent. Such problems require new approaches, that combine multidisciplinary, multinational teams of investigators, employing advanced technologies that can generate and analyze data sets not previously available to the scientific community. EOS and the EOS program have this potential. Yet if we are to exploit fully the potential of EOS, it must be done within an information systems context, linking scientists together with required facilities, data, and each other. Such an approach can improve the global science community's access to data sources and processing capabilities. The science of the biosphere is a data-intensive activity and, in its broadest sense, EOS, as an information system, can provide a tool for improved understanding of our planet.[25]

EOS is a complex system. It is currently planned to fly on the polar orbiting platform as a part of the total United States space station effort. The space station complex offers the global science community great potential, but a number of problems as well. There are still unanswered questions concerning the operational and commercial uses of the sensor systems on polar platforms. What will the United States National Oceanic and Atmospheric Administration's role be? Will the commercial Landsat vendor or other commercial entities be a major factor in sensor decisions? These and other technical problems must be carefully weighed. International scientific and technical cooperation and the role of the European Space Agency, SPOT Image Corporation, and the Japanese Earth and Marine Observing Systems must also be evaluated.

The challenge before the international scientific community is to continue to develop the infrastructure and expertise that will allow the EOS information system to work effectively. On the one hand, we must continue to develop the science and technology of remote sensing. Beyond the space station, this includes improved communications and advanced processing techniques, natural language interfaces, and advanced scientific workstations, as well as new, more directed sensor technology. On the other hand, we must embrace the concept of global biology and work toward a quantitative science of the biosphere and an understanding of the applications of such a science. We must put more stress on accuracy assessment and the qualification of the results of our studies. For only if we do this will we truly begin to understand the nature, limits, and variety of uses of the only known closed system capable of sustaining life for more than a few decades. It is the combination of these types of studies that will move us out of the realm of research toward global monitoring into operational data acquisition and analysis.

In this period, as we come to understand the uses of these new technologies, for monitoring and updating trends and the status of given Earth-based processes and phenomena, special-purpose sensor platforms and systems may be developed by private enterprise and the public sector. These systems could range from satellites with associated information systems directed solely towards specialized disciplines or industries (agricultural crop type determination and yield modeling, ocean current and storm monitoring, monitoring of global atmospheric ozone and carbon dioxide are but a few examples). The choice of systems and the infrastructure needed to derive the data required for a given basic or applied research on operational application will be complex. It will require the development of more sophisticated models of technology adoption, as discussed by Caswell in the following section of this volume.

NOTES

[1] D.B. Botkin (ed.), 1986, *Remote Sensing of the Biosphere*, National Academy of Science Press, Washington, DC.

[2] NASA, 1984a, *Earth Observing System*, Vol. I, Part I, p. 58, Working Group Report, NASA Tech. Memorandum 86129, NASA Goddard Space Flight Center; NASA, 1984b, *Earth Observing System*, Volume I, Part II, p. 59, Working Group Report, NASA Tech. Memorandum 86129, NASA Goddard Space Flight Center.

[3] D.B. Botkin, J.E. Estes, R. McDonald, and M. Wilson, 1984, "Studying the Earth's Vegetation from Space," *BioScience*, Vol. 34, No. 8, pp. 508–514; L. Mann, 1985, *Large Area Mapping Design for Multiple-Use Management in the Superior National Forest Minnesota*, 74 pp., Unpublished Masters Thesis, Santa Barbara, California.

[4] J.E. Estes, 1984, "Improved Information Systems: A Critical Need," *Proceedings 1984 Machine Processing of Remotely Sensed Data Symposium*, pp. 2–8, Lab for Apps. of Remote Sensing, Purdue University.

[5] NASA, 1984a, *op. cit.*

[6] J. Naisbitt, 1984, *Megatrends*, Warner Books, Inc. New York.

[7] J.E. Estes and G.A. Thorley (eds.), 1983, *Manual of Remote Sensing*, 2nd ed., Vol. II, pp. 1233–2315, Am. Soc. of Photogrammetry, Falls Church, Virginia.

[8] NASA, 1983a, *Land-Related Global Habitability Science Issues*, NASA Office of Space Science and Applications, Washington, DC, NASA Tech. Memorandum 85629; NASA, 1983b, *Global Biology Research Program*, M.B. Rambler (ed.), Program Plan, NASA Office of Space and Applications, Washington, DC, NASA Tech. Memorandum 85629; NASA, 1984a, *op. cit.;* NASA, 1984b, *op. cit.*

[9] J.E. Estes, 1984, op. cit.

[10] J.E. Estes, J.R. Jensen, and D.S. Simonett, 1980, "Impacts of Remote Sensing on U.S. Geography," Remote Sensing of Environment, Vol. 10, No. 1, pp. 43–80.

[11] J.E. Estes, 1985, "Geographic Applications of Remotely Sensed Data," Proceedings of the IEEE, Vol. 73, No. 6, pp. 1097–1107.

[12] J.R. Jensen, 1983, "Biophysical Remote Sensing," Annals of the American Assoc. Geography, Vol. 73, No. 1, pp. 111–132.

[13] J.E. Estes, E.J. Hajic, and L.R. Tinney (authors, ed.), 1983, "Fundamentals of Image Analysis: Analysis of Visible and Thermal Infrared Data," in Manual of Remote Sensing, 2nd ed., J.E. Estes and G.A. Thorley (eds.), Am. Soc. of Photogrammetry, Falls Church, Virginia, pp 987–1124.

[14] G.L. Atjay, P. Ketner, and P. Durignand, 1979, "Terrestrial Primary Production and Phytomass," in The Global Carbon Cycle, B. Bolin et al. (eds.), SCOPE 13, New York, pp. 129–181.

[15] D.S. Simonett, (ed.), 1976, "Applications Review for a Space Program Imaging Radar," Santa Barbara, Remote Sensing Unit, Tech. Rep. No. 1, NASA Contract No. NAS9-14816, Johnson Space Center, Houston, Texas, pp. 6–18.

[16] J.E. Estes, J. Scepan, L. Ritter, and H.M. Borella, 1983, Evaluation of Low-Altitude Remote Sensing for Obtaining Site Characteristic Information, 67 pp., Nuclear Regulatory Commission NUREG/CR-3583, S-762-R, RE, Washington, DC; G.H. Rosenfeld, K. Fitzpatrick-Lins, and H.S. Ling, 1981, "Sampling for Thematic Map Accuracy Testing," Photogramm. Eng. Remote Sensing, Vol. 48, No. 1, pp. 131–137.

[17] J.R. Anderson, 1977, "Land Use and Land Cover Changes–A Framework for Monitoring," J. Res. U.S. Geology Survey, Vol. 5, pp. 142–152.

[18] M.M. Waldrop, 1984, "An Inquiry into the State of the Earth," Science, Vol. 226, pp. 33–35; NASA, 1983a, op. cit.; NASA, 1983b, op. cit.

[19] J.W. Christenson and H.M. Lachowski, 1977, "Urban Area Delineation and Detection of Change Along the Urban-rural Boundary as Derived from Landsat Digital Data," presented at Fall Meeting of the American Soc. Photogramm., Falls Church, Virginia; S.Z. Friedman, 1978, "Change Detection and Urban Monitoring Research," National Aeronautics and Space Administration, Jet Propulsion Lab., Rep., Pasadena, California; J.L. Place, 1987, "Monitoring Change in Land Use Over Large Regions," A New Window on Our Planet, Geological Survey Professional Paper, Washington, DC, U.S. Government Printing Office; Computer Systems Corporation, 1979, "A Critical Review of the Change Detection and Urban Classification Literature," Goddard Space Flight Center, National Aeronautics and Space Administration Rep., Greenbelt, Maryland, 26 pp.

[20] J.E. Estes, D.S. Stow, and J.R. Jensen, 1982, "Monitoring Landuse Landcover Changes," in Remote Sensing for Resource Management, Johansen and Sanders (eds.), Ankeny, Iowa, pp. 100–120; J.E. Estes, 1985, op. cit.

[21] J.R. Anderson, 1977, op. cit.

[22] J.E. Estes, 1980, op. cit.

[23] K. Lulla, 1981, "Remote Sensing in Ecology," Can. J. Remote Sensing, Vol. 7, No. 2, pp. 97–107; G.R. Barker, 1983, "Forest Resource Information System: FRIS," Final Project Rep., St. Regis Paper Co., Jacksonville, Florida, 93 pp.; K. Lulla, 1983, "The Landsat Satellites and Selected Aspects of Physical Geography," Progr. Phys. Geography, Vol. 7, No. 1, pp. 1–45.

[24] W.L. Thomas (ed.), 1956, Man's Role in Changing the Face of the Earth, University of Chicago Press, Chicago.

[25] NASA, 1984a, op. cit.

SECTION III
SOCIAL AND ECONOMIC POLICY ISSUES

In the chapters included in Section I, some new advances in our scientific understanding of the biosphere were discussed. The chapters in Section II discussed new, impressive techniques that enable us to monitor and analyze the environment with greater accuracy than ever before. The question remains, however: Even if we have a greater understanding of the biosphere and have the technological capabilities to improve environmental quality, will we actually manage our global environment in a better way in the future? This question is addressed in Section III within the context of economic and social issues.

Some of the chapters in this section stress the need to clarify and understand economic and social choices. In the first chapter of this section, Dr. Pearce explains how the goal of sustainability, which has been recently advocated by the World Commission on Environment and Development, could be attained. Dr. Pearce stresses that environmental and economic growth can be harmonious objectives but changes will be required to reach the goal of sustainability. This chapter gives an excellent overview of the evolution of many environmental problems that we face and how we can move toward their solution.

While Dr. Pearce argues that a *sustainable future* may be possible, in Chapter 21 Dr. d'Arge questions whether it is probable. By comparing the level of CO_2 emissions that might be obtained if different ethical systems were dominant, he shows how certain economic systems (e. g., industrial, less developed) would benefit from pollution control. The ethical implications of environmental policies have seldom been discussed even though perceived fairness may have a greater impact on the decision process than scientific accuracy or economic efficiency. Dr. d'Arge points out that the methods used to evaluate environmental policies often mask an ethical bias.

Some global environmental problems are caused by the cumulative impact of local resource management decisions. For example, many local cases of estuarian and coastal water pollution can lead to a loss of productivity in the oceans. The deforestation of many areas will limit species diversity and possibly change global weather patterns. It is essential for resource management to be conducted efficiently at all levels—local, regional, and global. In Chapter 22, Drs. Brokensha and Riley report on the evolution of our understanding about which technologies and practices would be "appropriate" for solving some local problems. They show us that often the best of modern knowledge and technology have failed when applied to local environmental problems in developing nations. Their chapter provides two warnings to those who become overly optimistic after reading Sections I and II. First, it is important to understand local cultures and indigenous solutions if we hope to solve any environmental problem; second, we cannot solve global environmental problems unless we build a firm foundation on an appreciation of cultures and of natural resource management systems that have worked.

In Chapter 23, Mr. Bingham stresses the need to recognize that many of the planet's resources do not have prices assigned to them in economic markets. Although the timber value of a tree can be assessed, its value as part of a wildlife habitat or recreation experience is harder to determine. Such value is real, however, and must be incorporated into the calculations of benefits and costs. Mr. Bingham describes the most promising techniques for placing a monetary value on nonmarket goods.

In the following chapter, Dr. Bugliarello provides a novel conceptual approach, describing the complex interaction between natural systems, society, and technology, and the way that the balance between each of these components has evolved over time. He offers an optimistic view that we have the opportunity and the ability to use technology to have a "civilized" society while "avoiding destructive economic upheavals."

We have not been successful in the past in accurately assessing the impact of new technologies on society and the environment. Dr. Mar, in Chapter 25, suggests that such failures have been caused by our inexperience in doing interdisciplinary studies. Sometimes it is only a "language" barrier that causes the problem, but in other cases the cause may be more fundamental. His discussion echoes the main theme of this volume—now that we recognize that there are scientific, technological, and human components to environmental problems, how can we work together to find solutions? Dr. Mar offers a framework for evaluating the impact of new technologies that transcends disciplinary boundaries and ensures that all assumptions and methods are made clear.

Dr. Mar's chapter describes a procedure for assessing the impact of new technologies *if* they are adopted. The following chapter by Dr. Caswell addresses the question of *whether* a technology will be used. As is expressed in the chapter by Drs. Brokensha and Riley, some technologies may not be best for all circumstances. Using the example of modern remote sensing technology, Dr. Caswell shows that the appropriate technology will be defined by the environmental and social characteristics associated with those who might adopt it.

In the final chapter in this section, Dr. Endres describes several policies designed to improve environmental quality, and he compares them with respect to their effectiveness, economic efficiency, and political feasibility. This chapter returns to a theme introduced by Dr. Pearce when Dr. Endres concludes that a greater use of market-based (economic) incentives in environmental policy would lead to benefits for both the economy and environmental quality.

EDITORS' INTRODUCTION TO:
D. PEARCE
Sustainable Futures: Some Economic Issues

The first chapter in this section presents an overview of many of the crucial issues that are discussed in greater depth in subsequent chapters. Dr. Pearce is a distinguished economist from University College in London who began working on international environmental issues during the "negative" period of environmentalism and is well aware of the limitations of that approach. Here he offers a constructive approach, which he contrasts with the *Limits to Growth* and "steady-state" concepts that gained popularity in the 1970s. He assures us that a sustainable future is possible and that environmental quality and economic growth are harmonious objectives. He clarifies the role of economic theory in achieving these goals. The concept of sustainability of renewable and exhaustible resource use within the context of the laws of thermodynamics is clearly defined in this chapter, and Dr. Pearce discusses the interaction among changes in technology, social values, and economic conditions in attaining sustainability. Technological changes can lead to more efficient resource use in manufacturing or to more efficient pollution control; social changes can lead to more conservation and recycling of resources; economic changes can result in increased resource prices as relative scarcity increases. These changes will serve to moderate the depletion of resources and permit economic growth to occur in both industrial and less-developed countries.

20 SUSTAINABLE FUTURES: SOME ECONOMIC ISSUES

DAVID PEARCE
University College
London, England

INTRODUCTION

Sustainability is the fashionable buzzword for environmentalists in the 1980s. We now speak of *sustaining* the use of the environment, seeking *sustainable* rates of resource use, and making these environmental objectives consistent with the *sustainable* development of the economy. Environmental quality and economic growth are seen as harmonious objectives, not conflicting ones. Unfortunately, while widely used as a rallying cry for the new environmentalism, sustainability as a concept and a workable social objective has been subject to limited analysis. This chapter raises some of the issues that arise when thinking about sustainable futures.

THE EMERGENCE OF THE SUSTAINABILITY THEME

Sustainability has emerged as an integrative theme because the environmentalism of the 1970s grew from the idea that care of the environment and pursuit of economic growth are incompatible objectives. Notable but unscientific documents such as *The Limits to Growth*[1] and *The Blueprint for Survival*[2] were hallmarks of this period. While the environmental consciousness that came with that movement has been tremendously influential, the central theme, of conflict between social objectives, failed as a political selling point, despite the continued existence of so-called "green" radicalism. Growth-oriented societies could not be expected to reverse centuries of Protestant ethic. Poor communities could hardly be expected to sacrifice, even at the margin, what little they had. Sustainability offers an attractive alternative. Environmentalism becomes respectable and embraceable both by the industrial managerial community and the direct conservationists. Environmentalism need no longer be confused with green radicalism, although those who make that simple-minded equation as a historical observation are themselves guilty of ignorance of how environmentalism has progressed. The new environmentalism is thus sellable, in concept at least, and that alone partly explains its emergence.

While vaguely stated, sustainability appears as the underlying theme of the *World Conservation Strategy*[3] and some of the individual country responses to that document. A conspicuous feature of the World Conservation Strategy is the absence of any input from economists, an oddity given that resource and environmental economics (1) is a vast and growing subject with a history of some four decades, (2) contains a substantial literature on concepts of sustainability of resource use and environmental stability, and (3) is central to any analysis of making economic and environmental objectives compatible. The explanation for this neglect of economics lies in part in the continued ignorance of noneconomists about the concerns of economics. It would advance cooperative science no end if we could rid ourselves of the image that economics is about money or commerce, and that all economists automatically favor economic growth. It would help if the emphasis could be shifted to the concern of economists with *social prices,* the values that people place on goods and assets *regardless of whether they are sold in a market.* The relevance of social prices should become a little clearer as we consider some of the problems of affording sustainability a central role in our thinking about development and the environment.

THE TWO DIMENSIONS OF SUSTAINABILITY

There are two fundamental dimensions of sustainability. The first is *sustainable economic development,* (SED) which, for working purposes, we can take to mean the sustainable growth of per capita real incomes over time—the traditional economic growth objective.[4] (A point for discussion might be whether development and economic growth are the same thing, but we will not discuss this here.) The second feature is the *sustainable use of resources and environment* (SURE). Just as sustainable economic development implies some reasonably constant rate of growth in per capita real incomes without depleting the nation's capital stock, so the sustainable use of resources and the environment implies some rate of use of the environment that does not deplete its capital value. These definitions at least provide some rigor in the way we think about the SED and SURE principles, although they raise formidable problems in themselves—for example, how we measure the stock of natural assets. The essential features of sustainability are: (1) that we live only off the "income" generated by the wealth-creating process, where wealth has now to be defined to be inclusive of the stock of capital equipment *and* the stock of natural assets, together with the stock of knowledge; (2) as a corollary of (1), that we do not run down either the capital stock in its conventional accounting sense or in the sense of the stock of natural environmental assets; (3) that we may actually seek to increase the stock of environmental assets in the same way as the capitalist ethic teaches us to accumulate manufactured capital; and (4) that the pursuit of SED is *consistent* with the pursuit of SURE. In the economist's language, the critical feature of the SEDSURE philosophy is that we can be on an equilibrium growth path in the two broad sectors, the economy and the environment.

IS SUSTAINABILITY AN OBJECT OF FAITH OR LOGIC?

A major observation is that economic growth has a link to the amount of waste emitted to the environment. The reasoning here is familiar from the intellectual foundations of the zero growth or "steady-state" literature of the 1970s. The two laws we cannot avoid are the first two laws of themodynamics.

One implication of these laws is that, apart from materials "stored" in capital equipment, the mass and energy extracted as natural resources must reappear as waste in residuals flows. Whether this waste results in environmental damage through the accumulation of residuals or through impairment of capacity of the environment to absorb wastes depends upon the empirical magnitudes involved. Thus there is a link, but not a necessary one, between economic activity, waste emissions of energy and matter, and the scale of damage to the environment. Whether this link is stronger the higher the economic growth depends in turn upon the nature of the growth process. If growth is accompanied by a shift in the structure of gross national product towards payment for informational services (negentropy), then growth need not result in additional physical waste. But if growth is accompanied by increases in manufactured output, then increased physical waste will occur. In turn, the ability to prevent this waste from reaching the environment is limited by the second law, which tells us that recycling will require additonal available energy that can only be obtained from the environment.[5] The thermodynamic laws tell us that the resource problem and the pollution problem are obverse sides of the same coin.

As we have noted, one mechanism for reducing waste emissions is to change the *structure* of economic output. Technological change may reduce the inputs of material or energy per unit of output (but not limitlessly). Social change itself may act to do this through, for example, environmental consciousness and the conservation movement. Simple economic forces will also operate: few mechanisms are more powerful than the market price as an incentive to reduce input/output ratios, and market price is something we can manipulate through taxes or by fiat. The second mechanism, much embraced in the defense of economic growth during the 1970s, is that we can invest in environmental preservation to increase its capacity to deal with the increased wastes that economic growth is likely to generate. Basically, we take some percentage of the proceeds of growth and use them to correct the damage done by growth.[6]

Endless questions arise. At what speed are we using up resources and impairing the assimilative capacity of the environment? If the thermodynamic laws are in operation but bring about an unlivable society in 20,000 years, can it be a matter of concern even for an intergenerationally caring society today? To put it another way, is the issue of sustainability about living forever (in the sense of persistence of some social organization)? If so, is the philosophy indifferent to the quality of life in the sustained societies? Views about facts relating to rates of resource use and environmental destruction vary widely, as exemplified by looking at the *World Conservation Strategy* or its more structured form in Robert Allen's book,[7] and comparing these to the pervasive optimism of writers such as Julian Simon and Herman Kahn.[8]

What if the damage done by growth is irreversible? Once we have lost a species, we cannot re-create it. When this occurs, the philosophy of using the proceeds of growth to correct the damage as we go along is inapplicable. Yet irreversibility defines many of the environmental problems we are now encountering: wildlife loss, habitat loss, storage of radioactive and other hazardous waste, use of materials that have no counterpart degrader capacity in natural environments, destruction of the built heritage, the carbon dioxide problem, ozone depletion, and so on. Hope for sustainable growth advocates might lie in the demonstration that, as the stock of the environment *in fact* diminishes and as the demand for its preservation actually grows, so its social price rises rapidly.

This is one of the points where environmental economics comes in, for in the last decade we have seen impressive demonstrations that *if* markets existed in natural assets, the prices that people would pay for them are often very high.[9] This demonstrates that the social value of the environment is much underplayed. But this is not the "fault" of economics—it represents an historical evolution of the way in which rights to property have developed. Game birds survive because there is a profit motive in survival of the species. Wildlife parks are run as businesses. Kenya receives income from preserving wildlife in natural (or near-natural) habitats. What has happened here is that markets have developed in preservation. We still do not have markets in clean air, only limited ones in clean water, little or none in preserving whales, and so on. Nor is it worth waiting for history to solve the problem by evolving property rights that will produce those markets—the process is too slow and uncertain. Instead, what we need is the use of social prices—the prices that would emerge from such a hypothetical market—to demonstrate what our economies would look like if natural assets were incorporated into our national accounting systems in the same way as conventional capital assets. After all, if it is historical accident that one set of goods enters the accounts while the others do not, then we have no logical basis for ascribing positive values to those that do enter and implicit values of zero to those that do not. What we are after is an inventory and valuation of those things that people care about, and it is evident that they do care about the environment. What environmental economics has done is to demonstrate the impressive *magnitude* of some of these values.

To be clear, this is not an advocacy of markets in environmental assets, although it is not clear why Friends of the Earth should not bid in the market for *not* killing whales in the same way as the Japanese and Russians continue to bid for their slaughter. Nor is it advocacy of modifying our national accounts in some meticulous fashion to reflect the values of market and nonmarket assets, although there is virtue in such efforts. It is simply a request that we take what economic research we have and develop it if we can in order to show our political masters that we can *measure* the concern for the environment in many cases. By extrapolation at least, the value to be placed on assets not yet susceptible to economic value measurement is likely to be high. If we are to behave rationally under conditions of uncertainty, it is better then to avoid, as far as we can, irreversible damage. If that means increased social controls or higher prices for polluting products or heavily resource using products, so be it. SED and SURE offer no logical support for *free market* economic growth any more than they logically imply centralized coercion of economic activity. Finding the "middle way" is very much what SEDSURE advocates should be concerned with.

The outcome of our first observation—that physical laws present awkward problems for sustainable growth advocates—is complex. We must ask whether there are features of the changing economic and social system that will postpone their impact to some "acceptable" time horizon. Conventional beliefs in the use of some of our income growth to correct damage as we go along are naive and ignore the nature of environmental damage now being done. Policy has to be anticipatory, not retrospective. To turn conventional beliefs into workable policy, however, is extraordinarily complex: what incentive can we devise to *avoid* damage that has not yet occurred and that those who will suffer from it do not yet perceive? Again, this is a challenge to the middle way that should define the SEDSURE philosophy: how to devise incentives that preserve our freedom from coercion and yet secure our freedom and that of our children from irreversible damage. In other words, how do we devise incentives to secure livable and sustainable futures? If we can convert what we are now discovering about the social prices that people are placing on environmental preservation into actual incentives, we shall have gone a long way.

THE SURE PRINCIPLE

The SURE principle has its clearest interpretation with respect to *renewable resources*. The simplest rule for the sustainable use of a renewable resource is that its rate of use, or harvest (H), should not exceed the regeneration rate, or yield (Y). We can quickly see the sense of such a rule for resources such as fish, forests, soil fertility, and water. In the case of an aquifer, the rate of regeneration is simply the rate of natural recharge. The SURE rule thus becomes:

$$H \leq Y$$

There are two immediate problems with even this simple rule. The first is that Y will be stock-dependent. That is, yields will vary according to the absolute size of the stock of the renewable resource. A typical growth function for a renewable resource is shown in Figure 20–1. One tempting rule is to select the maximum possible sustainable yield (MSY),

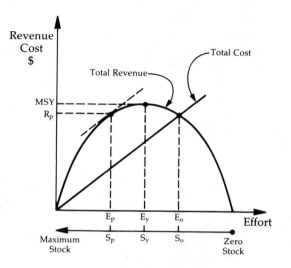

FIGURE 20-1 Total revenue is yield multiplied by price of the harvested resource. Total cost is the amount of effort (labor, capital etc.) multiplied by the price of effort (e.g., the wage). The optimal harvest is R_p rather than *MSY*, and the optimal stock is S_p rather than S_y. Note that if the resource is not owned by anyone (it is an "open access" resource) the tendency is for a solution such as S_o to come about, in which all profit is competed away.

that is, the greatest level of harvest, catch, or off-take that can occur without the stock declining. This is indeed how many life scientists continue to view the "optimal" rate of harvest for a renewable resource. But economists have long pointed out the *MSY* will only converge on the socially desirable rate of harvest by accident. If we draw in a function relating the amount of effort required to secure a given yield, we see that the maximum *net return*, defined as yield minus effort in Figure 20-1, is not *MSY*. That is, once we allow for the amount of labor and time involved in harvesting the resource (the "effort"), *MSY* is unlikely to be the desirable outcome. Figure 20-1 shows the optimal point (*P*) as being at a higher stock and lower yield and effort than *MSY*. The basic point is simple: there are many sustainable yields, and what we seek is the best one. That optimal yield is unlikely to coincide with the *MSY*, although as we shall see, the MSY still has attraction for other reasons. Note that an equilibrium such as that shown in Figure 20-1 is an equilibrium for the harvesting *industry* and for the ecosystem—it is a *bioeconomic equilibrium*, demonstrating that it is possible in the context of renewable resources to honor both economic and ecological objectives.

The second problem is that, as stated, our SURE rule as applied to renewable resources says nothing about growth in demand. The point about the simultaneous achievement of economic growth and sustainable resource use is that the consumption of the resource will tend to increase over time. We can imagine two responses. One would be for the yield to increase, and this can only happen if the carrying capacity of the ecosystem is altered so that larger stocks become feasible or there is a move toward maximum sustainable yield. Once *MSY* is reached, however, the system cannot meet any further increase in demand without running the stocks down. If this happens, ecological equilibrium is endangered, the stock will tend to extinction, and we have broken the basic rules of sustainability. Does this suggest that there is some "natural" limit to the growth process and hence to the feasibility of SEDSURE? The literature suggests not, provided the second means of adjustment is allowed to come into play, namely, that resource prices rise as demand increases. That is, there is a dynamic ecological equilibrium path of prices and harvest rates in the context of rising demand.

The SURE principle thus has an intuitive application to renewable resources. Before considering what it means in the context of non-renewable (exhaustible) resources, there is a need to consider the other major economic function of the natural environment—its role as a waste sink. Here we can define a rule that is entirely analogous to that given for renewable resources. The harvest rate is replaced with the amount of waste disposed of to the environment, and the yield is replaced with the rate of assimilation achievable by the environment. The SURE rule then becomes:

$$W \leq A$$

where W is waste emitted per unit time and A is the rate of assimilation.

Again, there are problems involved in ensuring that the rule is maintained. First, W is a multidimensional concept. It is not just the quantity of waste that matters but also its quality: 1 kilogram of cyanide is not to be compared to 1 kilogram of urban sewage. Second, for a large number of waste products the assimilative capacity of the environment is in fact zero. The environment has little or no degradation capacity for wastes such as cadmium and mercury. A strict application of the SURE principle would suggest that such wastes should be emitted to the environment in zero quantities—that is, either the products from which they come should not be produced at all or the flow of waste must not actually enter ecosystems. Believers in sustainability may therefore have to think in terms of foregoing certain products altogether or of a sealed encapsulation process for their associated wastes. Indeed, exactly this debate goes on, for example, with respect to radioactive waste, with some arguing that sustainablity implies no nuclear power and others arguing that nuclear waste must not enter ecosystems at all. In practice, a substantial part of the SURE principle's case on nuclear waste is actually honored by the nuclear industry insofar as it does conceive of plans to entomb nuclear waste. Yet we rarely hear of similar suggestions for, say, dioxin.

The issue of economic growth is the third problem for the application of SEDSURE to the environmental capacity issue. As demand grows, so the throughput of waste in the economy grows, and hence the flow of waste to the environment increases. The environment has a finite capacity for degrading waste. For example, a river can receive sewerage, break it down, and "purify" it, converting it to nutrients. But if assimilative capacity is fixed, waste flows must eventually exceed that capacity, bringing about the destruction of the counterpart degradation capacity of the environment. This interactive process is, of course, exactly that which ecologists have long told us about. The ways to escape from this problem, however, are less obvious and less likely than those for the analogous problem with renewable resources. We can increase the "stock" of the environment by investing in assimilative capacity—an example would be to augment river flows so as to increase the oxygen content of rivers. Similarly, just as there will be technological change that will increase the efficiency with which we use renewable resources (lowering the unit value of resource input per unit of output from the resource), so we can imagine technological change that will alter the nature of waste in terms of its damage to the environment. The hope would be for so-called "low waste" technologies to substitute for existing "high waste" technologies.[10] But whereas many renewable resources have an associated price, the assimilative capacity of the environment rarely carries such a price tag. Because of this, there is no particular incentive for anyone to adopt low waste technologies unless obliged to do so by a command or an incentive. Essentially, assimilative capacity is an *open access* resource, and it is this that leads to the danger of ecological instability through overuse. (Note that open access is not to be confused with common property resources. Common property resources are often managed so as to secure sustainability. To this end, Garret Hardin's famous "Tragedy of the Commons" is a fundamental misreading of the issue.[11])

While the problems for the application of the SURE principle to the environment have parallels in the renewable resource case (e.g., many renewable resources (fisheries, forests) are open access resources), the preceding discussion suggests that it is the waste disposal aspects of SURE that will cause most problems for believers in sustainable futures.

We may look briefly at the implications of sustainability for non-renewable resources. The first observation is self-evident. There can be no such thing as a sustainable use of a nonrenewable resource. Any rate of use that is positive will exhaust the resource in finite time. The argument that resources are infinite in supply will not be discussed here. This view can be found in recent works by Julian Simon.[12]

How is this problem to be overcome in order to secure consistent faith in sustainability? There are various suggestions. It may well be that the distinction between renewables and nonrenewables will have to be regarded as not applying to a strict interpretation of sustainability. Essentially, what would be argued is that sustainability applies to the *set* of all resources and not to the constituent parts. Sustainable use of renewables would then have to be supplemented by a second condition, namely, the substitution of renewables for nonrenewables at some rate consistent with the optimal use of nonrenewables. A schematic picture is shown in Figure 20–2, where we see that the demand for the totality of resources rises over time (in accordance with SED), that the time profile of use of nonrenewables is predetermined by some optimal plan, and that the profile for renewable resource use is given as a residual. Note, however, that the optimum path for nonrenewables would be partly determined by substitution possibilities. Moreover, the *stock* of renewables must rise through time in such a way that the harvest rate can become greater at an increasing rate.

The other sources of salvation for the nonrenewable resource problems lie in the much-discussed options of: 1) increased materials recycling, 2) technological change that reduces the ratio of materials and energy inputs to product outputs, and 3) the role of scarcity in causing real prices to rise, which will, in turn, induce conservation and substitution. But recycling rates are limited by the entropy of the economic system and, while the price mechanism works in such a way as to induce conservation, care must be taken in using observed market prices as signals of future scarcity.

SEDSURE IN DEVELOPING COUNTRIES

The preceding discussion has hinted at a few of the wide range of fundamental problems that must be faced by any advocate of SEDSURE. Without their resolution, it is uncertain that the clarion calls of docu-

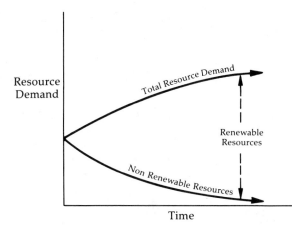

FIGURE 20-2 The planned time profile of the use of nonrenewable and renewable natural resources.

ments such as the *World Conservation Strategy* have much chance of being successfully implemented.

We can, however, illustrate some contexts in which the SEDSURE principle illuminates the growth problems and holds promise as a workable social objective. In a number of developing countries, it is now recognized that the process of ecological destruction has led to highly damaging impacts on economc growth. For these countries—those in the Sahel and Sudanian regions are cases in point—there is evidence to show that only by restoring ecological equilibrium can the growth process be restarted.

In highly simplified terms, such a disequilibrium has arisen because of a combination of factors that has led populations who, by and large, operated sustainable systems of agriculture to abandon those practices. As governments seek the commercialization of agriculture, there are incentives to move onto marginal lands and to increase the supply of labor by increasing population. The collective impact of these factors is not "internalized" in individual decisions because each individual acts in isolation as traditional communal thinking is forcefully replaced by centralized objectives and because events rapidly move out of the control of individuals.

As population expands, the trend towards urbanization grows. The demand for wood fuel, combined with forest clearance for agriculture and industrial purposes, leads to deforestation. As the forest cover is removed, the "sponge" and shelterbelt functions of the forest disappear. Average rainfall takes on the proportions of floods and topsoil is washed away. As the soil is denuded, so agricultural output declines; rivers and hydropower reservoirs are silted up. The deforestation thus produces ecological impacts that have costs attached to them. These "externalities"—to use the economist's language— can take on disaster proportions as previously damaging but generally manageable droughts become magnified in their effect. The outcome is well known. What were "natural" disasters become a combination of natural events and gross man-made damage through resource overuse.

This highly stylized account of what has happened, is happening, and will happen in poor countries is simplistic. Yet the evidence for its basic truth is mounting. What relevance then is SEDSURE to these countries? It is paramount. Short of wholesale relocation of populations and economically infeasible injections of foreign currency to purchase substitutes for existing resources (petroleum products for wood fuel, etc.), the only solution is regeneration of the renewable resources that have been depleted—forests, soil fertility, water supplies. In some cases, the cost of such programs is dramatically large and far beyond the resources currently available through international agencies and bilateral aid. But what the experience of these countries suggests is that sustainability *was* the defining quality of earlier agricultural systems and that only by restoring ecological equilibrium can the dramatic costs of resource depletion be contained. If sustainability can be restored, perhaps at a subsistence level of output in the worst cases, then the next stage can be a managed attempt at sustainable growth. SEDSURE would thus be in two stages—the restoration of ecological equilibrium first and sustainable growth thereafter. The alternative is a set of infeasible policies to provide an alternative to ecological equilibrium or no policy at all. The former must fail for many of the economies in question; this leaves the second. Just what that outcome looks like is already being experienced in the Sahel and elsewhere.[13]

SUMMARY

There can be no such thing as a workable policy of environmental and economic development without a clarification of concepts. Single-word objectives such as sustainability are deceptive and dangerous without rigorous thinking. Thinking about sustainablity as a social objective does not imply a large-scale standstill in policy formation while we work out what our aims are, even if such a "policy pause" were possible. We can afford to investigate further while at least avoiding major irreversible decisions. There can be no pretence that this chapter provides the rigor needed to lay the foundations for workable concepts of sustainability. We have, however, begun to scratch the surface of the problem, and that seems to take us a step further than has so far been the case.

NOTES

[1] D. Meadows et al., 1972, *The Limits of Growth*, Universe Books, New York.

[2] Ecologist Magazine, 1972, *Blueprint for Survival*, Penguin Books, Harmondsworth, U.K.

[3] International Union for the Conservation of Nature (IUCN), 1980, *World Conservation Strategy*, Gland, Switzerland; for an exposition see also R. Allen, 1980, *How to Save the World*, Kogan Page, London.

[4] Notice that this differs from Boulding's concept of *wealth* (of all kinds) as the source of well-being and income as the throughput to be minimized. See K. Boulding, 1966, "The Economics of the Coming Spaceship Earth," in H. Jarrett (ed.), *Environmental Quality in a Growing Economy*, Johns Hopkins University Press, Baltimore.

[5] See N. Georgescu-Roegen, 1973, "The Entropy Law and the Economic Problem," in H. Daly (ed.), *Toward a Steady State Economy*, W.H. Freeman, San Francisco; H. Daly 1987, "The Economic Growth Debate: What Some Economists Have Learned But Many Have Not," *Journal of Environmental Economics and Management*, Vol. 14, No. 4, pp. 323–336.

[6] A clear statement of the view can be found in W. Beckerman, 1974, *In Defence of Economic Growth*, Cape, London.

[7] Allen, 1980, *op. cit.*

[8] See J. Simon, 1981, *The Ultimate Resource*, Martin Robertson, Oxford; J. Simon and H. Kahn (eds.), 1984, *The Resourceful Earth*, Blackwell, Oxford.

[9] An overview of some recent work on such valuation techniques can be found in A.V. Kneese, 1984, *Measuring the Benefits of Clean Air and Water*, Resources for the Future, Inc., Washington, DC.

[10] See R.K. Turner and D.W. Pearce, 1984, "The Economic Evaluation of Low and Non-Waste Technologies," *Resources and Conservation*, Vol. 11.

[11] G. Hardin, 1968, "The Tragedy of the Commons," *Science*, Vol. 162, pp. 1243–1248.

[12] Simon, 1981, *op. cit.;* Simon and Kahn, 1984, *op. cit.*

[13] See D.W. Pearce and A. Markandya, 1985, "The Costs of Resource Depletion in Low Income Developing Countries," University College London, Department of Economics Discussion Paper 85–23.

EDITORS' INTRODUCTION TO:
R. C. d'ARGE
Ethical and Economic Systems for Managing the Global Commons

Dr. d'Arge has been recognized as a leading environmental economist since the early 1970s when he stressed that economic growth might be constrained by the environment's capacity to assimilate wastes (i.e., that resource *quality* rather than quantity may be the limiting factor). In this chapter, he opens up a new field for discussion. He uses the stratospheric problem of increased CO_2 emissions as an example of a global issue with intergenerational implications that will require cooperation among nations to solve. Whether actions to control CO_2 emissions will be considered "fair and/or efficient" will depend on the ethical criterion by which the action is judged. d'Arge compares four ethical systems in this context: Pareto, egalitarian, elitist, and utilitarian. He explains the elements that would charactize an "ideal" ethical system and then shows how the "diversity of ethical and religious beliefs within and between countries" may constrain us from attaining the ideal system. He explains how the dominance of any of the current systems would affect the relative well-being of four economic systems: Western developed, Middle Eastern, Socialist, and developing. This thought-provoking chapter emphasizes that no single solution to our pollution problems may be right for all.

21 ETHICAL AND ECONOMIC SYSTEMS FOR MANAGING THE GLOBAL COMMONS

RALPH C. d'ARGE
University of Wyoming
Laramie, Wyoming

INTRODUCTION

For some obscure reason, unsolvable policy problems often become the province and responsibility of economists. Certainly one of the most difficult to emerge in the past two decades is management of global common property resources, namely, resources that are not owned, controlled, or regulated by any one country but which provide services (and potentially disservices) to all, and thereby are of concern to every country. More than a decade ago, as was pointed out by d'Arge and Kneese, such global resources began to have a distinctly different set of institutional considerations of sovereignty of use and perception of national wealth than the typical set of common property resources examined by economists in the past.[1] Since then, there has been a growing interest in defining and measuring optimal use (or grazing) rates, identifying mitigating strategies, and re-examining economic principles associated with information costs and ethical considerations.[2] Using optimal control methods, Nordhaus has argued that the "shadow price" of CO_2 emissions into the stratosphere should increase through time in some relation to the "real interest rate on goods plus the CO_2 disappearance rate." If "the steady-state shadow price is $50 per ton CO_2, if the CO_2 discount rate is 5 percent and saturation occurs in 2052, today's shadow price would be $1.64 per ton."[3] This computation implies an increasing "shadow price" through time for carbon dioxide emissions not unlike the traditional Hotelling model for resource extraction where resource prices rise through time in some relation to the market rate of interest.[4] Intuitively, the price must rise through time to reflect the growing scarcity of the resource that is used up, whether it is a nonrenewable resource such as coal or the stratosphere. The Nordhaus model for CO_2 is identical in most respects to classical extraction models based on the maximization of *discounted* profits or utility through time. Future generations are only considered via their discounted utility of consumption and future catastrophic costs, which act as constraints to future emissions. Using a more heuristic and intuitive approach, several researchers have noted that if one calculated the cost today of a planetary disaster in 100 years using a moderate discount rate (e.g., 10 percent), one would discover that the value to society was negligible.[5] Lave pointed out that, given the great uncertainties of climate modeling and built-in institutional barriers for early mitigation, there is likely to be a substantial climatic change prior to the implementation of global control policies.[6] He suggested that the appropriate policy now is to look at how human adaptation to climate change might be increased and how such changes can be mitigated through adaptation and investment in accelerating human adjustment processes.

A radically different view of the problem has been proposed by d'Arge et al.[7] First, it is suggested that intergenerational choice ethically requires either compensation between generations or some ethical mechanism that does not discriminate against future generations. Rawls and Page have offered such a criterion where each generation is viewed as equal in value or utility terms.[8] That is, if individuals were placed in an "original position," not knowing what generation they would be part of, they would logically opt for equal treatment among generations. If this ethical criterion is adopted, then the utility discount rate across generations must be zero, that is future generations' utility is not reduced or given lesser weight. Given this assumption and the fact that the rate that the stratosphere regenerates itself is small or zero over the planning interval of relevance, then the "shadow price" for CO_2 emissions is constant or *decreases* over time. That is, its price is at its highest at the onset of planning and not at the end. A detailed proof of this result is given in d'Arge and Kogiku.[9]

Perhaps a heuristic argument will be convincing here. If a future generation suffers because of climatic change, it can only be compensated by increased consumption. In order for this compensation to be available, current generations must consume or pollute less. To do this implies a higher current "shadow price" on consumption and, therefore, emissions of CO_2. Note that consumption and emission of CO_2 are presumed to be strictly complemental and joint products. Thus, we have three distinct and possibly opposing policy recommendations regarding management of the stratosphere:

- Nordhaus: The optimal shadow price of CO_2 is low to start with and there is substantial uncertainty. By implication, we can delay making fundamental decisions until greater certainty exists.
- d'Arge: The optimal shadow price of CO_2, when intergenerational equity is considered, remains constant or decreases through time. There is substantial uncertainty. By implication, decisions need to be made now, particularly if they are of low cost and irreversible for reducing stratospheric emissions.
- Lave: Institutional forces preclude early controls of CO_2 emissions without a substantial and observed climatic change. Thus, current or early efforts at mitigation should concentrate on adaptation to the climatic changes.

It is the purpose of this chapter to examine the implications of these three paradoxical policy recommendations for management of the global commons, specifically, the stratosphere, in light of some recent ethical analyses. Also, a brief examination will be made of the political systems that are likely to influence rates of emissions of CO_2 to the stratosphere.

ETHICAL SYSTEMS

An ideal ethical system would be one that because of its fairness, efficiency, and completeness would select an optimal course of action over space, time, and institutional structures. That is, it would be both decisive in choice and best by unanimity across generations and societies. Unfortunately, such an all-inclusive ethical system is unlikely to emerge, if for no other reason than the diversity of ethical and religious beliefs within and between countries. Perhaps the most acceptable universal ethical system in a global context is the Pareto criterion where countries can take beneficial actions only if other countries are not made worse off than the status quo. Such a criterion may be highly restrictive because it may mean no action can be taken by countries that may harm present or future generations of other countries. A variation of this principle has already been adopted by the United Nations Stockholm Conference. It states that countries have "the responsibility to ensure that activities within their jurisdiction or control do not cause damage to the environment of other States or of areas beyond the limits of national jurisdiction."[10] The United Nations General Assembly adopted a less stringent resolution suggesting "due account must be taken of the need to avoid producing harmful effects on other countries."[11] Thus the Pareto criterion, with compensation implicitly ruled out or at least not explicitly identified, is embedded in accepted United Nations policy documents. Whether it can be implemented in practice remains doubtful for the case of carbon dioxide, where every country is both a polluter and recipient of the effects.

It has been suggested that the only efficient way to implement a Pareto type solution is for an autonomous international agency to be established to provide information and enforcement.[12] Yet, doubts have been raised about this possibility on the practical and realistic grounds that no nation will "voluntarily relinquish allocative choice-making for the stratosphere and its services over the country."[13] The Pareto ethical system has further disadvantages that only a subset of possibilities (without compensation) can be considered. For example, compensation could be paid to a slightly harmed country and all other nations be made substantially better off, and it makes permanent the current *status quo* in holdings of monetary assets, wealth, and environmental assets. Every change is judged at a point measuring a single distribution of income. That is, the worse off/better off Pareto criterion takes as a given where a country is today, rather than where it may aspire to be or where it incorrectly perceives itself to be.

On a more positive note, the stratospheric problem is one that can be characterized as a case of multiple reciprocal externalities, where each nation may harm (or benefit) others but also harms (or benefits) itself. In this case, there is a positive incentive for each nation to act unilaterally on CO_2 control. Since a country can reduce harm to itself through its own actions, even if other countries do not exert any control, there is an incentive, although a limited one, for each country to exert control unilaterally.[14]

A further problem with application of the Pareto criterion to the CO_2 problem is that some countries will be harmed while others will derive benefits, particularly from a global increase in temperature and greater variability of rainfall.[15] While some nations may have an incentive for international cooperation or at least limited unilateral action, others will have no such incentive. Thus, stratospheric CO_2 is a difficult policy problem since the gainers and losers all have sovereignty, there is substantial uncertainty as to who will gain and lose, and the current costs of substantially reduced energy conversion to developed and developing countries would be high.

A second simplified ethical criterion is the egalitarian ethic that indicates that the welfare of a society can be gauged by how well off is the individual who is worst off.[16] In a global context, welfare might be evaluated as to how well off the poorest nation is. Public policy in an international context would then specify that CO_2 policies make the least well off nation better off or at least be subjected to a minimum change toward being worse off.

Another way to state this ethic is that current and future policies of regulating CO_2 emissions place the least restrictions on controls in the developing nations or provide them with the greatest compensation, other factors being equal, for reducing emissions. With no international agreements, the developing countries are likely to suffer most, in that current generations are not substantially benefiting from current global CO_2 emissions and future generations will suffer from increased temperatures and climatic variability. Thus, the developing countries are likely to be in a long-term state where the worst off are made even worse off without international agreement, compensation, or intervention. The more developed nations alternatively are better off currently by not having to pay the price for CO_2 emissions and because of location; they are likely to be better off or, at least, less worse off from climatic change. Thus, from an egalitarian perspective, no controls on the developed nations' current CO_2 emissions is perhaps the worst ethical outcome.

Egalitarianism presupposes that somehow the worst off can be made better off through policy choices or compensation. This may not be the case for the least developed nations, where modern investments have often yielded low or zero net benefits per unit of investment. What this suggests is that, under the egalitarian ethical doctrine or other ethical systems, it might be impossible to compensate the losers from CO_2-induced climatic change.

The opposite of an egalitarian system might be classified as an elitist system where society welfare is judged by actions that make those already best off even better off. Without international agreement and regulation on the stratosphere, with each nation taking an independent course of action, this will be the likely outcome. The developed nations who are made better off by climatic change will not attempt to reduce emissions. It is presumed the developing nations in their pursuit of economic development will have little or no incentive to take unilateral actions. In consequence, a type of Nietzschean outcome emerges where some of the rich are made better off and some of the poor made worse off given current predictions as to CO_2-induced climate change.

Another ethical system or rule perhaps that is worth examining is a utilitarian system predicated on the idea that actions be taken that maximize the welfare of the entire global society. This implies that some countries would be consciously made worse off (but hopefully with compensation) for the good of the whole. A utilitarian approach presumes the existence of some type of measuring rod common to various countries or individuals to assess which actions lead to net improvements in welfare. Benefit-cost analysis yields one such form of measuring rod but may not be ethically acceptable across different cultures. For example, at a 1979 conference, more than 40 distinct objections were raised by representatives from various countries to the use of benefit-cost analyses to assess the impact of climatic change.[17]

There are undoubtedly a large number of distinct ethical systems by which to judge whether actions to control CO_2 emissions or evaluation of the CO_2 problem are fair or efficient. The four ethical system analyzed briefly here provide a glimpse of the rich possibilities for evaluating the global CO_2 problem. However, some tentative conclusions emerge. If no concerted actions are taken by any country except for self-interest focused on unilateral action, the CO_2 problem will yield a most undesirable ethical outcome by default. This ethical outcome will be to harm the poor and selectively aid the rich countries. It will be antiegalitarian, anti–Rawlsian, and is unlikely to be efficient (or utilitarian) from an economic perspective. The implied "shadow price" for CO_2 emissions from a global perspective will be positive but will be ignored, thus implying a zero price. Lave's emphasis on adaptation may be the only means that individual countries can take to reduce the costs or offset the impacts of climatic change.

The United Nations has already proposed a limited form of the Pareto ethical system, so perhaps this is the one most appropriate for guiding multilateral efforts at regulation. To carry this one step further, a rather simple proposal comes to mind. If all societies at all times are to be valued equally and the CO_2 problem exhibits differential effects by harming the poor and benefiting the rich countries, then in the spirit of a Pareto criterion, the developed nations might consider reducing current emissions or subsidizing non–CO_2 emissions-intensive growth currently in developing countries. Both these efforts would offset potential future losses in the poor countries and be compensated for by future benefits to the developed countries. Such a policy would tend also to be egalitarian and antielitist by aiding the least well off first. Finally, substantial restrictions on current emissions of CO_2 would imply a relatively high present "shadow price" for CO_2 emissions, which is consistent with the d'Arge and Kogiku condition for intertemporal global efficiency.[18]

POLITICAL SYSTEMS

The major long-term contributor to anticipated rising levels of stratospheric CO_2 is the continued and increasing combustion of fossil fuels and relatively slow rate of absorption of CO_2 in the oceans. In terms of a Western economic perspective, the sale and use of fossil fuels derives from the theory of mining extractible resources, where the owner of the resource sells as long as price exceeds marginal costs. The amount extracted and sold through time must decrease to induce a price rise sufficient for the present value of future years' profits to be equalized across years.[19] If each supplier reduces supply through time, then aggregate supply should also decrease over time as a result of profit-maximizing behavior. If demand is expanding rapidly enough, price increases will occur that maintain the present value of expected profits and also sustain or increase supply. However, the projected time profile, excluding considerations of increasing demand or reduced costs, is for the use of fossil fuels to be maximized as soon as possible.

The recent efforts of OPEC have been to restrict supply in order to sustain price. Thus, OPEC inadvertently has been beneficial to the CO_2 emissions problem by providing a substantive implicit early tax, consistent with a "shadow price" of 25 to 40 U. S. cents per gallon of gasoline or 8 to 20 U. S. cents for a typical pound of emissions of CO_2. This can be contrasted with the Nordhaus range of an optimal current tax of from 1.5 to 3.5 cents per pound of CO_2 emitted.[20] Unfortunately, the cartel does not include other fossil fuels, where both demand and use have expanded in response to higher oil prices.

Increased regulations on nuclear power production and a lessening of subsidies for solar power development in the United States have had opposing effects in that they have, on balance, accelerated CO_2 emissions. The important observation, however, is that in private market economies there is a continuing incentive to discover and extract fossil fuels as rapidly as market factors allow. This is due both to the necessity of increasing profits through time and to exploiting fully common pools where private rights are not adequately specified.

In the Middle Eastern economies, such rapid long-term exploitation of extractible resources would not be anticipated for several reasons. First, much of the extractible resource base and transport facilities are not privately held and, thereby, the exploitation of common pools can be expected to be more rational (i.e., slower). Second, the profit motive may not be the only rationale in decision making on extraction or sales. Islam has its own unique tenets of saving, accumulation of wealth, and consumption that imply a slower rate of extraction than the pure profit motive.[21] The central idea is that each individual owes part of profits, oil, or other wealth to others inclusive of future generations (called ZAKAH). Saving extractible resources for future generations leads to positive utility but not current profits. A more complex case can be made that the Muslim implied interest rate is lower than in market economies and therefore there is less emphasis on early extraction—that is, a lower interest rate implies that future extraction is more valuable compared with the present, so less will be extracted now. These factors suggest that Middle Eastern countries may have an institutionally constrained lower rate of extraction and thereby CO_2 emissions than the Western profit-maximizing model would predict.

A third paradigm on the supply side of CO_2 emissions are the socialist economies with substantial reserves of oil, coal, and natural gas. These economies tend to operate with high implicit interest rates and an emphasis on maximizing production rates, which suggests a high rate of current extraction and use, but not necessarily one that maximizes extraction and thereby combustion from each mine site. Finally, the developing economies are just commencing intensive use of fossil fuels in their development processes but will become increasing users from industrialization regardless whether they are net importers or exporters.

In Table 21–1 are recorded some qualitative observations on past, present, and future rates of extraction and combustion for these countries. Also included are the most and least preferred ethic based on which ethical principle yields the best economic outcome. There is obviously no dominant ethical principle based on individual economic interest. The Pareto principle is both the most and least preferred across political systems.

SUMMARY

There appears to be no dominant ethical principle for resolving the problem of CO_2 emissions into the stratosphere across various political systems. The most appealing one, the Pareto principle (with compensation within and between nations over time) would be the least preferred by the Western developed and socialist economies. The elitist principle, unfortunately, is the outcome of each nation pursuing its own interests and, in the long term, may be the most damaging for the global environment.

TABLE 21-1 SUMMARY OF QUALITATIVE ASSESSMENT—CO_2 EMISSIONS

Type of Economy	Current Rate of Extraction, Use, and CO_2 Emissions	Magnitude of Past Contributions	Magnitude of Future Emissions	Future CO_2 Related Costs	Most Preferred Ethical[a] Principle	Least Preferred Ethical Principle
Western developed economies	Rapid	High	High	Negative	Elitist or utilitarian	Pareto
Middle Eastern economies	Slower	Medium	Medium	Positive	Rawls[b]	Utilitarian
Socialist economies	Rapid	Low	Medium	Negative	Elitist or utilitarian	Pareto
Developing economies	Slow	Low	High	Highly positive	Pareto	Elitist

[a] Preference from the economic losses or gains as a result of the ethic being imposed.
[b] Presumes that a Muslim ethic overrides economic considerations.

NOTES

[1] R.C. d'Arge and A.V. Kneese, 1980, "State Liability for International Environmental Degradation," *Natural Resources Journal*, Vol. 20, pp. 427–450. Note: This paper was written and circulated in 1972 but did not appear in print until eight years later due to publication problems beyond the control of the authors.

[2] L. Lave, 1982, "Mitigating Strategies for Carbon Dioxide Problems," *American Economic Review*, Vol. 72, No. 2, pp. 157–261; W. Nordhaus, 1982, "How Fast Should We Graze the Global Commons?" *American Economic Review*, Vol. 72, No. 2, pp. 242–246; R.C. d'Arge, W.D. Schulze, and D.S. Brookshire, 1982, "Carbon Dioxide and Intergenerational Choice," *American Economic Review*, Vol. 72, No. 2, pp. 251–256; A. Scott, 1976, "Transfrontier Pollution and Institutional Choice," in *Studies in International Environmental Economics*, I. Walter (ed.), John Wiley & Sons, New York, pp. 303–318; and the references therein.

[3] W. Nordhaus, 1982, *op. cit.*

[4] H. Hotelling, 1931, "The Economics of Exhaustible Resources," *Journal of Political Economy*, Vol. 39, pp. 137–175.

[5] See L. Lave, 1982, *op. cit.*; d'Arge et al., 1982, *op. cit.*

[6] Lave, 1982, *op. cit.*

[7] d'Arge et al., 1982, *op. cit.*

[8] J. Rawls, 1971, *A Theory of Justice*, Harvard University Press, Cambridge, Massachusetts; T. Page, 1977, *Conservation and Economic Efficiency*, Resources for the Future, Johns Hopkins Press, Baltimore, Maryland.

[9] R.C. d'Arge, and K.C. Kogiku, 1973, "Economic Growth and the Environment," *Review of Economic Studies*, Vol. 40, pp. 61–77.

[10] U.N. Stockholm Conference on the Human Environment, 1972, "Declaration on the Human Environment," U.N. Document A/Conf. 48/4.

[11] U.N. Document, 1971, A/Res. 2849, XXVI.

[12] d'Arge and Kneese, 1980, *op. cit.*

[13] For an early yet very enlightening discussion of institutional problems see, A. Scott, 1976, *op. cit.*

[14] See R. C. d'Arge, 1976, "Transfrontier Pollution: Some Issues on Regulation," in *Studies in International Environmental Economics*, I. Walter (ed.), John Wiley & Sons, New York, pp. 257–278; R. C. d'Arge, 1975, "On the Economics of Transnational Environmental Externalities," in *Economic Analysis of Environmental Problems*, E. Mills (ed.), National Bureau of Economic Research, Columbia University Press, New York.

[15] R.C. d'Arge and V. Kerry Smith, 1982, "Uncertainty, Information, and Benefit-Cost Evaluation," in *The Economics of Managing Chlorofluorocarbons: Stratospheric Ozone and Climate Issues*, J.H. Cumberland, J.R. Hibbs, and I. Hoch (eds.), Resources for the Future, Johns Hopkins Press, Baltimore, Maryland, pp. 157–189; W.W. Kellogg, 1979, "Influences of Mankind on Climate," *Annual Review of Earth and Planetary Sciences;* Hotelling, 1931, *op. cit.*

[16] The discussion that follows draws heavily on the pioneering work of William Schulze, Allen Kneese, and W.D. Schulze, 1983, "Ethics and Environmental Economics," Resources for the Future, Washington, DC, Disc. paper QE85-01.

[17] R.C. d'Arge et al., 1975, *Economic and Social Measures of Biologic and Climatic Change*, U. S. Department of Transportation, Washington, DC, pp. 1–1241.

[18] d'Arge and Kogiku, 1973, *op. cit.*

[19] H. Hotelling, 1931, *op. cit.;* and W.D. Schulze, 1974, "The Optimal Use of Non-Renewable Resources: The Theory of Extraction," *Journal of Environmental Economics and Management*, Vol. 1, No. 1, pp. 53–73.

[20] Nordhaus, 1982, *op. cit.*

[21] M. Fahim Khan, 1984, "Macro Consumption Function in an Islamic Framework," *Journal of Research in Islamic Economics*, Vol. 1, No. 2, Winter 1404, pp. 1–24.

EDITORS' INTRODUCTION TO:
D. BROKENSHA AND B.W. RILEY
Managing Natural Resources: The Local Level

Drs. Brokensha and Riley, who have their primary training in anthropology and geography, respectively, have spent many years studying natural resource management techniques practiced in developing nations, with an emphasis on African communities. In this chapter, they report on the evolution of our understanding about the appropriate scale of technologies used in developing countries. In a well-meaning attempt to accelerate economic development in these areas, some industrialized nations and international agencies have promoted large-scale projects for fishery, forest, and water management. Many of these projects have had disastrous socioeconomic and environmental consequences. Drs. Brokensha and Riley argue that there are real advantages to involving local people in the planning and management of natural resources. They give examples of several types of projects from Asia, Africa, and South America that illustrate how local natural resource management can be effective and sustainable in economic, social, and ecological terms. They suggest that the lessons learned from these problems in local natural resource management must be heeded by those who hope to solve global environmentl problems.

22 MANAGING NATURAL RESOURCES: THE LOCAL LEVEL

DAVID BROKENSHA AND BERNARD W. RILEY
University of California
Santa Barbara, California

INTRODUCTION

George Perkins Marsh's *Man and Nature* (1864) has been described as "The fountainhead of conservation consciousness." Marsh was concerned with "how human activities had reshaped—largely unintentionally, often disastrously—the habitable Earth."[1] Not surprisingly, he was mainly concerned with European and other "civilized" societies. At that time, scarcely any systematic inquiries had been made about the relationship of small-scale, non-Western societies to their environment.

Coming forward nearly a century to the publication of *Man's Role in Changing the Face of the Earth* (1956), we find some significant statements made by the two anthropologists present, Alexander Spoehr and Sol Tax. We cite two statements as being especially relevant.

"So-called 'primitive' peoples do not exist in a state of ignorance of the natural world about them."[2] This may seem an obvious assertion, but, unfortunately, it needs to be repeated, and its truth demonstrated. A surprising number of people, especially those in the development community, need to be educated on this point. Sol Tax made the second significant statement when he summarized a session by saying "the consensus [of the session] seemed to be that a substantial part of the learning has to be done by those who would do the teaching of techniques to others."[3] This is the main point of this chapter—that *people who are involved in any aspect of natural resource management, global or local, should look first at what local societies do,* examining their existing practices and perceptions before recommending any intervention or innovation. What we describe concerns small-scale societies in developing countries in Africa, Asia, and Latin America, but the principle often applies to communities in economically more developed states.

Local Natural Resource Management (NRM) is important, both for those who plan for develoment in the Third World and also for those whose local resources are affected. This perspective is relevant for this book. Not only are some of our global environmental problems centered in regions of the Earth where there are many developing nations, but also plans for the study of the biosphere tend to be made in the developed nations with the assumption that all other nations will follow. Third World environments are generally deteriorating. During the last decade, most international and national agencies have realized that new approaches are necessary and that attempts must be made to find new institutions that will incorporate local people in development planning. During most of human history, the local community has been the basic unit for production and consumption of useful resources. Local systems of NRM (despite their variety) have been characterized by a high degree of social integration, low capital, and high labor inputs. They have also been dynamic, diversified, adaptive, and generally appropriate for their specific social and physical environments. As long as populations were small and technology limited, there was a balance. In the last century this was distorted by many forces, especially by rapid population increase. In Kenya, for example, an annual population increase of nearly 4 percent puts great pressure on the natural resource base. Introduction of new technologies, as well as development projects generally, have also altered the social and environmental balance, inducing a dependence on the state and erosion of local initiative. During the colonial period (which most Third World countries experienced), there was a tendency towards centralization. Government claimed ownership of basic natural resources such as land, forest, fish, and game. These claims further discouraged local responsibility, even though the state was seldom able to exercise effective control. Foreign interference was not the major problem since independent governments have further increased the level of centralization, and they too have often lacked effective management skills.

In recent years there have been modest signs in some countries and resource areas of a trend toward decentralization. Sometimes units chosen to control resources have been too large, such as the district in

Africa or the *panchayat* in South Asia. *Local* NRM—if it is be effective—should mean administration by a small group of people familiar with the area and with each other, who are united by historical and social bonds.

Another critical factor has been exploitation of resources for commercial purposes, when previous use had been mainly for subsistence. In the case of forestry, fisheries, and game, outsiders (either exotic ethnic groups or people of European decent) with sophisticated technology (such as shotguns, nylon nets, outboard motors, chain saws) engaged in wholesale resource base destruction.

One of the best-known statements on common property resources is Garrett Hardin's 1968 article, "The Tragedy of the Commons," in which he states that "freedom in a commons brings ruin to all." Hardin gives the example of pastoralists who have pasturage open to all, which works well until the carrying capacity of the commons is exceeded. At this point, he claims, herdsmen, acting rationally in their own interests, seek to maximize their gain, which they do by increasing their own herd. Hardin continues, applying the principle to population growth, "Freedom to breed is intolerable," because parents do not have the direct responsibility of caring for their children. He asserts that "as the human population has increased, the commons has had to be abandoned in one aspect after another"—in agriculture, grazing, hunting, and fishing.[4]

This doctrine finds a number of receptive listeners today, where many nations (especially the industrialized states, but also some of the less-developed countries) have enshrined a belief in "privatization" and "market forces," and have built up a resistance to and suspicion of any communal enterprise. Whatever the merits of this argument at national or international (IMF) levels, it should not be indiscriminantly applied to local communities. One major objection to Hardin's thesis is that he is describing an open access system, which is different from a common property regime. "Open access regimes do not recognize anyone's claim as a right. Consequently, there are no duties on the part of others."[5] So Hardin is correct insofar as he is describing an open access system but incorrect in the implication that this is a general way of managing common property. In fact, pastoralists (to look at the group that Hardin selects) have highly developed systems of rangeland management, with mutually recognized rights and duties.

The Maasai, probably the best known of Africa's pastoralists, had "elaborate grazing sequences, grazing flushes to create hay in dry-season reserves; regular use of donkeys to carry water . . . to permit camps to stay away from their dry season reserves as long as possible . . . and regular social rebuke and avoidance of families or camps that fail to adhere to good management practices."[6] This is clearly no free-for-all open access, but a carefully regulated system involving defined and enforceable rights and duties evolved to meet social and environmental needs.[7] We found it necessary to go into detail to refute Hardin's thesis, because it does have, unfortunately, a widespread and uncritical acceptance in academic and other circles.

We show below how societies evolved workable rules for managing their natural resources, taking examples from water management, fisheries, forestry, and wildlife. In all cases, we are concerned with rights and duties, that are specified for individuals or groups and that are enforced.[8] In contrast to common property regimes, many government-sponsored "top-down" projects leave local people unclear about their rights and responsibilities regarding maintenance, distribution, enforcement, and other matters. People need to be certain about costs and benefits (including, of course, social costs and benefits, often regulated by technicians and planners) before they will invest time or money in a project.

In the thirty years since the publication of *Man's Role in Changing the Face of the Earth*, there have been detailed, systematic studies of indigenous knowledge systems, with specific reference to resource management.[9] We now know enough about local NRM to make confident assertions about its value and the necessity for taking it seriously.

The 1956 publication of *Man's Role in Changing the Face of the Earth* came at a time when development symbolized large capital-intensive "high tech" projects, the prototype being a large hydroelectric power dam. Since those confident days of the 1950s, so many grandiose development projects failed to meet their objectives that officials now are inclined to seek alternatives, one being to start by looking at what local people actually do. We do not advocate in any way an abandonment of Western technology—small is not always beautiful—nor do we romanticize local NRM, recognizing that many of these systems are no longer effective because of inexorable distorting influences already mentioned. For example, one recent model examines "long-term persistence of communal tenure systems in Amazonia and recent dissolution of those systems." Government policy provided a direct incentive—granting individual title to land—to Indians who "improved"parcels of forest by establishing *permanent* fields or pasture. This was a clear instance of government intervention—attempting to increase resource extraction and promote agricultural colonization—destroying the previously sustainable system. New systems of land tenure mean that some Indians adopt destructive slash-and-burn techniques, abandoning their old sustaining agriculture.[10] Many other examples could be cited. We are trying to demonstrate that the local people do not live in a state of ignorance about their natural world; unless they had a detailed and systematic knowledge of their environment, they would not survive. It is still necessary to demonstrate this because many officials and planners, both expatriates and developing country people, need to learn its truth.

IRRIGATION

The hazards of huge hydroelectric projects, like the Aswan Dam in Egypt, are well known; problems regarding health, effects on downstream fisheries, resettlement of displaced people, siltation, and weeds have been well described, as have the disappointing results of irrigation.[11] Rather than reiterate the problems of big dams, we consider briefly the Kano River Project, a large-scale, capital-intensive project that was designed to irrigate 58,000 acres in northern Nigeria. According to Wallace,[12] the Nigerian development planning process was generally "technocratic, economic, and authoritarian," human and social aspects being ignored. Some 13,000 people were relocated, and this was done in a way that soon reduced them from self-sufficient peasant farmers to government dependents. Most large-scale irrigation projects have been costly (up to $20,000 per hectare) and seldom produced anticipated benefits (the value of increased crop production). Further, benefits have been unequally distributed, with higher income farmers capturing a disproportionate share.

The promise of irrigation to increase agricultural productivity leads governments and international donors to continue promoting irrigation. Because of problems associated with large-scale projects, some planners are beginning to look at possibilities for small-scale irrigation, building on existing local systems. Another reason for looking at small-scale irrigation is that the easiest locations for large projects have already been used. Most examples of successful small-scale irrigation projects come from south and southeast Asia,[13] but there are a few cases from other continents.

Siy emphasizes officials' need to deal with *existing* irrigation organizations, even if planners regard them, at first glance, as "inefficient, primitive and far below engineering standards."[14] In many cases they should be studied and improved, which may prove less expensive than dismantling them and building new systems. A consideration of the existing organization offers local people an opportunity to take part and encourages them to contribute labor and materials.

Small-scale irrigation projects that we are considering are farmer-managed systems, mostly covering not more than 100 hectares although one system in Nepal exceeds 10,000 hectares. These "small" systems account for 80 percent of the irrigated area in Nepal, 60 percent in the Philippines, and substantial percentages in India, Bangladesh, Sri Lanka, Thailand, Korea, and Peru. Even in Africa, the U.N. Food and Agriculture Organization (FAO) estimated that 49 percent of the total irrigated area is accounted for by "small-scale traditional systems."

Many problems are encountered when governments try to use existing local systems as a base for wider irrigation projects and thus for increased food production. First, there is no certainty about potentials for enlarging small local systems. In some circumstances, such systems may already be at an optimum size in terms of organization and available technology, and considering socioeconomic and topographical factors. Engineers, seeing rudimentary technologies involved, usually underestimate potentials, while social scientists, impressed by the particular culture, frequently overestimate potentials; each may inadvertently distort a modest but effective system.

These are some of the questions that should be asked in considering any upgrading to local irrigation systems. What will be the effects on equity? Will poorer people benefit? How will water be allocated? How will conflicts over water allocation be resolved? Can labor be mobilized when, for example, canal repairs are needed? How will maintenance be handled? Will there be a problem of "free-riders?" ("Free-riders" refers to people who reap benefits of a common property regime—in this case, the irrigation water—while not meeting their obligations to contribute money or labor.) How will water use be coordinated with crop schedules? Who will impose—and enforce—regulations? How will management be chosen? We do have instances where local water users' associations have evolved satisfactory solutions to these questions, but we cannot be certain that all solutions will remain valid with an increase in the scale of operations. Local water users do, however, have several advantages, notably their knowledge of local conditions and ability to use social sanctions to enforce rules, with a capacity to be flexible and to show initiative. Even with goodwill, governments may destroy this initiative if too much is done for the people.

Some small projects start from scratch, with no existing organization; here it is essential to organize users' associations before any construction takes place, so that users can actively participate in all phases and can work with the engineers and planners. It is important that users do not leave all decisions and actions to the government. The National Irrigation Association (NIA) in the Philippines is often cited as a model of an enabling and encouraging, rather than an authoritarian, organization. Precise institutional arrangements are crucial with irrigation or any other NRM system.

IRRIGATION IN THE PHILIPPINES

A study by Korten[15] noted several important characteristics of the Philippines National Irrigation Association. Local farmers wanted to participate in the development of physical systems and their participation strengthened their sense of ownership and concern with operations and maintenance. Also, farmers had knowledge that could contribute to system design and layout. In the early stage of the project, the engineers proposed to build a semipermanent type of dam (*gambion*), which they claimed was the most cost-effective design. Based on their knowledge of the area, the farmers insisted that the proposed structure would not withstand the force of local floods. The *gambion* dam was built but shortly after completion was washed out by a typhoon.

Farmers also understood landholding patterns and whether it would be possible to obtain rights of way in various areas. They pointed out land ownership boundaries and proposed that canals follow such boundaries where possible to avoid taking too much land from any single farmer. They knew rainy season conditions (engineers' surveys often being done during the dry season) such as which creeks would swell after heavy rains and which areas would become water-logged. They also knew soil conditions, pointing out where sandy soil might induce significant water loss from the canal. The task of designing a workable system was not easy. Engineers and administrators are not used to communicating with the local users of the system, and community organizers often mistrust "outsiders." Many planners, who are based in major cities, are often unaware that conditions may vary radically between areas and, even for a single area, between seasons.

Several points deserve elaboration. Chambers[16] pointed out "dry season bias" as a handicap for "rural development tourists" who make their visits in the dry season when travel is easier and conditions are generally better for the local people, the harvest having been gathered. Such naivete shows the value of local knowledge, which can be an essential complement to outside technical knowledge. Finally, Chambers emphasizes again the crucial importance of learning on everybody's part: of farmers, community organizers, and engineers—anyone who "would do the teaching of techniques." Philippine National Irrigation Association engineers were, apparently, unusual. They were "open and honest in examining their agency's work" and were prepared to be flexible. The NIA was persuaded that the participatory approach offered benefits; they set up a reciprocal program using resources of the local irrigation authority.

Philippine experience indicates that much planning and care is required; specific conditions, rules, and regulations must be made unequivocally clear. The NIA has been unusually imaginative and open. Sometimes government bodies lack this ability; they would do better to engage NGOs (nongovernment organizations) to act for them. (In Kenya, some rural churches play an active role as effective brokers between the people and the department of water development.)

IRRIGATION IN INDONESIA

In Bali, irrigation is centrally organized by a system of water temples; they manage regional terrace ecosystems. Such temples have been detached from the traditional political order. Consequently, they survived the demise of the latter. They work subtly, so have often been "invisible" to foreign consultants, who emphasize *subaks,* the more visible local-level farmer associations.

Lansing[17] documents the problems that followed the initial success of the green revolution a decade ago. Traditional temple scheduling systems produced two bountiful crops per year—without resort to dangerous chemical controls. They had for eight centuries functioned successfully as efficient ecologically balanced mechanisms for avoiding serious paddy (rice) cultivation pests. Government instructions to ignore these traditional practices from the start of the green revolution resulted in widespread environmental damage to flora and fauna alike— disturbing vital balances to disrupt established productivity for *subak* members. These changes produced divisive dual irrigation management systems, so fundamentally dissimilar as to be "all but invisible to each other."[18] One was concerned with macroscopic interpretations of aerial photography, the other with microscopic ground truth of information that the peasants conveyed.

IRRIGATION IN KENYA

Compared to Asia, tropical African societies have mostly had rudimentary systems of water management, although there are some exceptions, such as the Taita and Marakwet people in Kenya, who have technologically efficient irrigation systems.

The Taita have had irrigation works for at least a century. Some peoples of neighboring Tanzania also practiced irrigation, especially the Sonjo,[19] and also the Shambaa, Chaga, Pare, and Arusha. The Taita irrigation system has 16 canals (some several kilometers in length) leading off a 2.5 km stretch of the Mwatate River.

The structure of the irrigation canals, the system of distribuion of water to fields, and the overall management of irrigation all depend on the local social organization and the people's values and perceptions. Sometimes engineers regard irrigation solely as a technical institution, while it is always closely related to the social structure.

The Taita system survived throughout both German and British colonial periods, despite British administrators who regarded it as "unscientific" and "wasteful." Their modern successors hold similar views: that is, engineers and planners are often too ready to dismiss indigenous resource management systems as "primitive" without taking the trouble to examine them. Ironically, several of the "scientific" irrigation projects that were developed in the Taita Hills in the 1950s, (the last colonial decade) failed. "It proved difficult to better or even match the performance of the indigenous system."[20]

Disputes between users—usually downstream users complaining that their upstream neighbors use too much water—are settled before the assistant chief of the area. "Access to water is regulated by the same mechanisms that regulate access to land, livestock, and marriageable women: that is, relations embodied within the dense web of affinal and agnatic ties that is the principal feature of Taita social organization. Management of the canal is seemingly effortless, almost incidental, because water relations are governed by, or are perhaps even epiphenomena of, social relations. Problems may still arise in sorting out water relations, as indeed they may in sorting out social relations, but solutions may usually be found through application of the universally-accepted logic of kinship, as contained in rules of descent, inheritance, marriage, and residence."[21]

Informal canal committees arrange for maintaining the system, cleaning and rebuilding canals or distribution furrows. This work is done effectively—because users rely on irrigation for subsistence needs and cash income during certain seasons; they cannot afford to neglect maintenance. This organizational reliability has proved difficult for introduced systems to achieve. "There is a puzzling imperative to development, namely that new, special-purpose organizations are often brought in with new technologies, to disseminate and control them. The Taita case shows that irrigation can be managed by multi-purpose organizations, including indigenous modes of organization that merge almost imperceptibly with modern organizations."[22]

Taita people, like most poor people in developing countries, cannot afford (in terms of time and money) "elaborate, specialized organizational super-structures"; and it would obviously be appropriate for government to ascertain what steps are needed to ensure the continuity of the present NRM system.

IRRIGATION SUMMARY

Government irrigation agencies need to learn to work with local people, to use existing organizations, or to encourage new bodies, and to do this before beginning actual construction. The planners and engineers should try to understand the daily lives, problems, and opportunities of the people. Then, together they can develop appropriate ways of tackling problems and of enhancing long-term benefits of the project. In most cases, it is preferable to work with existing organizations than to create new ones, even if the distribution of benefits is imperfect. For example, cooperative societies for production, marketing, and services have been promoted—in irrigation and in other domains—as a solution, but they have seldom worked well, partly because inappropriate models from Europe or North America have been introduced to Third World communities without regard to local socioeconomic conditions.

Drawing on experience in the Philippines, Alfonso[23] describes the basic requirements of local participation (which he regards as essential) in irrigation projects:

- Technical specialists and social scientists need to cooperate and to work as a team.
- Farmers must be involved at all stages of the development process.
- The organizers must know the salient characteristics of individual communities and develop appropriate intervention strategies.
- The technical designs have to "fit," that is, to be appropriate to specific conditions and capabilities.

Like many other conclusions in rural development, these requirements seem so obvious as hardly to be worth elaborating, but they are *not* obvious to everyone. Social aspects are as important as the technical aspects, and the two must be integrated if the project is to succeed. This means that cost-benefit calculations must include the farmer's perspective (i.e., social costs and benefits) as well as the strictly technical viewpoint. Also, outside experts must recognize the heterogeneity among local communities and not assume that what worked in one locality will necessarily work in another. Planners should not only recognize that farmers should be involved in the decision making about *their* irrigation systems—they will be the users—but also that when local knowledge is incorporated, a better system may be built. This calls for engineers to develop new skills—listening to farmers and imparting technical information in a comprehensible manner. (We should point out that we have dealt only with irrigation, whereas water management includes many other activities.)

FISHERIES

Fish are almost universally a common-property resource, at least as long as they are in rivers or oceans. Fish, like wildlife, are a "fugitive" resource, which poses special problems in their "management." There may be a problem of open access, as among the Bemba of Zambia, who "catch whatever they can during the breeding season, or before the fish are full grown."[24] But "many tribal and peasant societies have customs and traditions which, in effect, control the freedom of the marine commons, and which, thereby, protect marine resources against excessive exploitation." Moerman divides these mechanisms into two basic types, "social specialization" and "internal control." The former "limits the number of people who have access to marine resources," and the latter includes restricting access to specific classes of people.[25] These mechanisms illustrate that local people have understood the population dynamics of fish and managed the resource accordingly.

Artisanal, or small-scale fisheries, account for one-third of the total catch of fish worldwide and for 90 percent of the fishing population in Africa and Asia. The global figure of those engaged in these fisheries is 30 million, according to FAO. Fisheries represent a major source of employment and income. Fish are also an important source of protein. Generally labor-intensive with relatively little imported technology, small-scale fishing has been often overlooked or dismissed by "experts" because of its "primitive" equipment and techniques. Yet these fishermen often operate at a more favorable cost-benefit ratio than do the larger fisheries, which cannot operate in all areas where small-scale fishing is carried out. Despite the large numbers involved, small-scale fisheries receive less than one fifth of the total technical international assistance given to the fishing industry.[26]

Our contention, which we will support with selected examples, is that—as with small-scale irrigation—it would be in the national and local interest to pay more attention to artisanal (small-scale) fishing. What is needed is an understanding, first, of what local fishermen do. (We use the masculine case, as few women fish, although they have important roles as processors, distributors, sellers and even—as in Ghana—as boat or net owners.) We recognize that not all existing practices are desirable: dynamiting, for example, is common and is a particularly grave threat to fragile coral reefs, as in northeastern Tanzania. Dynamite also has been used extensively in fishing by the Gurkhas in Nepal.

There should be a comprehensive approach to resource management that includes actual fishing, processing, and marketing, and integrates social and economic aspects. Relatively small investments in motor maintenance, landing facilities, and equipment could produce good dividends—much better, in fact, than some of the expensive "high tech" interventions have done.

The introduction of large trawlers put local fishermen out of business, and led to severe overfishing in many areas; reduced anchovy catches off the Peruvian coast are the best-known example. Similar social and environmental disasters have occurred in south India, the Philippines, and Malaysia. In Malawi, the World Bank financed an ambitious project with large "improved" vessels, huge docks, and refrigeration facilities, all of which proved unsuitable for local conditions. In Sri Lanka, fish production doubled from 1960 to 1967 because of the introduction of nylon nets and 3.5-ton vessels with inboard engines. At the same time, there were fewer employment opportunities, social inequality increased, and there were signs of overfishing. If the aim of development is to help people achieve sustainable livelihoods, this was hardly "development."

There have been some expressions of international interest in small-scale fisheries, as evidenced by two recent studies commissioned by the Fisheries Department of the Food and Agriculture Organization; these studies provide many examples of effective local managment of fisheries.[27]

The general trends in local natural resource management are also evident in fishing: population increase leads to overexploitation of the resource; increased commercialization; abrogation of traditional methods of conservation, due in part to the government taking over control of the resource; disastrous results of some new technologies on the resource. Fishermen, traditionally often among the poorest members of their communities, have become poorer as a result of these changes and have been unable to compete with the aggressive, technologically sophisticated international fishing fleets—from Japan, the USSR, the United States, Poland, Spain, the Netherlands, South Korea, Italy, Rumania, and the People's Republic of China (PRC).

The 1978 Law of the Sea treaty should have helped smaller countries protect their marine resources by recognizing the 200-mile Exclusive Economic Zone (EEZ), but Senegal, for example, does not have the force to patrol its zone, so its rich fishing grounds are easily exploited. Some small Pacific nations have entered into agreements with Russia, China (PRC), and others, granting these countries exclusive fishing rights. Few national governments are making determined efforts to help their own small-scale fishermen or to conserve their supplies of fish. In 1982, a United States "superseiner" that was illegally fishing 35 miles inside the EEZ was seized in Papua, New Guinea. The United States claimed that any island state had jurisdiction over only 12 nautical miles. Later, the United States agreed to begin negotiating treaties with 16 Pacific Island countries.[28]

FISHERIES IN INDONESIA

Nicholas Polunin, writing of "Traditional Marine Practices in Indonesia," emphasizes that traditional "limited-entry" areas exist in order to protect marine resources, especially in the outer, less-populated parts of the country. But some of these areas are now threatened, partly because government officials are unaware of their existence. Polunin advocates traditional reserves be maintained, even though they "may not be consistent with the purist view that in reserves nature should ideally be left entirely to its own desires." He also recognizes that local organizations may not be able to cope with the effects of a rapidly expanding population, especially increases resulting from immigration. Although traditional reserve systems may lose their effectiveness due to outside pressures, they do "provide a basis for local responsiblity in resource management." Polunin stresses—as do many others writing on similar topics—the urgent need for more research about such traditional conservation methods.[29]

Another Indonesian example concerns flying fish (*Cypselurus* spp.), which has been a staple of the diet for fishing people on the Strait of Makassar for 200 years. Until the early 1970s, the Mandarese fishermen used to catch enough flying fish for their needs in the April–September season using catamarans and barrel-shaped traps. Flying fish roe was a local delicacy either freely available or sold for a nominal charge. All this changed due to the Japanese market for flying fish roe, which caused the price of roe to rise to $10 per kilogram, an increase of 10,000 percent. This meant higher profits and a higher standard of living for the fishermen, but it led to overfishing. This is a dramatic example of the rapid changes caused by commercialization of a natural resource in an international marketplace—the flying fish moved from being a commonly held resource, important in both subsistence and ritual, to a source of private profit.[30]

FISHERIES IN BRAZIL

A detailed study from Bahia examines poor fishermen who evolved, over at least a century, an effective system to control fishing space, especially regarding net-shooting of seine nets. The system is based on *respeito* (social respect) and a complex system of community rewards and punishments. Many believe that the poor cannot afford the luxury of conservation, but in Bahia the system is marked by cooperation and generosity, and also by effective sanctions on offenders. It also ensures that fishing stocks are not depleted. This example underlines the need to examine local natural resource management in its total social context, which government planners have often failed to do. The Bahia system is now threatened by market expansion and government-sponsored improvements, with resultant overfishing and an undermining of the *respeito*. What is needed in Bahia is neither government (public) nor private control, but some sort of joint, shared or collective tenure. "The real managerial strengths in Third World coastal fisheries are indigenous: the power is clearly not vested in the state or its bureaucracies but in fishermen's own informal institutions, norms and co-operative organizations."[31]

FISHERIES IN THE PHILIPPINES

With an estimated one million people engaged in small-scale fishing, the Philippines provides many examples of their problems and some solutions associated with artisanal fisheries. Of 3,500 fishing households near San Miguel Bay, few have any rights in agricultural land; there are few economic alternatives to fishing. Although the small-scale fishermen have relatively inexpensive boats and do not use expensive technology, they have been affected by increases in costs, especially for fuel, while fish prices have increased only slightly. Most fishermen reported that their catch has declined over the past two years.

In the early 1970s, many trawlers started fishing in the bay and soon there was "a significant concentration of fishing power controlled by a few individuals, employing but a small fraction of the work force." Some of the trawlers are legally defined as small-scale fishing units because they are under three tons, so they may operate legally in most of the bay. The problem here is political and organizational. If the local municipalities, whose jurisdiction includes the shoreline of the bay, changed their regulations, they could ensure that the *true* small-scale fishermen—not the commercial "small" trawlers—had the right to fish in the bay. Such regulations would require enforcement and community management of the fisheries.

"If properly organized, the small-scale fishermen could themselves take part in monitoring and reporting illegal fishing. In the course of their daily activities, these fishermen witness what happens at sea. They know who is responsible for illegal fishing, but individually lack the confidence and political influence necessary to turn their complaints into effective administrative or judicial action. With the support of an organization of their peers, small-scale fishermen could be integrated into legitimate local political processes and be able to lobby effectively for their interests. Only by acting in concert can effective pressure be exerted on political leaders and enforcement agencies."[32]

Other attempts to establish organizations among small-scale fishermen have seldom been successful, partly because these organizations have not been concerned with the fishermen's real needs. For example, marketing cooperatives have been a favorite solution put forward by officials. "The focus of organizing efforts in fishing communities should be directed to the central problem, the collective management of the commons according to rules established by the local community, to the equitable benefit of all the community's members."[33]

FISHERIES IN PAPUA, NEW GUINEA

A recent volume on *Traditional Conservation in Papua, New Guinea* included a chapter on inland fisheries management, which stressed that the first step for any subsistence fishery development strategy is to examine what is already present in order to get baseline data about production skills, attitudes about control mechanisms, and fishermen's interests and ideas. Additional necessary surveys should ensure that there is a sufficient sustainable resource to allow economically viable development acceptable to local people without necessarily adopting outside "high technology" changes. Successful self-reliant projects are those where outside expertise is minimal and simple technology comprehensibly incorporated into traditional skills, easily taught and not disruptive of established patterns, even while traditional systems' limitations are recognized. Fish stocks are everywhere under pressure, and all strategies need to take this into account.[34]

FISHERIES IN AFRICA

When studying African fisheries, we once again encounter a lack of systematic documentation of indigenous fishing practices.[35] But we do see that familiar scene of community-based systems that operated effectively in the past on a mainly subsistence basis until disturbed by population growth, commercialization, technology, and government interventions. In many areas, chiefs or other traditional authorities used to have control over fishing, exercising various restrictions that had the effect of promoting conservation. When colonial governments started charging fees for fishing, anyone could pay the fee and fish; old sanctions no longer had force. "Strangers" came in, both other Africans and non–Africans: in the 1950s there was a community of 50 Greek fishermen from the Dodecanese Islands in Lake Mweru, Zambia, where their superior vessels and technology ensured that they had a disproportionate share of the catch, to the detriment of the locals. The introduction of nylon thread nets in Lake Nyasa (now Lake Malawi) in 1958 led to the emergence of a new class of "big" fishermen, in this case, Africans.

Most interventions emphasized production and ignored sociocultural aspects, often with disastrous results. "Tradition" has been seen by fisheries experts as a handicap or, at best, a primitive transitional stage that should be "developed" and "modernized" without delay. A more beneficial approach might be to treat local fishermen and their knowledge as a valuable resource and a basis from which to make improvements.

Writing of Africa, one observer states: "Peasant fisheries created employment, used modest local investment, exploited abundant resources, produced high-quality fresh fish for local markets, consumed little energy and caused little pollution. The industrial fisheries, on the other hand, created unemployment, relied on foreign investment, exploited poor offshore resources, produced fish for export, consumed much energy and caused substantial pollution."[36] This observation has proven to be true in many parts of the world.

FISHERIES SUMMARY

To summarize, local management of fisheries is still, despite the distortions of the systems, deserving of study because of the detailed knowledge of fish and the waters that traditional fishermen have. In addition, upgrading what is already there is likely to be of more lasting benefit than attempting a radical transformation using what may well be an inappropriate model. Upgrading can focus on equipment (crafts, nets), processing, distribution, and training of men and women in the fishing industry. Local people must be integrated into management of their resources from the design stage of any project, which must be suitable for local institutions and beliefs. Access to the fishery may have to be restricted to ensure a sustainable system. Institutional development is vital to any project; it demands imagination, with ability to experiment, since existing institutions, whether modern or traditional, are seldom adequate. A special Fisheries Development Unit should be set up to deal with the needs of small-scale fisheries; progress should be step by step in accordance with long-term plans, and local people must be consulted at every step.[37] One possible solution is to set up "Community Participatory Shareholders Organizations," which are outlined in Scudder and Conelly.[38] The point here is that experiments must be made (even if some inevitably fail) until workable institutions are created.[39]

FORESTRY

From the mid-1970s partly as a result of the discovery of what Erik Eckholm termed "The Other Energy Crisis—Firewood," increased attention has been paid to problems of deforestation. The original response was to encourage communal woodlots, which have seldom been successful (with few exceptions, such as Korea, and, less so, Gujurat) because they were imposed from outside and were based on false assumptions about the local community with regard to common property regimes. By the 1980s, it was clear most rural communities had a far greater degree of individualism and private property, in relation to forests and trees, than was appreciated by outsiders. In most cases, locals were not at all clear about communal woodlot costs and benefits—who would do the hard work of guarding and weeding, for example, and who would be able to harvest the trees? Unless these basic questions about rights and obligations are spelled out precisely, it is unreasonable to expect local participation.

Colonial forestry departments initiated a series of policies that were continued by their successors. These included the setting aside of large areas of forest reserves, keeping all local people out, and emphasizing exotic trees, often in monoculture stands, for single uses (usually commercial timber). This is in stark contrast to most local ideas of forest managment, where indigenous species grow in heterogeneous diversity and where favored trees have multiple uses. Many locals perceive foresters as police officers whose main function—so it seems to them—was to prosecute people for using their own resources.

During our joint enquiries (1970–1987) into social and ecological change in Mbeere, Kenya, we studied changes in incidence and use of vegetation. We were most impressed by the wealth of knowledge that the local people had—it was common for men, women, and children to be able to identify over 200 tree species and to know a great deal about their properties. Considering that, at least until the 1960s, 90 percent of Mbeere material culture came from vegetation, this extensive knowledge is not surprising: it was essential for survival in a marginal area. We have seen an encouraging increase in tree planting by individual farmers and also by groups. The old groups—generation sets, lineages, and neighborhood groups—have largely lost their corporate functions, but here are new groups to take over: women, church members, and schoolchildren are all busy planting trees, with the active help of the Forestry Department.[40]

In Nepal, in response to heavy deforestation, all forests were put under government control in the 1950s. But the government was unable to exercise adequate supervision and, from the late 1970s, there has been a move to decentralization, with Panchayat Protected Forests being handed over to local communities. These communities have had to appoint guards, build fences, and solve disputes. A common cause of conflict has been the rights that livestock have to graze in forests. The government of Nepal is actively seeking an appropriate framework for programs of local forestry management systems that are suited to the needs of the local communities. Some groups have little representation from women, which is unfortunate in that women are primary collectors of forest produce. They were more adequately represented in the old system. Also, equity is not always easy to achieve in an hierarchical society, at least in the short term. Management groups have regulated forest use along traditional lines in making regulations restricting collection according to quantity, season, person (e.g., no outsiders), or state of product (e.g., only dead wood). They have levied fees and made decisions about closing areas. The rules adopted by them are easier to enforce than those that are imposed from outside the community, without the understanding and consent of local people.[41]

Some Indian communities in the Himalayas were so incensed by the logging concession given to outside timber merchants over *their* forests that they started the *Chipko* movement; local men and women literally "hugged the trees," forcing the government to alter its policies and give local people greater participation in the management of their forests. Women were, and still are, prominent in this movement, which has spread to other parts of northern India. "Chipko's search for a strategy for survival has global implications. What Chipko is trying to conserve is not merely local forest resources but the entire life-support system . . . [its emphasis is on] an ecologically sustainable path for development."[42]

WILDLIFE

Game, perhaps more than any other resource, has traditionally been viewed as a free resource. (This was not the case in Western societies, of course, where all sorts of royal and landowner prerogatives restricted the right to hunt.) Much hunting was communal, which was often essential given the simple weapons and traps available. Despite claims to the contrary, not all hunters "lived in harmony with nature." This myth is based on examples, usually from "tribal" peoples who did respect animals and who killed only what they needed for food and clothing. One well-documented account of the Ba-Ila in what is now Zambia describes how zebra were driven into boggy marshes, the tails cut for fly-whisks, and the animals left to perish. Other examples could be cited, the point being that, again, we should be careful not to romanticize traditional systems, for it is certain that not all followed conservation ethics: some mined, rather than harvested, wildlife.

Again we see the same forces at work—overpopulation (or poverty) leading hungry people to kill for sale, now that game has been commercialized. Because of affluent Western—and Oriental—desire for trophies (leopard skins, rhinoceros horns, elephant ivory, etc.), many poor farmers in Africa and elsewhere are willing to undertake killing, even if they receive a fraction of the final price that the trophies finally fetch. New technology, too, has caused dramatic change, as exemplified by the introduction of the shotgun to Amazonian Indians.[43] Off-road vehicles, flashlights, and helicopters have all caused changes.

Government intervention usually took the form of declaring Game Reserves, which excluded people, and provided protection for certain species, which usually meant that only wealthy foreigners could kill them. The response of many indigenous groups has been to protest and claim that they possess inalienable or "natural" rights to hunt and fish in their territory.

The special issue of *Cultural Survival Quarterly*, "Parks and People," has invaluable information on actual incidents, and discusses prospects for incorporating local people into game parks. Recently, the Third World Council on Indigenous Peoples passed a resolution supporting the rights of indigenous peoples to hunt and gather. This resolution states that economies of such peoples would be severely damaged if there were restrictions in their harvesting that were based on "false ecologic and so-called conservationist premises" and if such restrictions were implemented.

Yet many archeological studies suggest subsistence hunting in the past occurred at nonsustainable rates and that within traditional hunting methods "resource extinction" practices once did, and could yet again, manifest themselves. As we now have an imprecise knowledge of most reproductive biology criteria for commercially valuable wildlife species, we cannot yet regard them as a potentially viable resource or guarantee their future. As researchers interested in the practices of conservation, we do not wish to deny indigenous people the right to harvest their resources, but we wish to present the facts on which they may, we hope, base future decisions on the harvesting of natural resources.

We now present some specific examples, beginning with the most complete African hunting ethnography on the Valley Bisa of Zambia.[44] The Bisa are surrounded on three sides by National Parks, which they are forbidden to enter. They believe that government is more concerned with animals than with people—a view shared by other African peasants. The Bisa are arrested as poachers if they hunt. "[T]he imposition and enforcement of an alien game code and restrictions pertaining to weapons do have an effect on Valley Bisa hunting behavior. The fear of detection by resident game officials explains to some extent why hunters walk almost continuously rather than use a more cautious approach in areas where they expect game. Hunters realize that investing longer periods of their time in an area increases their chances of detection by itinerant officials. The greater firing range of bow and arrow, together with its lack of noise and disturbance, makes it a more appropriate weapon for current tactics than the muzzle-loading guns now in use. Furthermore, the recovery rate of hunters of wildlife wounded with poison is greater than that with muzzle-loading guns."[45]

Marks recommends that the Bisa be allowed to hunt, subject to stringent controls, which would increase the supply of animal protein for local use and meet the people's requests. But most governments are wary of incorporating local hunters into their projects.

The Maasai pastoralists who have lived for years in what is now Ngorongoro National Park, Tanzania, were at one time promised that they and their herds would be able to stay in the park, except in the crater. The original plan for the park was to combine protection of wildlife with promotion of the well-being of the inhabitants, but cattle were soon perceived as being in competition for grazing with wildebeests and zebra. The government is attempting to resettle the Maasai.[46]

Scudder describes two programs in Zimbabwe run by government officials for the benefit of local people. "Operation Windfall" provides for proceeds from culling of surplus elephants (i.e., sale of ivory and meat) to go to local councils. But one problem here is that the most remote villagers, the ones who are most affected by elephants destroying their crops, see few benefits. "Operation Campfire" sets up "Community Land Councils" with local people as shareholders. Here, land is fenced off, game kept out, and local herdsmen allowed a stated number of livestock. Zimbabwe is a pioneer in attempting to involve local people in wildlife management to see that they get some benefits. In time, locals should become directly involved in the profitable tourist and safari business.[47]

Most examples of "wildlife anthropology" come from Africa, but we do have examples from Asia. Katy Moran, writing of elephant management in Sri Lanka, says, "traditional cultures have profound and detailed knowledge of animal species in their environment and traditional practices for animal management that can be useful in conservation and appropriate to the values of local cultures." She is specifically concerned with the multiple conservation strategies—keeper-elephant relationships, domestication of wild elephants, protection for Mahaweli farmers—that are needed to preserve the endangered elephants.[48]

One of the most successful examples of local people managing their own wildlife comes from Panama, where Kuna Indians (numbering 30,000) corporately manage 50,000 hectares as a reserve that allows for "scientific tourism" and effectively keeps out commercial developments (unless the Kuna agree to these). The reserve contains a rich variety of flora and fauna in different ecosystems—"a wide variety of pristine marine and terrestrial associations including coral reefs, islands, mangroves, coastal lagoons, gallery forest, mixed agricultural plots and evergreen hardwood forest."[49]

The Kuna recognize the power of the Panamanian government, and have joined the national political structure, while maintaining management authority over the forest, marine, and mineral resources. "Kuna traditional belief teaches that the primary forest is the sacred home of the spirits. As a result, they have maintained vast tracts of unaltered forest. The dense tropical forest cover of Kuna Yala contrasts sharply with the denuded hillsides in the immediately adjacent province of Panama, where the increasing deforestation by slash and burn agriculturalists followed by the introduction of cattle is rapidly degrading the natural resources base." The Kuna hunt occasionally, but they are aware of the need to conserve their resources and are actively engaged in teaching the young people the cultural significance and economic value of their forest resources. However, they still face threats from colonists eager to farm in the forests, as has happened on a wide scale in Central and South America. The Kuna example could serve as a model to show that local natural resource management can work and to demonstrate the need for "political action to protect the land and rights of other indigenous peoples."[50]

In summary, while population pressure may cause local people to destroy their ecological base, there are many ways in which governments, scientists, and indigenous people can work together to devise plans that will conserve wildlife *and* help the people achieve a sustainable livelihood. It is essential that local people be actively involved as participants. Simply to exclude people from protected areas does not necessarily have the desired conservational effects. "Protected areas could ensure the survival of habitats as well as indigenous inhabitants. Reserves can either preserve traditional life-styles or slow the rate of change to levels more acceptable to and controlled by local residents. Indigenuous inhabitants can benefit from the protection of their rights to traditional areas as well as the sale of goods or income generated from tourism."[51]

SUMMARY

Concern with global environmental problems is leading to proposals for international programs and actions. In general, plans for these are made in the developed nations for all nations. In addition, some key global environmental issues focus on parts of the world that are predominately Third World nations. For example, much of the tropical rainforests that receive so much interest are in Third World nations. In dealing with global environmental problems and in proposing the use of advanced technology, we must pay attention to local indigenous practices if we hope to avoid the traps and failures of many development projects of the past.

Before any planning is started that involves developing nations, it is essential first to examine existing institutions and practices, and to see how these can be improved, rather than simply introducing a radical new institution. In some cases (the *mwethya* labor group among the Kamba of Kenya or the *Kgotla* assembly of cattle-owners in Botswana), an existing institution can serve modern purposes or can be adapted.[52]

Those concerned about global environmental issues who seek the participation of Third World governments must realize that government and local people have different priorities in research use, different objectives regarding that use, and different ways of resolving conflicts. For example, in forestry, government foresters think in terms of yields, inventories, and harvesting schedules, whereas locals place more importance on qualitative ecological aspects. It is necessary to take all viewpoints into account. Local national resource management has a strong social component, which needs to be understood before changes in resource management are proposed.

We want to emphasize that no one approach will be appropriate for all situations, for much depends on specific socioeconomic, biophysical, and cultural historical circumstances; on traditions of collective action; and on the quality of local leadership, as well as on the role of government. In most cases, effective natural resource management will neither be a resuscitated traditional form, nor will it be imposed from outside; rather, it will be a blend of old and new institutions and will represent a collaborative venture between government and people. Because of the pressures for change, few traditional natural resource management systems are viable in their present form, but they can be modified or incorporated into broader systems, with adequate controls.

Good local natural resource management can mean management that is more informed, more appropriate, and more finely tuned; it may be both cheaper and more effective than imported forms would be. Local contributions can help in the monitoring of resource use and also in setting up early warning systems for environmental change. Those studying global environmental issues would also benefit from observing the local management regimes that have been successful.

In many parts of the world, local natural resource management systems have been replaced by large-scale projects that are often accompanied by a centralized bureaucracy. Many of these projects have failed to achieve their social and economic objectives and have had many deleterious environmental consequences as well. Although the rural poor in the developing nations suffer the immediate negative impacts of such resource planning errors, there is a cumulative effect on the global environment that may have a severe long-term impact. For example, the overexploitation of many coastal fisheries may lead to the extinction of some species. The same is true for forestry management. By recognizing the positive aspects of a local natural resources management system, development planners may be able to achieve their economic goals without degrading either the local or the global environment.

NOTES

[1] D. Lowenthal, 1985, *The Past Is a Foreign Country,* Cambridge University Press, Cambridge, UK, p. xvii.

[2] A. Spoehr, 1956, in *Man's Role in Changing the Face of the Earth,* A. Thomas (ed.), University of Chicago Press, Chicago.

[3] S. Tax, 1956, in *Man's Role in Changing the Face of the Earth,* A. Thomas (ed.), University of Chicago Press, Chicago.

[4] G. Hardin, 1968, "The Tragedy of the Commons," *Science,* Vol. 162, pp 1243–1248.

[5] A.A. Dani, C. Gibbs, and D. W. Bromley, 1987, *Institutional Development for Local Management of Rural Resources,* East-West Environment and Policy Institute, Honolulu, Hawaii.

[6] A. Jacobs, 1980, "Pastoral Maasai and Tropical Rural Development," in *Agricultural Development in Tropical Africa,* R. Bates and M. Lofchie (eds.), Preager, New York.

[7] N. Dyson-Hudson, 1985, "Pastoral Production Systems and Livestock Development Projects: An East African Perspective," in *Putting People First,* M. Cernea (ed.), World Bank, Washington, DC; M.M. Horowitz, 1986, "Ideology, Policy and Praxis in Pastoral Livestock Development," in *Anthropology of Development in West Africa,* M. Horowitz and T. Painter (eds.), Westview, Boulder, Colorado; P. Little and D. Brokensha, 1987, "Local Institutions, Tenure and Resource Management in East Africa," in *Conservation in Africa,* D. Anderson and R. Grove, (eds), Cambridge University Press, Cambridge, UK.

[8] See National Research Council, 1986, *Proceedings of the Conference on Common Property Resource Management,* National Academy Press, Washington, DC, for a most comprehensive survey, which includes Oakerson's useful methodology for analyzing Common Property Regimes.

[9] For bibliographies, see D. Brokensha, D.M. Warren, and O. Werner (eds.), 1980, *Indigenous Knowledge Systems and Development,* Lanham, University Press of America, Maryland; H.C. Conklin, 1972, *Folk Classification,* Yale University, Department of Anthropology, New Haven, Connecticut; G. Klee (ed.), 1980, *World Systems of Traditional Resource Management,* Winston and Sons, New York.

[10] D.D. Southgate and C.F. Runge, 1986, "Toward an Economic Model of Deforestation and Social Change in Amazonia," *The Science of the Total Environment,* Vol. 55, pp. 121–126.

[11] E. Goldsmith and N. Hildyard, 1985, *The Social and Environmental Effects of Large Dams* (3 vols.), Ecosystems Ltd., Camelford, Cornwall, UK.

[12] T. Wallace, 1981, "The Kano River Project in Nigeria: The Impact of an Irrigation Scheme on Productivity and Welfare," in *Rural Development in Tropical Africa,* J. Heyer et al., (eds.), St. Martin's Press, New York.

[13] See E.W. Coward (ed.), 1980, *Irrigation and Agricultural Development in Asia: Perspectives from the Social Sciences*, Cornell University Press, Ithaca, New York; E. Martin, R. Yoder, and D. Groenfeldt, 1986, "Farmer Managed Irrigation: Research Issues," (ODI/IIMI Paper 86/3c), Overseas Devt. Inst., London, for an overview.

[14] R.Y. Siy, 1982, *Community Resource Management: Lessons from the Zanjera*, University of Phillippines Press, Quezon City.

[15] F.F. Korten, 1982, "Building National Capacity to Develop Water Users' Associations: Experience from the Phillipines," (Working Paper No 528), World Bank, Washington, DC.

[16] R. Chambers, 1983, *Rural Development: Putting the Last First*, Longman, London, pp. 20–21.

[17] J.S. Lansing, 1987, "Balinese 'Water Temples' and the Management of Irrigation," *Amer. Anthropologist*, Vol. 89, No. 2, pp. 326–341.

[18] *IBID*, p. 339.

[19] R.F. Gray, 1963, *The Sonjo of Tanganyika*, for the International African Institute, Oxford University Press, London.

[20] P. Fleuret, 1985, "The Social Organization of Water Control in the Taita Hills, Kenya," *American Ethnologist*, pp. 103–118.

[21] *Ibid.*

[22] *Ibid.*

[23] F.B. Alfonso, 1983, "Assisting Farmer Controlled Development of Communal Irrigation Systems," in *Bureaucracy and the Poor: Closing the Gap*, D.C. Korten and F.B. Alfonso (eds.), Kumarian Press, West Hartford, Connecticut.

[24] A.I. Richards, 1939, *Land, Labour and Diet in Northern Rhodesia*, Oxford University Press, London, p. 336.

[25] D.E. Moerman, 1984, "Common Property and the Common Good: Ecological Factors among Peasant and Tribal Fishermen," in *The Fishing Culture of the World*, B. Gunda (ed.), Akademiai Kiado, Budapest.

[26] FAO (Food and Agriculture Organization of the United Nations), 1985, *Community Fishery Centres: Guidelines for Establishment and Operations*, FAO, Rome.

[27] T. Scudder and T. Conelly, 1985, *Management Systems for Riverine Fisheries* (Fish Tech. Pap. 263) FAO, Rome; B. W. Riley and D. Brokensha, Unpublished, "Part-time Subsistence and Traditional Inland Fisheries," (Paper prepared for FAO, Fisheries.)

[28] B. Nietschmann, 1987, "The New Pacific," *Cultural Survival Quarterly*, Vol. 11, pp. 7–9.

[29] N.V.C. Polunin, 1985, "Traditional Marine Practices in Indonesia and Their Bearing on Conservation," in *Culture and Conservation: The Human Dimension in Environmental Planning*, J.A. McNeely and D. Pitt (eds.), Croom Helm, London.

[30] C. Zerner, 1987, "The Flying Fishermen of Mandar," *Cultural Survival Quarterly*, Vol. 11, No. 2, pp. 18–22.

[31] J. Cordell, 1985, "Sea Tenure in Bahia" in *Common Property Resources*, National Academy of Sciences, Washington, DC.

[32] C. Bailey, 1984, "Managing an Open-Access Resource: The Case of Coastal Fisheries," in *People Centered Development*, D. Korten and R. Klauss (eds.), Kumerian Press, West Hartford, Connecticut, pp. 88–100.

[33] *Ibid.*, pp. 102–103.

[34] A.K. Haines, 1982, "Traditional Concepts and Practices and Inland Fisheries Management," in *Traditional Conservation in Papua, New Guinea: Implications for Today*, L. Morauta et al., (eds.), Boroko, PNG: Inst. App. Soc. & Econ. Research; R.E. Johannes, 1982, "Traditional Conservation Methods and Protected Marine Areas in Oceania," *Ambio*, Vol. 11, No. 5, pp. 258–261.

[35] See B.W. Riley and D. Brokensha, Unpublished, *op. cit.*, for a review of the literature, and also Lars Sundstrom, 1972, *Ecology and Symbiosis: Niger Water Folk*, Stud. Ethnographica 35, Uppsala, for an impressive exception to the dearth of good studies.

[36] L. Timberlake, 1985, *Africa in Crisis: The Causes, the Cures of Environmental Bankruptcy*, Earthscan, London, p. 7.

[37] M. Ben-Yami and A.M. Anderson, 1985, *Community Fishery Centres: Guidelines for Establishment and Operation*, (FAO Fish. Tech. Paper, 264), FAO, Rome, p. 5.

[38] Scudder and Conelly, 1985, *op. cit.*, pp. 56–58.

[39] *Cultural Survival Quarterly*, 1987, special issue "Fishing Communities," Vol. 11, No. 2.

[40] B.W. Riley and D. Brokensha, 1988, *The Mbeere of Kenya: Changing Rural Ecology*, University Press of America, Lanham, Maryland.

[41] D.A. Messerschmidt, 1985, "People and Resources in Nepal: Customary Resource Management Systems of the Upper Kali Gandaki," paper prepared for the Conference on Common Property Management, Washington, DC, Board on Science and Technology for International Development, National Research Council.

[42] V. Shiva and J. Bandyopadhyay, 1987, "Chipko: Rekindling India's Forest Culture," *Ecologist*, Vol. 17, No. 1, pp. 26–34.

[43] R.B. Hames, 1979, "A Comparison of the Efficiencies of the Shotgun and the Bow in Neotropical Forest Hunting," *Human Ecology*, Vol. 7, No. 3, pp. 219–252.

[44] S. Marks, 1976, *Large Mammals and a Brave People: Subsistence Hunters in Zambia*, University of Washington Press, Seattle; S. Marks, 1977, "Hunting Behavior and Strategies of the Valley Bisa in Zambia," *Human Ecology*, Vol. 5, No. 1, pp. 1–36.

[45] Marks, 1977, *op. cit.*

[46] K. Arhem, 1985, *Pastoral Man in the Garden of Eden: The Maasai of the Ngorongoro Conservation Area, Tanzania*, Uppsala Research Reports in Cultural Anthropology, University of Uppsala and the Scandinavian Institute of African Studies, Uppsala, SEK 70.–, 91–7106–232–7, 123 pp.

[47] T. Scudder, 1982, *Regional Planning for People, Parks and Wildlife in Sebungwe (Zimbabwe)*, Department of Land Management, University of Zimbabwe, Harare.

[48] K. Moran, 1987, "Traditional Elephant Management in Sri Lanka," *Cultural Survival Quarterly*, Vol. 11, No. 1, pp. 23–26.

[49] B. Houseal et al., 1985, "Indigenous Cultures and Protected Areas in Central America," *Cultural Survival Quarterly*, Vol. 9, No. 1, pp. 10–19.

[50] *Ibid.*

[51] *Cultural Survival Quarterly*, 1985, special issue, "Parks and People," Vol. 9, No. 1, pp. 2–56.

[52] Brokensha, Warren, and Werner (eds.), 1980, *op. cit.*; M. Cernea (ed.), 1985, *Putting People First: Sociological Variables in Rural Development*, Oxford University Press (for the World Bank), New York.

EDITORS' INTRODUCTION TO:
T.H. BINGHAM
Social Values and Environmental Quality

In this chapter, Mr. Bingham expands on an issue mentioned by Pearce in the first chapter of this section: that it has become increasingly important to identify the value of some aspects of the environment. He points out that such valuations are necessary because in many cases it is necessary to make trade-offs between environmental quality and the goods and services that can only be produced by degrading the environment. To make such choices, the relative values of the alternatives must be estimated and compared. Our best objective value is individuals' willingness to pay for environmental dimensions. It is relatively easy to evaluate those components for which we have a market price, but placing a value on elements such as air or water quality are much more difficult. Mr. Bingham describes the most recent developments that have been made in developing a methodology to estimate individuals' valuation for environmental quality. Mr. Bingham provides a foundation for the subsequent chapters by Drs. Caswell and Endres, who develop specific methodologies for evaluating choices.

23 SOCIAL VALUES AND ENVIRONMENTAL QUALITY

TAYLER H. BINGHAM*
Center for Economics Research
Research Triangle Institute
Research Triangle Park, North Carolina

INTRODUCTION

George Perkins Marsh's book *Man and Nature* published in 1864 signaled the beginning of an awareness of the profound effect of industrialized human beings on the global environment.[1] His early observations and insights provide a useful backdrop to current issues concerning society's choices and the quality of the environment.

Marsh was, among many other things, a naturalist and a geographer. He recognized the interrelationships in the natural environment and saw human beings as disturbers of nature and destroyers of the natural equilibrium in the environment. He found perfection in these natural balances and urged people to restore "the disturbed harmonies of nature, whose well balanced influences are so propitious to all her organic offspring, of repaying to our great mother the debt which the prodigality and thriftlessness of former generations have imposed upon their successors."[2]

Marsh reserved his criticism for modern society. "[P]urely untutored humanity, it is true, interferes comparatively little with the arrange-

* Appreciation is extended to William H. Desvousges, Hall B. Ashmore, Richard W. Dunford, and Margriet F. Caswell for their review of an earlier version of this chapter. The usual caveat applies.

ments of nature, and the destructive agency of man becomes more and more energetic and unsparing as he advances in civilization."[3] He had almost a reverence for primitive cultures, seeing them in harmony with nature.

Marsh's specific concerns about human effects on the environment were many. However, his major attention centered on the tremendous changes industrialized society was making on the face of the Earth, especially deforestation. He described it as "man's first physical conquest, his first violation of the harmonies of inanimate nature."[4] He saw the forest as the matrix that bound the environment. Marsh chronicled the effects of deforestation on soil erosion and the contours of the Earth, on plant and animal life, on climate, and on water quality and availability. His other concerns included drainage of lakes and marshes; local or global extinction of plant or animal species; purposeful or accidental introduction of nonindigenous plants, animals, insects; destruction of the natural contours of the land; and effects of aqueducts, reservoirs, canals, and irrigation.

In the early 1860s, Marsh could not foresee that the primary environmental issue a century later would be the more insidious effects of air, water, and soil pollution that result from production and consumption residuals, especially chemicals, intruding into the environment. These residuals, as identified by Ayres and Kneese,[5] include the unwanted by-products of energy conversion (e.g., carbon dioxide, fly ash, sulfur dioxide, carbon monoxide, radioactive wastes, and waste heat), of materials processing (e.g., slag, particulates, inorganic wastes, processing losses and organic wastes, chemicals dissipated in processing, and scrap, trash), and of final consumption (e.g., garbage, sewage, trash, and junk). They persuasively argue that such residuals are an inevitable part of virtually all production and consumption processes.

Marsh did see, however, the importance of the law of conservation of matter to the environment, writing "[N]ature has provided against the absolute destruction of any of her elementary matter, the raw materials of her works. But she left it within the power of man irreparably to derange the combinations of inorganic matter and of organic life."[6] This observation finds modern expression in the concept of materials balance—that the "weight of basic fuels, food, and raw materials entering the processing and production system, plus oxygen taken from the atmosphere" must approximately equal the weight of residuals discharged to the environment.[7]

In Marsh's view, ignorance is the main reason for human being's destructiveness to the environment. He sought to educate his readers as to the "comprehensive mutations man has produced, and is producing, in earth, sea, and sky, sometimes, indeed, with conscious purpose, but for the most part, as unforeseen though natural consequences of acts performed for narrower and more immediate ends."[8]

The economist finds explanation for humanity's destruction of the environment primarily in the concept of externalities as developed by the English economist Pigou.[9] This concept is straightforward. An economic choice has, by definition, costs. These costs are the value of the foregone opportunities as a result of the choice. They may be incurred by the decision maker (private costs) or by others (external costs). To produce steel, for example, coal, iron ore, labor, capital, etc., are employed and, therefore, are not available for other uses. These sacrifices are the private costs of production. However, when the residuals from steel production (e.g., sulfur oxides) are discarded to the environment, thus reducing environmental (e.g., air) quality, affected individuals also experience a loss in their welfare or a cost from others' decision to manage the inevitable residuals in an environmentally degrading manner.

Since Adam Smith's insights over two centuries ago, economists have typically characterized human beings as self-interested individuals attempting to maximize their well-being.[10] They do this by judiciously balancing the benefits and costs of their choices. Individuals would not knowingly foul their own nests. However, Pigou argued that Smith's economic individuals would not be adverse to fouling the nest of others if in so doing they bettered themselves. The failure of markets to address this behavior effectively is a frequent rationale for justifying government intervention in a market-based economy. Specifically, there exists a class of goods that are collectively consumed—public goods. In contrast to private goods, their consumption is characterized by nonrivalry. Everyone can simultaneously enjoy (consume) a smog-free day, a public good, whereas a loaf of bread, a private good, when consumed by one person is not available to others. Public goods will not be produced without some form of collective action.

We make choices daily that create externalities and thereby affect the global environment. We make these choices individually and collectively through governments and other organizations. When we engage in a consumption activity, for example, by driving polluting automobiles, we may directly affect the quality of the environment. Such consumption activities may also indirectly affect the environment (as estimated by Bingham et al.) through production-consumption interrelationships.[11] To continue with the example, to make automobiles requires steel, among other inputs. In our choice of such intermediate products and of production methods, we affect the environment since residuals are generated as unwanted by-products of steel production. We may make environmentally degrading choices out of ignorance or out of a failure to have institutions that cause decisionmakers to bear the costs of their actions. Marsh urged that man "become a co-worker with nature in the reconstruction of the damaged fabric which the negligence of the wantonness of former lodgers has rendered untenantable." He believed our choice was relatively simple—reform or risk reducing the globe "to such a condition of impoverished productiveness, of shattered surface, of climatic excess, as to threaten the depravation, barbarism, and perhaps even extinction of the species."[12]

Clearly there are benefits to our current life-style that many share—the quality of modern health care and longevity, the opportunities for material comfort, the surplus that allows us to pursue the arts and gives us the leisure time that permits opportunities for self-expression. It is doubtful that many of us would choose to exchange places with Marsh's "wandering savage." But to secure these benefits may, and indeed frequently will, involve some degree of degradation of the natural environment. Two important questions are: 1) What are the relative values of environmental quality and the goods and services that can only be produced by degrading the environment? and 2) How can we ensure that individuals' values for environmental services are incorporated into private and public decision making?

The methods for discovering these values are discussed later in this chapter.

MEASURES OF VALUE

Someone has corrupted Oscar Wilde by saying that "an economist is someone who knows the price of everything and the value of nothing." But our best objective measure of value is the maximum sacrifice one is willing to make to obtain a desired result—willingness to pay—or, alternatively, the minimum compensation required to accept a loss—willingness to accept. Money is a useful measure because relative prices and the way individuals allocate their budgets reflect values. Thus, if one is willing to pay one dollar for a loaf of bread, it must be worth at least that amount to the buyer. Alternatively, if the owner of a loaf of bread is willing to accept one dollar for it, its value to him or her must be no more than a dollar.

Because of the public good nature of the environment and its finite assimilative capacity, governments throughout the world have attempted to limit access to the environment for those looking for a place to deposit production and consumption residuals. However, such restrictions have costs since other methods for managing residuals must be employed that have higher private costs. A critical question that must enter into every decision on these limits, even if only implicitly, is "Is the protection or improvement in environmental quality worth the costs?" This is ultimately a question of society's values. In a democratic society, the values of all affected individuals should be considered in answering such a question.

To determine society's value for a public good such as environmental quality, one of two values will be added up for all affected individuals. The value will be either the sum of each person's maximum willingness to pay for the improvement in environmental quality or the sum of each person's minimum amount they would be willing to accept in order to forego the improvement. It is interesting to contrast this with the private good case. Because there is rivalry with private goods, the value of a private good is the amount that the single most desirous person would pay or accept, as appropriate, for the commodity. The value of one more loaf of bread annually might, therefore, be one dollar per loaf; the value of one more smog-free day annually might be a million dollars, even if one million similar people were only willing to pay one dollar each for the day.

Without knowledge of the preferences of individuals for the services provided by the environment, public decision making that affects the environment must be made on a whim, hunch, or the preferences of the decision maker. Economists have attempted to measure individuals' preferences for a number of environmentally related commodities. Measurements have been attempted for animals such as geese, elk, grizzly bear, and big horn sheep for sport hunting; use of the natural environment for hiking, boating, fishing, and swimming; and aesthetic and productive dimensions of the environment, such as scenic beauty, visibility, water quality, "woods," air quality, and odors.

Most attention has been directed to the effects of air, water, and, more recently, soil pollution from the release of chemical residuals into the environment. Marsh's concern over deforestation, drainage of lakes and marshes, extinction of species of plant and animal life, and the revision of the shape of our planet has received much less attention from economists in the academic and research communities, probably because governments have been less willing to support research in these areas.

Unfortunately, ordinary exchange does not as readily disclose how people value environmental quality as it does how they value a loaf of bread. There is no market where, for example, one may purchase "clean air." As identified by Freeman, other approaches must be used to discover these values.[13] These approaches include the results of public referenda on an environmentally related choice; examining the effect of environmental quality on the value of commodities exchanged in the market—the revealed preference approach; and directly asking individuals their willingness to pay or accept amount for a postulated change in the environment—the expressed preference approach.

The first approach identified—voting—has not found wide application because of the high cost of public referenda, the zero-one nature of the choices typically offered, and its failure to reveal the intensity of preferences. For example, suppose one thousand voting citizens were given the opportunity to cast ballots on a public mosquito control policy for their community. The policy will cost $2,000, to be paid in higher taxes equally distributed over all citizens. Under the simple majority rule principle, if 501 people vote against the policy, it will not be enacted. However, the average willingness of the 499 proponents to pay could be, say $3.50, while that of the 501 opponents might be $1.90. The 499 would enjoy a surplus of $1.50 ($3.50 − $2.00) each, or $748.50 in total, if the policy were enacted; the majority would lose $.10 ($1.90 − $2.00) each, or $50.10 in total. The strength of the preferences of the minority, as measured by their willingness to pay, exceeds that of the majority, yet the policy will not be implemented if referenda are used to make the decision.

The revealed and expressed preference approaches have been extensively employed. They are briefly described below.

REVEALED PREFERENCE APPROACH

The revealed preference approach uses an indirect method to measure individuals' values for changes in environmental quality. The traditional approach developed by Clawson and Knetsch is the travel cost method.[14] Visitors to a recreation site typically come from diverse origins. Since the trip requires travel time and cost, as well as, perhaps, a user fee, there is an implicit price, or cost, for each trip. The optimizing traveler will select the frequency of visits that equates this cost with the marginal value of the visit. All else being equal, visitors from further away would make fewer visits to the site than those closer to it. Site characteristics— including such environmental dimensions as water quality or quantity, scenic beauty, and visibility—will affect individuals' marginal valuation of a visit. Where variations in these characteristics are observed either over time for a particular site or across sites and where data are available, differences in the number of visits and cost per visit can be used to infer indirectly the value of the characteristics to the visitor.

The travel cost method is site-specific and measures only the visitors' values. The values, if any, of nonusers are not revealed since they make no trips. Two important methodological issues with the travel cost method that have not been resolved include the valuation of time spent at the site and the assignment of the travel costs when a trip involves a visit to more than one site.

The second major method that uses a revealed preference approach is based on the hedonic price methodology developed by Rosen.[15] This method recognizes that many commodities have different attributes that are an intrinsic feature of the commodity and cannot be disassociated from it. For example, a house will be associated with neighborhood characteristics such as employment proximity, average smog levels, school quality, and distance from a toxic waste site. Individuals may value these attributes differently. When environmental quality is one of these attributes, the differences in the value of, for example, houses otherwise similar in location and features but differing in the level of environmental quality could be used to infer the value of the environmental differences.

Like the travel cost method, the hedonic method is also confined to estimating user values. Markets are assumed to be in equilibrium and adjustment costless with the hedonic method. The subjective observations of buyers and sellers regarding the quality of the environment and the objective measurements used in the estimation model are assumed to be highly correlated.

EXPRESSED PREFERENCE APPROACH

The expressed preference approach involves directly asking individuals for their preferences. The forerunner of this method as practiced by economists was the Gallup poll and other opinion surveys. Gallup and Louis Harris asked the public to rank environmental issues among other issues of national importance.[16] In 1965, "reducing pollution of air and water" ranked ninth among national problems. The ranking rose to third by Earth Day, five years later. Concern over the environment has since slipped from among the top issues.

A ranking is an ordinal measure of concern or preference. Attempts to provide cardinal measures lead the national opinion samplers to ask questions such as whether we are spending "too little," "about right," or "too much" on "improving and protecting the environment." The response to these and other similar questions indicate a higher level of citizen concern today than do the ranking questions and point out the problem with simple rankings. However, this question format still does not provide the desired information on the intensity of citizen preferences.

The expressed preference approach to measuring the value of changes in the environment—or, as it is more commonly known, the contingent valuation method—involves the use of a survey questionnaire to solicit information on respondents' values and on the respondent per se. The contingent commodity (e.g., an improvement in environmental quality) is described using one of several alternative valuation questionnaire formats to develop a data set for statistical analysis (Table 23–1).[17]

The contingent valuation method is flexible. Any commodity can be created and both user and nonuser values obtained. However, concerns have been raised regarding the potential for bias in expressed preference experiments. Respondents may purposefully or inadvertently give incorrect responses. There also are a number of potential instrument-related biases, including those associated with the starting point used in the bidding game format, the vehicle employed, and the effect of any information provided to the respondent.

SUMMARY OF VALUATION APPROACHES

The revealed preference approach is favored for estimating the value of changes in environmental quality because, with it, individuals actually

TABLE 23–1   QUESTIONNAIRE FORMATS

- With the *direct question* format, the interviewer directly asks individuals about their willingness to pay for or accept compensation for a postulated change in the amount of environmental services provided.
- With the *bidding game* format, an auction, in either the Dutch (top-down) or English (bottom-up) format is used.
- With the *payment card* method, respondents are asked to select the value from a card that most closely approximates their value for the change. In some cases, the card shows the share of an individual's taxes associated wth specific public programs such as education and defense (i.e., anchored).
- With the *bidding game with budget constraint* method, respondents are asked to identify how they would reallocate their fixed budget to "purchase" additional amounts of environmental quality.
- With the *ranked choice* method, the individual ranks alternative hypothetical market outcomes.

---

demonstrate their willingness to pay or accept rather than simply respond to a hypothetical question. There is continuing concern that respondents to a contingent valuation questionnaire can be influenced by the format of the questionnaire and value-solicitation method. There is also concern in some quarters that individuals may purposefully misrepresent their preferences to achieve a personal goal, although there is little, if any, evidence to support this position.

The revealed preference approach is, however, limited. First, it is limited to situations where circumstances have provided data for the measurement of values. It is also limited to user values. Values that are not revealed through behavior (e.g., individuals who are only potential users or individuals who gain utility from a sense of stewardship or from vicarious consumption) are not accounted for by this method. With the expressed preference approach, one can define hypothetical outcomes outside the current range of experience, for example, a smog-free year in Los Angeles. There are several direct comparisons of the revealed and expressed preference approaches. Although such comparisons can only be suggestive of the validity of either approach, they tend to find a high degree of similarity in the estimates obtained by both methods.

It is seldom possible to avoid making difficult interpersonal comparisons of the economic welfare changes from environmental policies. Consider, for example, a policy that affects the economic welfare of two individuals: Ted and Alice. Figure 23–1 shows their willingness to pay for the possible welfare changes relative to the status quo that is the center of each diagram. In panel (a), an environmental policy that places them in the third quadrant is clearly not preferred to the status quo by either person, as both have negative willingness to pay for the change. Therefore, the policy should not be implemented. In contrast, a policy outcome falling in quadrant I [see panel (b)], results in an improvement in economic welfare for both individuals. Outcomes along the positive region of either axis depict situations where one person gains and the other is unaffected. Policy outcomes in quadrant I are termed Pareto improvements, after the Italian economist and sociologist. He argued that only policies where at least one person is better off and no one is worse off can be objectively preferred.

However, most public policies will adversely impact some members of society. In such cases, Kaldor and Hicks argued that a policy should be implemented if the potential exists for gainers to compensate losers without exhausting all the gains (i.e., if aggregate benefits exceed aggregate costs).[18] For example, consider policy outcomes falling within the shaded area of quadrant II in panel (c). In this area, Ted's gain always exceeds Alice's loss. Ted could fully compensate Alice for her losses and

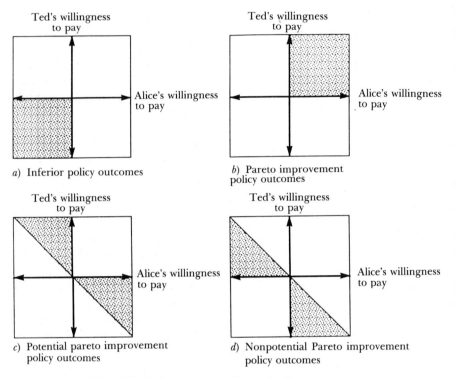

Figure 23-1   Interpersonal effects of policy outcomes.

still have part of the gain left for himself. In quadrant IV, the shaded area represents the corresponding set of outcomes where Alice's gain exceeds Ted's loss. The mosquito control example presented earlier was a policy that would fall in this area. Along the diagonal, Ted's (Alice's) gain equals Alice's (Ted's) loss. Thus, policy outcomes lying to the right of the diagonal up to the axes represent potential Pareto improvements in economic welfare.

In panel *d*), the losses experienced by one party (Alice in quadrant II, Ted in IV) exceed the benefits of the other (Ted in II, Alice in IV). There is no potential Pareto improvement in economic welfare with such policy outcomes. Selection of a policy with an outcome in either of the shaded areas of panel *d*) implies unequal weighting by the policy-maker of the willingness to pay of the two affected individuals.

Simon revised Smith's description of self-interested individuals to recognize that, if human beings attempt to maximize utility, they do so using limited computational skills and information.[19] Recognition of the effect of these additional constraints on choice is important for estimating individuals' value for the services of the environment. For example, in most cases, health scientists have an incomplete understanding of the effect of pollutants on human health since dose-response relationships exist for only a few pollutants. Even for these, low-dose or animal-human extrapolation is necessary because most epidemiological data pertain only to occupational exposures. These exposures are typically much higher than those found in the ambient environment. Another difficulty in valuing health risks is that large confidence intervals exist around these dose-response relationships. How does the general public evaluate health risks when the available information is incomplete? Further, the health effects of exposure to pollutants typically have long latency periods. Individuals' ability to solve this intertemporal expected utility maximization problem consistently must, obviously, be limited. The extremely wide range of value-of-life estimates found in the literature must, in part, be due to these problems.

The methods for measuring the health effects—disease, illness, death—are another issue related to the valuation of health risks. In empirical studies these effects have typically been measured in terms of lost work days due to illness or restricted activity days for unemployed individuals. Mortality is usually described in a zero-one context. Clearly, however, diseases, illnesses, and even the form of death differ in ways that are presumably important to the valuation question.

SUMMARY

This chapter has argued that environmental quality is a public good and that a society's choices regarding the level of environmental quality and other commodities are technically constrained. Therefore, economic choices have to be made by governments entrusted with custodianship of the environment. Further, we argue that: 1) in most cases, the preferences or values of the affected individuals should form the basis for public policy decisions related to the environment, 2) willingness to pay for environmental quality is our best objective measure of the intensity of an individual's preferences, and 3) empirical methods are available to estimate willingness to pay.

A substantial amount of progress has been made over the last decade in the valuation of nonmarket goods, especially environmental quality. However, even if one could secure agreement on the principles and observations discussed earlier, there are many unresolved issues on the incorporation of the willingness to pay values into public policy decisions.

Earlier, this chapter considered interpersonal comparison for individuals. Issues of interpersonal comparisons of welfare changes are substantially more complex when different governments represent different affected individuals. Many environmental issues such as the harvesting of whales, the deforestation of the Amazon jungle, poaching of endangered species, and effects of acid rain impact individuals in different nations. There is no omniscient government skillfully trading off the interests of these people. Typically, issues of international importance are bartered in bilateral agreements between countries. The countries are frequently contiguous and the bartering takes place between governments for environmental problems. However, a recent exchange is noteworthy. A nonprofit United States group, Conservation International, purchased $650,000 of Bolivia's external debt at a discount and retired it. In exchange, the Bolivian government agreed to protect 3.7 million acres of threatened tropical lowlands from deforestation.[20]

Intertemporal choices present difficult valuation problems, especially when one is dealing with unborn generations and where the choices we make today regarding the environment are irreversible. In such cases, the values to future generations should be incorporated into today's decisions. However, it is not possible to know with certainty their values and the technical constraints they will face. Consequently, current policy decisions will not be optimal to the extent that expectations regarding future conditions are not met.

Marsh articulated the key interaction between human beings and the environment: human beings act on and modify the environment. In turn, the environment acts on them and their "social life and social progress." But our knowledge of the linkages between human beings and the environment is still primitive despite the great technological advances that have been discussed elsewhere in this volume. We may, perhaps even with some precision, be able to identify the effects of a paper mill's effluent on objective measures of water quality such as biological oxygen demand (BOD). But our knowledge of how individuals perceive changes in characteristics of water quality and respond to these perceptions is much less complete. Further, there may be acclimatization or habituation to reduced levels of environmental services, raising questions about the stability of preferences over time.

Although Marsh argued that ignorance is the reason for our poor treatment of the environment, we may recognize two sources of ignorance. The first involves the effects of human beings on the environment, which Marsh described. The second is our incomplete knowledge regarding individuals' valuation of these effects.

Each decision that is made that affects natural resource use and environmental quality (local or global) will require that we assess the benefits and costs of such action. We cannot do so without including a measure of the value of environmental quality. Even though the methods of estimating the value of nonmarket goods that were presented here are still embryonic, they are an important first step in assessing the true damages of global environmental change and the benefits that would be gained by reallocating our resources. More work must be done in this field, but we agree with Marsh that one cannot wait for all facts to be resolved—even if we could look forward to such an unlikely prospect. In his words, "we are, even now, breaking up the floor and wainscoting and doors and window frames of our dwellings, . . . and the world cannot afford to wait till the slow and sure progress of exact science has taught it a better economy."[21]

NOTES

[1] G.P. Marsh, 1864, *Man and Nature*, Scribners, New York, reprinted 1965, Belknap Press of Harvard University Press, Cambridge, Massachusetts.

[2] *Ibid.*, p. 13.

[3] *Ibid.*, p. 40.

[4] *Ibid.*, p. 119.

[5] R.U. Ayres and A.V. Kneese, 1969, "Production, Consumption, and Externalities," *American Economic Review*, Vol. LIX, No. 3, pp. 282–297.

[6] Marsh, 1864, *op. cit.*, p. 36.

[7] Ayres and Kneese, 1969, *op. cit.*, p. 284.

[8] Marsh, 1864, *op. cit.*, p. 19.

[9] A.C. Pigou, 1932, *The Economics of Welfare*, 4th ed., London.

[10] A. Smith, 1776, *The Wealth of Nations*, various editions.

[11] T.H. Bingham, D.W. Anderson, and P.C. Cooley, 1987, "Distribution of the Generation of Air Pollution," *Journal of Environmental Economics and Management*, Vol. 14, No. 1, pp. 30–40.

[12] Marsh, 1864, *op. cit.*, pp. 35 and 43, respectively.

[13] A.M. Freeman III, 1979, *The Benefits of Environmental Improvement: Theory and Practice*, Johns Hopkins Press for Resources for the Future, Inc., Baltimore, Maryland.

[14] M. Clawson and J.L. Knetsch, 1966, *Economics of Outdoor Recreation*, Resources for the Future, Inc., Washington, DC.

[15] S. Rosen, 1974, "Hedonic Prices and Implicit Markets: Product Differentiation in Perfect Competition," *Journal of Political Economy*, Vol. 82, pp. 34–55.

[16] R.E. Dunlop, 1985, *EPA Journal*, Vol. 11, pp. 15–17.

[17] W.H. Desvousges and V.K. Smith, 1983, *Benefit-Cost Assessment Handbook for Water*

Programs, Volume I, prepared for U.S. Environmental Protection Agency, Research Triangle Institute, Research Triangle Park, North Carolina.

[18] N. Kaldor, 1939, "Welfare Propositions of Economics and Interpersonal Comparisons of Utility," *Economic Journal,* Vol. 39, pp. 549—551; J.R. Hicks, 1940, "The Valuation of Social Income," *Econometrica,* Vol. 7, May, pp. 105–124.

[19] H.A. Simon, 1955, "A Behavioral Model of Rational Choice," *Quarterly Journal of Economics,* Vol. 69, pp. 99–118.

[20] New York Times, 1987, "Bolivia to Protect Lands in Swap for Lower Debt," *New York Times,* July 14, p. C2.

[21] Marsh, 1864, *op. cit.*

EDITORS' INTRODUCTION TO:
G. BUGLIARELLO
Technology and the Environment

In this chapter, Dr. Bugliarello, president of Polytechnic University, provides a novel perspective about how technology can be viewed in an evolutionary context as an entity combining indissolubly biological organisms, society, and machines (biosoma). This concept, coupled with the definitions of three leitmotifs of this evolution—materials, energy, and information—provides an integrated view of the role of technology in the environment. Dr. Bugliarello then discusses how technology can be managed to create new environmental niches for human beings and to lead to a more rational use of existing environments.

24 TECHNOLOGY AND THE ENVIRONMENT

GEORGE BUGLIARELLO
Polytechnic University
Brooklyn, New York

INTRODUCTION

The complexity, variety, and magnitude of the effects, both direct and indirect, of technology on the environment are already large and growing at such a rapid rate as to make it imperative to create a framework to better understand and control them. This chapter is an attempt to propose such a framework and point out its implications for future environmental strategies.

TECHNOLOGY AND THE BIOSOMA

Ecologists have used the term *ecological niche* to refer to the set of all environmental conditions under which a species can persist. As suggested intriguingly by Kenneth Boulding, human beings are niche expanders because their population has steadily grown as a result of agriculture, industry, science, and the occupation of new territories.[1] For our purposes here, it is most useful to use the term *niche* in a complementary way, by referring to the environment in which a species operates. We can view human beings as intense, constant creators of new niches—as creators of technology and social organization to make life possible in environments that otherwise our unassisted physiology would deny to large numbers of us. The creation of new niches is the product of evolutionary processes that, in the case of human beings, have been partly biological and partly social and technological, that is, carried out through society and machines (the latter defined here as any artifact created by human beings). These two evolving entities—society and machines—extend extracorporeally our reach as biological organisms.

Fundamentally, society and machines are the product of human beings, but they also have affected modern humankind. Today, the combination of humans, society, and machines has developed to the point that it has acquired a dynamics of its own. It has become truly a new entity, greater than the sum of its parts, in the same way that a living organism acquires characteristics that go beyond those of the cells and diverse organs that compose it. I shall call this new entity, this indissoluble unit of *bio*logical organisms, *so*ciety, and *machines*, the *biosoma*. The biosoma, through the interactions and synergisms of its components, is the synthetic agent of the interaction between human beings and the environment, and the platform from which future evolution will occur.[2]

Since the appearance of *Homo sapiens,* the establishment of new human niches has been achieved through the synergism of artifacts and organization. The artifacts—machine components of the biosoma—are the primary instruments for carving out new niches, whereas the social components of the biosoma have been primarily niche organizers, coordinating and reinforcing the activity of individual human beings. The artifacts are both tangible and intangible—"hardware" and "software"—with the hardware ranging from tools to clothing, houses, irrigation works, factories, and spaceships, and the software (the "know-how") from computer programs to weather modification, animal husbandry, and surgery. In turn, the social components of the biosoma range from government to religion, management, customs, or the financial system.

Technology, as a process or system for the production and utilization of artifacts, is a biosomic entity combining social organizations (factories, work yards, design offices, engineering schools, research laboratories, financial institutions, markets, military organizations, etc.) and machines. Because of its biosomic nature, its impacts on the environment, both biological and inanimate, are basically twofold: the direct impact of machines and the technological systems that produce them and the indirect impact of social organizations and processes made possible by machines (because the way a society is organized and operates depends increasingly on the technology it has available).

The impact of machines—and of the pertinent technology that has produced them—on biological organisms and the inanimate environment is exemplified by the impact of the automobile. That impact is that of the factories manufacturing automobiles as well as the automobile itself. The factories affect the environment with noise, chemicals in the air, water, and ground, with the appropriation of space, with workers' residences, etc. The automobile, in addition to generating similar effects (think of the emission of particulate matter, hydrocarbons, and noise, and of the space occupied by highways), has also reshaped our way of life, created extended suburbs, and brought people as tourists in large numbers to previously undisturbed regions.

Since the 1955 conference on *Man's Role in Changing the Face of the Earth,* two aspects of technology of great environmental consequence have emerged: biotechnology and sociotechnology. These two aspects have been flanked by the accelerated growth in information technology, which will be discussed later. Biotechnology—the purposeful modification of biological organisms—can create new species and, hence, new environmental niches, with at times unforeseen environmental consequences. Sociotechnology is a multifaceted activity endeavoring to study and guide the interaction between technology and other social systems.[3] It includes science and technology policy, technology assessment, environmental impact assessments, and the sociology and philosophy of technology. Sociotechnology can have immense environmental consequences, both for good and bad. It can be beneficial when the powerful synergisms of technology and other social systems reinforce intelligent approaches to the environment. It can be devastating when it magnifies environmental injuries, either for lack of information or because it is guided by philosophical, religious, or legal concepts, or by social customs, that ignore the environment or, worse, actively promote an indiscriminate human-centric view of nature.

TECHNOLOGICAL LEITMOTIFS AND THE BIOSOMIC MATRIX

The impact of technology on the environment is extremely diverse and fragmented, given the existence of myriad specific technologies and machines, each with their peculiar interaction with other technologies, other social systems, and with the environment. A broad conceptual systematization of technology that appears particularly useful because it applies, either in a direct or indirect sense, to every element of the biosoma has three leitmotifs:

- Materials
- Energy
- Information

We can represent the biosoma as a matrix relating the three biosoma components to these leitmotifs (Figure 24–1). In considering such a matrix, we must not lose sight of the fact that the biosoma is a unit, and that the leitmotifs are only facets of a whole within each component of the biosoma. We must also be aware that each leitmotif has a different meaning for each component of the biosoma and represents an organizing principle rather than exactly comparable physical or logico-mathematical concepts. Thus, we can look at machines that are primarily focused on materials, machines that are primarily focused on energy, and machines that are primarily focused on information. Examples of the first are civil engineering structures, of the second, engines and vehicles, and of the third, computers and telephones.

In the biological domain, too, we can look at living organisms from the viewpoints of materials, energy, and information. For instance, material aspects have to do with the skeletal structure and with tissues; energy aspects with the energy that activates our physiological functions; and information aspects with a range of processes from genetic instruc-

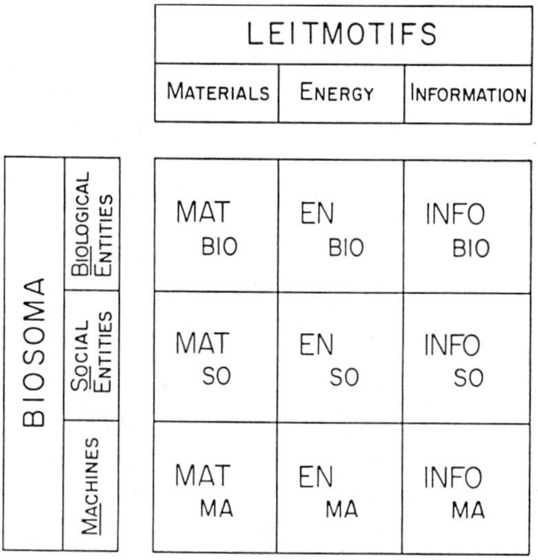

LEITMOTIFS		
MATERIALS	ENERGY	INFORMATION

BIOSOMA	BIOLOGICAL ENTITIES	MAT BIO	EN BIO	INFO BIO
	SOCIAL ENTITIES	MAT SO	EN SO	INFO SO
	MACHINES	MAT MA	EN MA	INFO MA

FIGURE 24–1   The biosomic matrix.

tions for reproduction to the transmission of messages humorally or by the nervous system, and the sensing by an organism of its environment and its own condition.

In the social domain, we can look at the organization of a society as it is affected by the technological leitmotifs. Ancient societies were primarily dominated by materials—walls and roads—and by a limited availability of energy, which was either biological (provided by plants, animals, and human beings) or solar (such as in wind or falling water).[4] Societies of the Industrial Revolution were made possible by the utilization of large amounts of energy, primarily from coal. Our contemporary societies, while utilizing both sophisticated materials and large amounts of energy of various forms, are uniquely characterized by their dependence on intense generation and utilization of information.

The biosomic matrix in Figure 24–1 in effect represents how human society, as an organized ensemble of individual biological organisms creating and using machines, operates on the environment and is in turn affected by it. The significance of the matrix in helping us understand and govern environmental impacts lies in its ability to schematize and clarify historical trends in the development of technology, to indicate the environmental impacts arising from the synergisms of its nine elements, and to suggest strategic trade-offs among such elements.

HISTORIC TRENDS IN THE EVOLUTION OF TECHNOLOGY

The evolution of technology can be described with the help of the rows and the columns of the biosomic matrix. Three major trends can be singled out—with the warning that they represent, obviously, broad generalizations.

The first major trend is the shift from biology to the machine. For instance, in the domain of materials utilized by technology, the utilization of biological materials such as wood or animal skins has given place increasingly to that of machine-made or machine-altered materials such as plastics or steel. (The utilization of materials such as stone and clay through the development of mortar, firing techniques, etc., represented an intermediate but still technologically primitive step. On the other hand, today's silicon wafers, fiber optics, and ceramics for high temperatures represent technically sophisticated machine-altered materials.)

This technological direction from the biological to the human-made (machine) manifests itself in the energy and information domains as well. In energy, it is represented by a shift from natural fuels and foods to synthetic ones (e.g., from wood-burning to nuclear energy, from harnessing animal energy to the internal-combustion engine, from food obtained through hunting and animal husbandry to synthetic proteins); in information, from a predominance of genetic storage and transmission of information and the use, extracorporeally, of human messengers, to an increasing societal use of books, computers, and telecommunications.

With the advent of biotechnology, however, the historical trend of technology from biology to the machine—from wood to plastics, from the horse to the automobile—is being reversed and made more complex. Biotechnology, as a new synergism of biology and machines, is both biological and artifact, the latter being the knowledge of how to manipulate genes and the capability to utilize such knowledge. Biotechnology makes it possible to modify genes (generating, in effect, artificial ones), so as to create new materials, new foods, new fuels, new information machines (biological computers), and, perhaps eventually, even new human beings. In the biosoma matrix, these developments represent a new emphasis on the biological component, but more properly the results of biotechnology are biological artifacts—truly biological machines.

The second major trend occurs along the rows of the matrix, as the focus of the leitmotifs of technological innovation shifts from materials and energy toward information. In the domain of the machine, computers, telecommunication, data banks, robots, etc., are all manifestations of this shift, made possible by electronics. A similar shift has occurred in the domain of the biological sciences, with the growing emphasis on information, exemplified by genetics and the study of the brain. (It is useful here to remind ourselves of the nature of science as a social activity focused on the various domains of the biosoma and its environment, and synergistic with technology.)

The third major trend manifests itself in the societal domain, with society's increasing tendency to assume some of the tasks that were once exclusively biological—as exemplified by learning or by social pressures for birth control—and to delegate in turn some of its activities to machines (e.g., teaching to computers). This has profound but seldom sufficiently perceived environmental implications, as the interaction between humans and environment ceases to be instinctive but becomes —by necessity—governed by a complex host of societal controls and conditioned by society's ideologies. Concurrently, moving away from the earlier themes of energy and material, the very organization of society is being increasingly shaped by the leitmotif of information, with its implications for knowledge and control of all of society's manifestations and actions, including environmental impacts and strategies.

CHANGES IN THE CENTER OF GRAVITY OF THE BIOSOMIC MATRIX

These historical trends in the evolution of the three components of the biosoma in interaction with the three leitmotifs can be further summarized through the bold generalization shown in Figure 24–2. In the figure, the light-shaded areas represent the portions of the biosomic matrix prevalently involved in a given kind of societal or historical period; the darker-shaded areas indicate the major focus or "center of gravity" of the biosomic activities.

As the result of the interactions of technology with the rest of society, the center of gravity of the shaded areas has been moving constantly in a counterclockwise direction in the matrix, describing what I shall call the *biosomic circle*. In preindustrial civilizations, technology as well as society in general utilized primarily biological elements. The societies spawned by the Industrial Revolution were shaped by the synergisms of energy and materials. For a brief period in the postindustrial societies, the focus

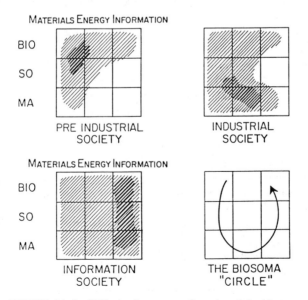

FIGURE 24–2 Shifts in the center of gravity of the biosoma.

of industry and other societal activities was primarily information in the machine domain (telecommunications and computers). Through biotechnology, that is, the ability to manipulate biological (genetic) information, biology again becomes of major significance; the postindustrial society is being replaced by a fuller information society, having as center of gravity information emphasized in all three components of the biosoma—biological, social, and machine. This is the first biosoma society in the full sense of the word—a society in which for the first time in human history the biological, social, and machine components begin to acquire comparable importance and complexity. Society and technology have thus come almost full circle in the matrix, with the potential, as we discuss next, of moving away from the environmentally most destructive synergisms that have characterized the industrial and the early postindustrial societies.

SYNERGISMS AND TRADE-OFFS

Interactions among the elements of the biosomic matrix give rise to synergisms and trade-offs of key environmental significance. For instance, in the machine domain, the synergism of materials and energy has led to the creation of totally new materials such as plastics or to the concentration of existing materials—creating in effect new substances—through distillation.

Each synergism has environmental implications, as it creates new technological products or manifestations, or it reinforces the impact of existing technological components and other biosomic components. The environmental impacts of the synergisms among the nine elements of the matrix are perhaps more clearly understood with reference to Figure 24–3. The vertical dimension of the matrix in that figure can be viewed as representing a spectrum from the *natural* (the biological) to the *natural multiplied* (the social component) to the *metanatural* (the intensification and at times replacement of the natural by artifacts, that is, machines). Simultaneously, the leitmotifs are in part *tangible* or producing tangible effects—materials and energy—and in part *metatangible*—as in the case of information.[5]

From the environmental viewpoint, the most undesirable synergisms are those in the tangible domain that combine the natural multiplied and the metanatural—that is, individuals in large numbers intensively utilizing materials and energy. Historically, these synergisms have also been the major niche creators and expanders.

The environmentally desirable synergisms are of two kinds. In the

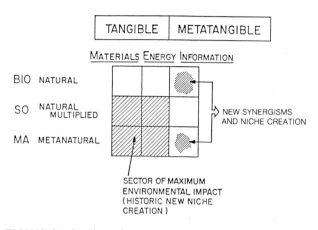

FIGURE 24–3   Biosomic synergisms and environmental impacts.

first place, there are the synergisms that organize existing niches by shifts from the tangible to the metatangible, in the context both of machines and of the organizing principles of society. As we have discussed, a society shaped in its organization by the use primarily of information rather than of materials and energy is much more conserving of the environment.

Second, there are the synergisms that make it possible to create new niches by the interactions of biology with the metanatural and the metatangible (machine-carried and machine-manipulated information). These synergisms are the great theme of the information society and create new niches by making it possible to control life, both human and nonhuman, and even to "manufacture" it, in activities ranging from agriculture and aquaculture to genetic engineering.

Trade-offs among the components of the biosomic matrix can occur vertically among the biosoma components, horizontally among the leitmotifs, or diagonally. Through vertical trade-offs we can replace, as already indicated, biological processes with social or machine ones. Through horizontal trade-offs we can reduce gross use of materials and energy with more precise information about materials or energy processes, or we can even totally replace such processes with information. In a diagonal trade-off we can replace a social organization requiring energy (fuel or food) with machines focused on information. A trivial but descriptive example is telecommunications replacing the physical delivery of mail by the postal service. As a further example of trade-offs, biotechnology makes it possible to replace in several basic industrial processes the traditional high levels of temperature or pressure that characterize smokestack industries with the much lower and thus environmentally far less damaging magnitudes characteristic of living environments. We thus have, in effect, the potential of reducing the scarring of the surface of the Earth by reducing our quest for minerals, of saving larger quantities of cooling water, of conquering the acid rain problem, and of mitigating the greenhouse effect.

FURTHER DISCUSSION OF ENVIRONMENTAL IMPLICATIONS

This overview of the nature and environmental significance of the biosomic matrix provides a useful background for some additional considerations on the relation of technology and the environment. Nine such considerations are included in this discussion. It must be underscored again, however, that often these considerations are delineations of megatrends, and thus need to be taken with a grain of salt.

A key but much neglected element of technology and its impact on the environment is the relation of technology to social dynamics. The environmental importance of that relation is exemplified by society's synergism with machines that make both technology and large and affluent (and therefore resource-consuming) populations possible. Such synergism makes it imperative that we approach the question of environmental impacts by starting "upstream" of the manifest impacts of specific technologies and population developments. We need to look, for instance, at patterns of employment, at the dynamics of consumer societies, at the sociology of technology (e.g., the reward mechanisms that generally are geared to success in construction and production rather than to the thoughtful refusal to carry out a project for sound environmental reasons or that reward the specialist rather than the rare and much needed generalist) or at the pathologies of the interactions between technology and other social systems. This is one of the key tasks of the emerging discipline of sociotechnology.[6]

Environmentally most devastating are the synergisms between society and tangible machines focused on material and energy processes. The environmental impact is the result of a spectrum of activities ranging in scale from the micro-level of individual machines such as cars, houses, spray cans, or plastic bags to the realm of megaprojects, that is large-scale projects typically involving enormous quantities of materials and energy. In societies that possess the necessary resources and skills, megaprojects are a favorite and at times indispensable response to challenges the societies face—be it the conquest of space, defense (the Maginot Line), protection from the elements (the Dutch Zuiderzee dams), economic development, or even preservation of some aspects of the environment (e.g., hydroelectric power plants as an alternative to thermal plants). Some of these projects have only limited environmental impact, but in most cases this is not so. Major water diversions, the Aswan dam, the Amazonian highway, or large-scale power plants, all made possible by the synergism of large amounts of energy and materials, can have profound and devastating environmental effects. This is exemplified by the disappearance of wetlands in California to make possible large irrigated farms, the salting of fresh-water marshes in Louisiana as a result of canals dug to facilitate large-scale oil drilling operations, and the impenetrable ecological barrier created by a highway in a wilderness area.

To reduce the environmental impacts of the machines that an increasingly developing world society needs, it is important that the machines become primarily information-oriented. Generally, the shift from materials and energy to information, in whatever biosomic domain, is beneficial in preserving the environment. This is so because information-based activities are carried out with relatively little use of materials and energy (compare, for instance, the space taken by a telephone line and the energy consumed in a telephone conversation with the space occupied by a highway and the energy consumed—and the pollution generated—by the vehicles that use it). The more precise the information we possess in a given activity or phenomena, the less can be the waste of energy and materials, and the concomitant environmental impact. Thus, the knowledge we have gained about aerodynamics, materials, and combustion has enabled us to design fuel-efficient lowweight cars. Telecommunications can save us many a trip, computer memories can take the place of massive physical archives, precisely controlled milling machines can reduce substantially the waste of materials in fabrication, and so on.

The favorable environmental implications of the shift from a concentration on energy to one on information are encountered in a host of other processes of substantial environmental significance. For example:

- In desalinization, the trend from evaporation processes, with their heavy energy requirements, to membrane ones. Evaporation is an energetically costly process removing water from salt ions, whereas membrane separation, based on more detailed understanding of the dynamic of ions in solutions, removes the ions from the bulk of the water with a technology requiring much less energy and capable of further improvements through more advanced membrane design.

- In irrigation, the trend from free-surface irrigation, again an energetically costly process because of its attendant heavy losses through evaporation and misdistribution, to low-volume systems that deliver water directly to the plant themselves (e.g., drip and trickle irrigation systems) in response to need as perceived by sensors at the plant or predicted by meteorological observations.

- In food production, the reduction of the environmentally damaging use of fertilizers and chemical pesticides through processes that directly or indirectly (by reducing waste) affect fertilizer and pesticide consumption. These processes include improved efficiency of fertilizer uptake by a plant (today no more than 50 percent in temperate climates and less in tropical ones); biological pest control practices; production of plants more resistant to stress (i.e., to hot, cold, dry, or wet conditions); and reduction of postharvest losses through better storage, handling, and distribution.[7] The key in each case is, again, information about the functioning of plant hormones and metabolites associated with growth and response to environmental stress, more effective application of fertilizer, insect signals, how to better handle postharvest storage and distribution of foods, and so on. Also key to this and all other aspects of natural resource management and utilization—although yet far from being widespread practice—are information-driven "command and control" systems[8] and the use of artificial intelligence.[9]

To be truly beneficial for the environment, the shift to the leitmotif of information must be a biosomic one, that is, one not limited to machines but involving the entire organization and functioning of society. This is at present not the case, as we have not yet developed to any sufficient extent the new social instruments required to take advantage of the new technologies. For example, work in the service sector of our economy is segregated from home, forcing us to have two abodes, one for work and one for habitation, and to undertake massive commuting, with far-reaching consequences for the environment. Yet information and telecommunications technologies are available to help us with the coordination and control functons that otherwise require the physical presence of the work force in central work places. For example, service activities such as health care and teaching are also unduly centralized today due to lack of mechanisms for the effective delivery of services at home.

As a further elaboration of the previous point, the environmentally favorable shift from the tangible to the metatangible cannot truly occur without a re-examination of today's consumer societies. Such societies, with a constant cycle of production and rapid throwaway of artifacts (cars, containers, etc.) are heavy consumers of materials and energy. Yet, they have been effective engines of economic expansion and achievement. Thus, the questions that require far more serious and dispassionate study than has been the case to date are whether a consumer society can be much more strongly based on the metatangible—on information—or whether a more radical shift involving the abandonment of the entire consumer society concept is required if the environment is to be protected. The move away from a consumer society that uses large quantities of energy and materials—a throwaway-oriented society—requires a deep reorganization of both its economic system and its way of life and habits.

In terms of the evolution of the human species—that is, in terms of the evolution of the biosoma and the enhancement of our biological reach—a consumer society is one that in many respects represents not as much a step forward as a lateral step. An immense amount of human energy and resources is involved in feeding the consumer machinery rather than in activities that can further enhance our reach. A fundamental question is whether we can succeed in separating the great advances in science and technology that have occurred in the consumer society from the environmentally negative characteristics of that society. Certainly, it is possible to have a perfectly civilized life by keeping machines operating for a much longer period of time, limiting the amount of throwaway objects, the consumption of paper, and so on. But can that occur through an orderly social process, avoiding destructive economic upheavals?

The vertical evolutionary direction in the biosomic matrix from biology to machines is transforming the natural environment into an artifact (i.e., an environment altered by us for our purposes). A simple example is the transition in aquaculture from the random process of catching to controlled environments (e.g., enclosures) with controlled feeding.

The making of the environment into an artifact, as we know, is occurring with great speed on a global scale. It changes land and water habitats to suit our immediate needs, transforming accordingly vegetation and animal populations. In the process, however, it also alters atmosphere and climate (e.g., by producing the greenhouse effect or acid rain it eliminates some species and interferes with geological processes through urbanization or seashore modification). A fundamental question (indeed *the* fundamental question) for our future is whether the environment as an artifact—as a necessary machine for the creation and expansion of human niches—can be steered by us in beneficial directions to reduce the rate of damage to the ecology of the Earth's crust while supplying us with our needs or whether the environment as an artifact will be a machine over which we are unable to exert control.

The ability to develop a controlled or *manufactured* environment (the latter being one of intensive artificiality) alters the natural environment in the areas of human intervention on it but can lead to preservation of the remaining portion of the natural environment. Greater geographical concentration of processes such as agriculture, energy production, mineral extractions, or human habitation can be to the advantage of other large segments of the environment that can then remain untouched. This is so thanks to the greater efficiency achievable in a controlled or manufactured environment in carrying out a number of human activities. An example is the economic geography of England, where high-density population concentration in large human habitats has preserved the countryside in its natural state to a much greater degree than in many other nations of similar size and population density.

The extension of the manufactured environment must be limited. Otherwise, its pressure on the rest of the environment may be impossible to resist, resulting in total and irrevocable alteration of the environment. This can have severe consequences for the human species. Suffice it to think of the risks associated with monocultures, which are characteristics of the manufactured environment because their productive efficiency makes the feeding of large populations possible. Thus it is important for the world to maintain a balance among manufactured, controlled, and natural environments. (The latter, however, are natural only in a narrow sense; in a broader sense they, too, (as we have indicated) have become artifacts by being affected by global human-provoked changes in climate, etc.) What the balance may be we do not know at this moment; we know even less how we could bring it about, given the geopolitical fragmentation of the Earth's surface and the pressures of population, agriculture, industry, and tourism. The past 25 years have not been kind to the environment and have permitted only limited attempts to formulate plans for maintaining an environmental balance in a strategic worldwide context.[10]

A strategy of providing a balance in the environment among manufactured, controlled, and natural or quasinatural components—even if we were to know what that balance should be—depends critically on our ability to plan and design systematically truly manufactured environments and on the creation of appropriate geopolitical structures. The intensely developed environments that constitute today's manufactured environments are still a haphazard phenomenon. We need to learn to achieve rationally planned manufactured environments that retain pleasing qualities for human life—the qualities we tend to seek in natural environments, such as variety, beauty, adventure—yet make possible heavy concentrations of dwellings and productive activities. As to the strategies for developing the necessary geopolitical structures, we must find a way to deal with the fact that the political and administrative jurisdictions exerting their authority over the Earth's surface—in the entire spectrum from national to municipal—reflect military or political history, rather than environmental considerations. This makes it that much more difficult to map a strategy of balance among the components of the environment.

A strategy aimed at providing a worldwide balance among manufactured, controlled, and natural environments must also take into account the fact that most of the environments that survive in a condition close to the natural state are asssociated primarily with developing countries. The issue is how to preserve as much as possible such environments and yet avoid penalizing the economies of developing countries. This cannot be achieved without a worldwide flow of economic support to the developing countries—far greater than is the case today—from the developed countries that have overmined the environment. It is as if a hypothetical global environmental administration were to cash IOUs on behalf of the developing countries that the developed countries have written in order to be allowed to alter profoundly their environment—and the entire global environment—in the pursuit of their economic development.

In general, each shift in the focus of the biosomic matrix can be a double-edged sword in terms of human ecology and the environment. For instance, the replacement of natural foods with manufactured ones (e.g. through large-scale manufacturing of proteins) can help create new niches by alleviating hunger, improving nutrition, and making human food supply less dependent on environment. On the other hand, to the extent that it makes larger human population possible and over broader geographical reaches, the shift is environmentally damaging. It can also be damaging to the human population, as in the case with the costly use of soft drinks of little nutritional value by financially poor populations.

SUMMARY

The process by which technology allows humans to carve for themselves new ecological niches by expanding in number, geographical reach, and material comfort is transforming the environment into an artifact. The sooner we adopt this viewpoint, the sooner will we be able to address the question of the extent to which we want or can allow that artifact to be different from its natural condition and the sooner also will we be able to develop mechanisms for controlling it, rather than leaving its future evolution to chance or careless acts.

The view of the environment as an artifact, to be sure, was often accepted in the past without much critical thought. The farm, mani-cured Italian garden, smoke-belching factory, highway, suburb, large urban conglomerate are all manifestations of the environment made into artifact—some pleasing, others not, but all generally seen and often enjoyed as indicators of human progress. In the second half of this century, however, recognition of the enormous and haphazard environ-mental impacts brought about by the explosive development of technol-ogy in the 180 years after the beginning of the Industrial Revolution have convinced us that, as the key agent of the transformation of the environment into an artifact, the technological process must be critically evaluated and planned in all its manifestations.

Until recently, we were handicapped in doing so by narrow characteri-zations of technology that ignored the environment. Technology was thought of in terms that were too simplistic ("the science of the application of knowledge to practical purposes"[11]), too mechanistic ("the totality of the means employed by a people to provide themselves with the objects of material culture"[12]), or too detailed (e.g., in terms of civil, mechanical, electrical, chemical, aeronautical or military engineering). These characterizations and our current lack of a rigorously conceived and widely shared philosophy of technology have led us to deal with the relation of technology with the environment on a piecemeal or episodic and, thus, far from effective, basis. They have allowed us to evade the rethinking of the sociopolitical structures within which technology operates and of the sweeping changes necessary if the impacts of technology on the environment are to be controlled. The most serious consequence, however, has been lack of informed consensus as to the main directions we need to pursue—and the means to pursue them. Namely, how can we establish a balance between our technological power to expand our hold on the globe and the degree to which the environment can be transformed into an artifact without ultimately ceasing to satisfy its essential function of buffer for our lives on the thin crust of the Earth?

The concept of the biosoma—the indissoluble unity of human beings with society and machines—and the viewing of the nature and evolution of technology in the framework of that concept offer us a new perspective on technology's impact on the environment and the nature of the environment itself as transformed by technology. There are four important components of that perspective.

In the first place, we have far greater options than we have ever had to satisfy our human needs by a wide range of trade-offs involving society and machines—delegating to either of these entities functions or needs that before could only be satisfied biologically. An example is the resort to procreation to provide a work force for a family engaged in farming or to perpetuate our business or our memory. Machines, corporations, and charitable foundations can now satisfy, at least in part, that need, with the far-reaching environmental consequences that stem from a reduction in rate of growth of population or from the increase in number and capability of machines.

Second, we are learning the importance of moving away from technological leitmotifs and synergisms highly taxing for the environment, such as the synergism of materials and energy technology, to leitmotifs and synergisms potentially much less taxing, such as the leitmotif of information or the synergism of information technology and biology.

In the third place, because of the new technological knowledge and capabilities we have developed in the past twenty years or so, technology and society are indeed evolving to a point where the dominant factor is information in all its manifestations. Although that evolution has not been spurred by environmental concerns any more than the Industrial Revolution was deterred by them, it has profoundly beneficial environmental implications. This is particularly so if it is accompanied by a massive reduction or transformation of the technological activities shaped by energy and material leitmotifs. Obviously, this cannot occur without relentlessly pressing for that transformation and without restructuring the way in which society operates and uses technology.

Fourth, when the environment, as a result of our interventions, becomes itself an artifact (a machine), we must think of how, as such, we can control it for our purposes without the destructiveness that characterizes our current interactions. That control must be based on a long and broad view of the essential purpose of the environment and can be exerted through an appropriate balance among its controlled, "manufactured," and untouched components.

As a result of these insights, trends, and potentialities, we must conclude that the future of our interaction with the environment will be qualitatively different from that of the past. Its direction will be determined by two factors: the constant creation of new human niches made possible by technology and the changes in the biosomic leitmotifs of technology and society. To the extent that we succeed in changing the leitmotifs in environmentally intelligent directions, we will be able to continue to open up new niches to realize fully our human potential— not by growth of numbers, but by expansion of our capacities, on land, in the oceans, in space.

NOTES

[1] K. Boulding, 1977, *Ecodynamics*, Sage, Beverly Hills.

[2] G. Bugliarello, 1984, "Tecnologia," *Enciclopedia del Novecento*, Instituto della Enciclopedia Italiana, Roma, pp. 381–414.

[3] G. Bugliarello, 1986, "Socio-Technology: The Quest for a New Discipline," *Science, Technology & Society*, No. 53, April, pp. 2–5.

[4] Of course, the energy of biological organisms ultimately comes also from the sun, through food, the photosynthetic process, etc.

[5] For the sake of brevity, I have labeled energy a tangible leitmotif, even if it is not a tangible entity but a concept evolved by humans to explain phenomena such as motion, heat, wind, water power, or electric current, most of which can be sensed.

[6] G. Bugliarello, 1986, *op. cit.*

[7] For example, see *Crop Productivity—Research Imperatives*, 1975, Michigan-Kettering International Conference, Charles F. Kettering Foundation, Yellow Springs, Ohio.

[8] G. Bugliarello, 1987, "Computers and Water Resources Education: A Projection," *Journal of Water Resources Planning and Management*, American Society of Civil Engineers, Vol. 113, No. 4, July, pp. 498–511.

[9] R.N. Coulson, L.J. Folsc, and D.K. Loh, 1987, "Artificial Intelligence and Natural Resource Management," *Science*, Vol. 237, July 17, pp. 262–267.

[10] J.V. Krutilla, 1972, *Natural Environments*, Johns Hopkins University Press, Baltimore.

[11] *Webster's Third New International Dictionary of the English Language*, 1966, G&C Merriam Co., Springfield, Massachusetts.

[12] *Ibid.*

EDITORS' INTRODUCTION TO:

B.W. MAR

Management of High Technology: A Cure or Cause of Global Environmental Changes?

Dr. Mar is a leading expert in systems engineering. In this chapter, he shows how current research efforts to study environmental problems are fragmented by discipline. This fragmentation creates the risk that the assumptions and abstractions introduced in each aspect of the research may result in surprises and unanticipated outcomes when these research results are used by decision makers. He shows how the process of system engineering may provide a systematic and thorough framework for the examination and management of actions that may impact the global environment.

25 MANAGEMENT OF HIGH TECHNOLOGY: A CURE OR CAUSE OF GLOBAL ENVIRONMENTAL CHANGES?

BRIAN W. MAR
University of Washington
Seattle, Washington

INTRODUCTION

High technology or new applications of science have been and will continue to be major forces that change the global environment. The development of building technologies created better structures and transportation systems but also created concentrations of people that exceeded the assimilative capacities of local waters and airsheds. The development of fossil-fueled machinery stimulated the Industrial Revolution and also created more air and water pollution. The development of nuclear power not only provided new weapons and energy, but also introduced radioactive waste disposal problems. The development of pesticides and synthetic fertilizers not only increased food production, but also created threats to the members of the food web. Many new technological advances seem to create another global environmental impact.

The need to manage the application of high technology to benefit humans and still avoid degradation of the global environment is well recognized, but the complexity of this management may create more problems than it can solve. In this chapter, the ability to manage the application of high technology to benefit the populations of the globe will be questioned.

Global environmental issues are characterized by complex interactions between societies and the environment. Any single discipline will have difficulty understanding the complex set of interactions encountered when studying or attempting to manage actions that can cause or reduce global environmental changes. One strategy used for the analysis of complex problems is to reduce the problems to a few critical variables and assume that the other interactions are essentially constant for the purposes of the analysis. Another strategy is to adopt a broad overview perspective where most of the interactions will not appear to be important and the few remaining interactions are those that may yield to the tools of the discipline performing the analysis. The first strategy permits the analyst to retain a desired level of detail while selecting the variables to study, while the second strategy generalizes the problem until only a few variables seem to be important.

This classification of analyses by their choice of the number of variables considered was noted long ago by Weaver.[1] He defined "two-variable" scientists as those who considered only the relationship between two variables. Other scientists were classified as those who studied many variables using statistical methods. Weaver noted the need for scientists who can study or understand the relationships between many variables simultaneously. This chapter synthesizes the results of four major research efforts of the author that explored whether those managing and developing high technology continue to use two-variable science or statistical analysis or whether these individuals have created new and more appropriate tools to address the level of complexity that exists in current global environmental problems.

The first research effort was a study of the management of high technology activities, where change and complexity are the basic characteristics of the process and interdisciplinary teams are commonly created to attack these issues.[2] The second research effort was an examination of 20 large-scale regional environmental studies programs sponsored by the National Science Foundation in the early 1970s.[3] The third research effort was a study of system engineering management concepts used in the aerospace industry in the United States. System engineering concepts address change and complexity by introducing a hierarchial, top-down, comprehensive definition and documentation of the requirements for each component and link in the system.[4] The final research effort to be discussed was an effort to integrate both expert opinion and quantitative knowledge into a computer-aided tool for the design of aquatic monitoring programs.[5]

DISCIPLINARY RESEARCH

The tendency of scientists to specialize rather than generalize may be one of the causes of the failure to recognize the global environmental implications of new technology. INTERSTUDY is an international association to foster the study of interdisciplinary problem-solving groups. It has held three international conferences and was originally dedicated to the examination of interdisciplinary problems in university settings, but has, under the efforts of the author, expanded its thrust to interdisciplinary problems in all organizational settings.[6] This strategy has provided the methods and tools to disaggregate problems into questions that can be examined in detail and to debate and validate new knowledge resulting from this process. The major weakness of this strategy is that the disaggregation process does not ensure that all aspects of any problem will be addressed. An individual researcher will select the problems for study based on various subjective criteria. There is no holistic force to ensure that all aspects of critical global issues will be examined.

At many supposedly interdisciplinary conferences, the experts are separated into disciplinary groups and the problems are also separated by topics such as water pollution, air pollution, desertification, resource exploitation, hunger, and wars. Members of these subgroups may represent a few disciplines, but members of each group do not communicate with members of the other groups. Most participants find it difficult to interact with more than one discipline or topic group, but these participants do acknowledge the complexity of the global issues. Individuals feel the need to isolate their problem to address it, but they do not always provide guidance for the integration of their findings into the larger global issues. When information is collected from other disciplines, it is taken from the literature without thorough examination of the limitations or assumptions associated with the findings.

The specialization by each discipline usually results in the development of a unique language and paradigm that creates communication problems when multidisciplinary or interdisciplinary activities are necessary. This communication problem increases the chance that information from other disciplines may be misinterpreted. The leader of an effort to coordinate disciplinary inputs often imposes his or her language and paradigm on the group. An alternative strategy is to encourage the group to create a language for the group, but this results in communication problems between the group members and others.

Most often, inputs from group members are just collected without integration. None of these alternatives provide an effective strategy for the orchestration of disciplinary contributions to address a given problem and reduce the risk of misinterpreting knowledge.

Many disciplines in both science and technology are organized using hierarchial structures that flow from the general to the specific. For example, the study of the environment of the Earth may be separated into the general areas of the water, land, and atmosphere (subsystems). Each subsystem may be separated into physical, chemical, and biological sectors, and each biologic sector could be divided into elements such as animals, plants, and birds. Each element could be further divided into skeletal, external, and internal parts, and so forth. In some fields of science, the generalists are identified as the "macro" level researchers and the specialists as the "micro" level researchers. These are often referred to as the "lumpers" and the "splitters."

Even within classical disciplines such as chemistry, physics, mathematics, and the biological sciences, the level of specialization has increased to the extent that subgroups within the disciplines have difficulty communicating with each other. New disciplines are continually being created to examine different components of existing disciplines or to integrate a component of a discipline that has been ignored. In the environmental area, environmental engineering evolved from the classic sanitary engineering because the sanitary engineers concentrated on treatment of water supply and wastewaters but not on the atmosphere and the receiving waters. Many other disciplines have spawned splinter groups that focus on environmental problems such as water chemistry, biostatistics, resource economics, ecotoxicology, and environmental health. These disciplines provide added knowledge for their specialty areas, but they do not provide mechanisms to integrate their knowledge with that of other disciplines. The integration process still relies on a search of the literature and the ability of individuals to integrate the diverse information.

It is almost impossible to use the knowledge about the behavior of the general levels of a subject and disaggregate these concepts to describe the behavior of the lower levels. It is just as difficult to aggregate the knowledge of the lower levels to predict the behavior of the upper levels. Each level of study of the same subject seems to evolve its own theories and methods, and these tools have not been effective to aggregate or disaggregate the knowledge generated at a given level of detail.

If disciplinary scientists maintain the posture that unless their particular level of resolution is understood the theories of higher or lower level behavior are questionable, then major interdisciplinary conflicts on research priorities and the value of particular research efforts will result. A counterargument is that it is not necessary to understand all the underlying phenomena in order to apply scientific knowledge for economic gain or societal benefit. This counterargument will be explored further in a later section.

INTERDISCIPLINARY INTERACTIONS

The first study to be discussed was an attempt to examine cases of interdisciplinary interactions outside the university setting to determine if disciplinary conflicts are limited to the university environment.[7] The cases were obtained from papers presented at the Third International Conference on Interdisciplinary Research—Managing High Technology.

High technology activities have been characterized as those efforts where all or part of the products or services are based on new innovations, when rapid technological change creates shortened product life-cycles, where resources are limited to meet compressed development schedules, where the human resources are not fully knowledgeable of the technologies needed to create the end product, and where the environment for the activities is extremely turbulent (changing customer needs, regulations, and new alignment of competitors).[8] These types of activities were selected for research since only interdisciplinary teams could address such programs.

Examination of this international set of high technology case studies revealed that communication and cooperation problems are encountered not only at the interface between individuals from different disciplines (as was observed in the university settings), but also at the interfaces between groups within an organization, between organizations, and between institutions. The case studies of disciplinary and interdisciplinary groups in industry revealed that when groups are organized by traditional disciplines found in the academic setting such as engineering, computer science, and chemistry, there are conflicts among groups on problem-solving approaches. When individuals from these different academic disciplines are grouped into problem-oriented units such as design, manufacturing, tests, or sales, there was little evidence of the disciplinary conflicts observed among units consisting of individuals having a common disciplinary training. The bond of common goals, common management, and a group performance-oriented award structure seems to negate the disciplinary conflicts noted in studies of university groups. The case studies revealed that the interface problems in industry are not limited to interdisciplinary conflicts, but that interfaces among organizations at any level also present major communication and cooperation problems. Differences among sexes, culture, age groups, and employers can also contribute to the perspective and assumptions adopted to address a problem.

It may be useful to employ an organization hierarchy to classify the various interfaces that may need orchestration in the conduct of global environmental research. Brown has proposed a useful point of departure to develop such a hierarchy.[9] He proposes a vertical and horizontal set of interfaces that present potential conflict. The vertical interfaces follow the typical organizational structure where an individual must interface with his or her superior and each superior must interface with his or her superiors. In a large organization, this vertical structure may begin at the top with the chairperson of the board of directors, the president of the company, the vice-presidents, the division managers, the supervisors, the group leaders, and, finally, the individual. The horizontal interfaces become much more complex because individuals must interface with their peers within their own group, but then face a series of other interfaces among individuals in other groups in their unit, among individuals in groups in other units, among individuals in groups in other companies, and so forth. The case study on high technology programs suggests that there are other interfaces at the group level with which an individual must cope. If the group consists of individuals from different disciplines, there can be the disciplinary interfaces discussed earlier, plus the age, sex, and culture interfaces that were observed in the high technology case studies. Each of the interfaces can present different problems; improving communication and cooperation at these interfaces requires formal attention.

The recognition that there is a matrix of potential interface conflicts that must be faced in the orchestration of major research or development activities is the first step in addressing global environmental problems. Independent research concentrating on fragments of the global environmental issues will not ensure that global environmental problems will be identified in a timely manner or that solutions for these problems can be developed. The management of these organizational and individual interfaces must be a conscientious part of the process of global environmental studies. If this type of management cannot be implemented, the second best action would be to have researchers include in their documentation a section describing the researchers' assumptions and perceptions of the problem. The unstated assumptions are usually the ones that can cause problems in integrating the new results.

ENVIRONMENTAL RESEARCH

Environmental researchers may feel that the conflicts identified earlier are not applicable to those engaged in large-scale environmental research programs. The next set of studies examined groups that conducted studies of regional environmental issues to determine if interface conflicts occur in environmental research programs. The term *region* included areas as small as a fraction of a large urban area to areas as large as the United States. No studies of global environmental problems were included in this set of studies.

In the 1970s, the National Science Foundation funded about 20 million dollars worth of regional environmental research programs under their Research Applied to National Needs (RANN) program. The author received funds to evaluate the interdisciplinary process used in these programs as well as to examine the role of large-scale modeling in facilitating the interdisciplinary interaction among group members.[10] These programs were funded for several years; involved disciplines from engineering, economic, and ecological fields; and received funds on the order of hundreds of thousands of dollars. There were about 20 programs studied. Each involved construction of a model to describe the impact of technological development on the regional environment. The research included the collection of data on the environment of the region, the formulation of models relating causes and effects, and the development, calibration, verification, and application of the models.

If program success is measured by the completion and use of the models, then programs that selected narrow issues and used higher levels of detail were more successful than those that did not. This suggests that efforts that followed disciplinary paths were more successful. (They actually developed a model that was used.) There were many programs studied in this group that had a difficult time developing a model. They allocated too much of their resources toward data collection. The disciplinary conflicts related to the type and quantity of data that were needed to develop a model were observed to be a major cause of excessive data collection. If all researchers must develop knowledge adequate to support the greatest level of resolution demanded by any team member, there may not be adequate resources or time to do this. Another problem observed in the environmental research programs involving individuals from many disciplines was that neither the project leader nor the team members could agree upon clear objectives or criteria for their models. When a submodel was identified by an individual, it was successfully completed and in many instances was adopted by disciplinary peers at other institutions, but the submodels were not well integrated into the program product. There were also submodels that could be defined, but the state of knowledge was inadequate to provide enough information to develop or calibrate the submodel. If the program waited to develop the required knowledge for that submodel, the rest of the program could not be completed.

This second set of studies indicated that problems encountered in the conduct of interdisciplinary environmental programs are similar to problems observed in most interdisciplinary efforts. Individuals seem to have trouble defining a common problem or defining their contribution to the common problem. Rather than using their knowledge to address a problem at the appropriate level of detail, the tendency is to solve the problem at the disciplinary level of resolution and hope that someone can translate this information to the problem level. This often results in some issues receiving too much study and other issues being ignored. There are few researchers that are willing to adjust their efforts to address the problem at the problem level.

A key to more effective studies of global environmental change may be the ability to identify the set of variables that control a global phenomenon of interest and to relate this subset of the global environment to the behavior of the total global environment. This suggests the need for a hierarchy of research efforts that can be disaggregated downward and aggregated upward for problem definition and knowledge synthesis activities.

SYSTEM ENGINEERING

The third set of studies is an examination of the concepts and process of system engineering to determine if these may provide a vehicle to organize and structure global environmental research. System engineering has been adopted by the aerospace industry in the United States as a management tool to cope with conflicts in program development under conditions of high complexity and rapid change. While system engineering has had success, there has not been adequate attention given to the education and development of new system engineers. Therefore, an increasing shortage of individuals with knowledge to practice system engineering is developing. Lockheed has published one of the few documents that describe the system engineering process.[11] Most texts on system engineering concentrate on the modeling, scheduling, and optimization concepts, but not the actual integration or management process. This set of studies of system engineering management will illustrate the practice of system engineering in the management of large, complex sets of activities.

System engineering is a detailed iterative process of defining a hierarchy of goals, preparing specifications describing how these goals can be met, and then verifying that the specifications have been satisfied. The process is characterized by a thorough examination of all feasible alternatives to meet each goal, optimization to select the best specification as well as the best solution, and a structured and disciplined documentation process of all actions and decisions. When system engineering has been effective in the integration and orchestration of a program, it has required at least 10 percent of the resources to provide these functions adequately. Since system engineering is in fact a discipline, with its own language and paradigm, system engineering is no better received than any other discipline. System engineers, however, recognize that they must develop communication and negotiation skills if they are to be effective.

In high technology programs where system engineering has been practiced, there are usually five major iterations of analysis to define goals and specifications. Each iteration provides an increasing level of resolution and focuses on a lower level of the system. These iterations are separated into 1) a concept/feasibility phase where the client's needs are defined, and general solution concepts are proposed, 2) a system definition phase where alternative concepts are explored and a system or solution is selected, 3) a development phase where technical development necessary to implement the solution is conducted, 4) a full-scale design and production phase where the solution is designed and produced, and 5) an operation where the system is used, maintained, or modified.

In each phase, the same sequence of steps is used to translate goals into specifications and solutions. The solutions for one level become the goals for the next lower level, creating a pyramid that can facilitate linkage between any two parts of the system. The steps used are:

1. Define the goals and objectives of this step
2. Identify the criteria to be used to judge when the goals have been met
3. Determine the constraints and resources available to perform this step
4. Given the information developed in the first three steps, develop alternatives that will meet the goals

5. Evaluate the alternatives and conduct sensitivity analyses to define the best solution
6. Translate the solution into goals for the next lower level of this system
7. Thoroughly document these steps and enter this information into the program data base

This stepwise, iterative process provides a rigorous audit trail of analysis and decisions that can be examined and traced to accommodate new information and changing goals. Each iteration develops more detail and requires more knowledge, but trade-offs must be made at each step to remain within resource constraints and still meet criteria for goal achievement. While the expense and effort required to implement the system engineering process is high, these costs have recently been drastically reduced by the application of personal computers for word processing, data management, and system analyses.

System engineering has provided a forum and structure for the orchestration of interdisciplinary and interorganizational programs. It requires that participants adopt a top-down approach to any program (general to specific) and features a process where goals must be clearly stated and as many alternatives identified and evaluated as possible before making a decision. In this process, participants can contribute to the goal definition and propose their contribution to the solution, but their concepts will be examined in relationship to the program goals rather than personal or system element goals. In this process, the performance of the overall program dominates over personal or system element goals, moderating the disciplinary tendency to place personal goals above the system goals.

System engineering has been able to reduce the fragmentation in the aerospace industry approaches to complex programs, but much still remains to be accomplished. The aerospace programs have addressed complex systems where each system component can be designed and tested. This is not the case in global environmental research, and one can question whether system engineering can be applied to environmental problems.

SYSTEM ENGINEERING OF ENVIRONMENTAL PROBLEMS

The final set of studies to be presented in this discussion is a description of an effort to apply system engineering to the management of a major research program to develop a method to design environmental monitoring programs for electric power generating utilities. This study is presented in support of the hypothesis that system engineering or some other form of formal integration is a needed component of any effective study of global environmental problems.

One of the thrusts of this work to improve environmental monitoring programs was a survey of almost 100 experts in the field of monitoring design.[12] This survey indicated that much of the information supporting this field was subjective and nonquantifiable. Many existing monitoring programs had ill-defined goals, and some lacked goals altogether. Hypotheses to be tested using the monitoring data were usually not formulated until after the data were acquired. The optimization of monitoring programs based on cost or the quality of information produced was not of prime interest; rather, the customs and practices of participating diciplines dictated the design. Engineers collected water quality and hydrologic data, biologists collected phytoplankton data, fishery biologists collected fish data, and botanists collected plant data. The survey results supported the hypothesis that the design of environmental monitoring programs was fragmented by disciplines and that the disciplines were not intergrating the programs or concentrating on program goals.

The application of system engineering to the task of cost-effective aquatic monitoring design was performed using an interdisciplinary team of administrators, practitioners, engineers, statisticians, and ecologists. The first step was to define the needs and the goals of environmental monitoring programs. Using the iterative system engineering process, the top level goals included needs to satisfy regulatory requirements, public concerns, and environmental and economic concerns. The next iteration defined the needs to establish baseline data, needs to determine trends, and needs to demonstrate that changes did or did not occur. The third level iteration defined the specific causes or target populations that needed to be observed. Once a specific hypothesis that must be tested could be specified, then alternative sampling methods could be examined and sampling programs optimized.

The next steps included the collection of available information to support the goal definition activities, and the program design optimizations. One of the surprising findings was that the costs of sampling and analysis are not documented for most programs, and little information about costs existed for optimization purposes. Most disciplines that were active in monitoring programs worked with a fixed budget and were not required to justify the cost-effectiveness of their efforts. The usual practice was to allocate the monitoring resources to the disciplines in a subjective manner; each discipline attempted to provide what they could. Another observation was that the statistical analysis of the data was not started until after the data collection was completed, and assumptions used in the design of the programs were not verified so correction could be made.

A top-down integration process to coordinate interdisciplinary team efforts is necessary to monitor the aquatic environments adjacent to large electric generation facilities in a cost-effective manner. Such a process fulfills the need to coordinate the knowledge of engineers, ecologists, economists, and statisticians in a unified effort and provides a mechanism to manage large-scale environmental research. It is but a single observation, but suggests that system engineering may provide a framework to improve the communication and reduce the interface conflict in efforts to study environmental issues. This research is an initial step in the process to demonstrate that system engineering may be a process that can facilitate communication and reduce conflict encountered in addressing complex interdisciplinary issues.

When discussing the application of system engineering to the study of environmental problems at any level, there is an immediate negative reaction that the environmental systems are so different from aerospace systems that the two situations are not comparable. While the goals of these two activities are quite different (one is to develop a high technology based piece of hardware that consists of parts that can be designed and fabricated and the other is to understand or predict the response of a natural system that is only partially understood), both activities share the common concept that each system is hierarchial in nature with complex feedback relationships. Further, both activities involve individuals from different disciplines and from different organizations who must eventually integrate their contributions to produce the desired outcomes.

SUMMARY

In summary, the incremental management of technology using disciplinary-based attempts to mitigate global environmental changes may increase rather than decrease the magnitude of the problems addressed. Will the reduction of solid and hazardous wastes using incineration increase air and water pollution? Will use of nuclear energy instead of fossil fuels reduce air pollution but increase the risk of cancer? Will the fertilization of crops to increase food production pollute the groundwaters? Too often the solution of one environmental problem creates another. Many case studies of interdisciplinary research in both university and industrial settings suggest that single objective problem solving is still common to disciplinary groups. The inability to create a structure that can integrate the disciplinary contributions and that will recognize the broader impacts of a proposed action can be traced to conflicts among disciplines, groups, and organizations. Disciplinary or organizational bias will introduce hidden agendas and narrow-minded solutions into recommended actions. The introduction of environmental impact assessment has been a necessary step to introduce views of many disciplines into the management of technology, but the methodology to synthesize and analyze these disciplinary inputs is weak.

An examination of high technology management has suggested that the system engineering process is one of the few techniques that can cope with the high degree of complexity and change encountered in global environmental issues. While system engineering has demonstrated a limited effectiveness in aerospace system development such as the development of complex space shuttle systems, it has not avoided major catastrophes. Global environmental systems are known to be more complex than even aerospace systems, and there is less known about these natural systems. Even so, the iterative top-down approach of system engineering may provide a tool for orchestrating the disciplinary inputs that are offered for the management and understanding of complex global environmental systems.

If new technologies are to help solve existing global environmental problems, improved mechanisms must be found to ensure that the broadest possible examinations of the impacts of these new technologies are articulated and evaluated prior to their widespread use.

NOTES

[1] W. Weaver, 1948, "Science and Complexity," *American Scientist*, Vol. 36, No. 1.

[2] B.W. Mar, W.T. Newell, and B.O. Saxberg (eds.), 1985a, *Managing High Technology—An Interdisciplinary Perspective*, North Holland Publishers.

[3] B.W. Mar, 1978, "Regional Environmental Systems—Assessment of RANN Project," University of Washington, Department of Civil Engineering for the National Science Foundation Project NSF/ENV. 76-04273.

[4] B.W. Mar, 1986, "Increasing the Supply of System Engineers," submitted to American Society for Engineering Education.

[5] B.W. Mar, D.P. Lettenmaier, R.R. Horner, J.S. Richey, R.N. Palmer, S.P. Millard, and M.C. MacKenzie, 1985b, "Sampling Design for Aquatic Ecological Monitoring," University of Washington, Department of Civil Engineering for Electric Power Research Institute, Palo Alto, California.

[6] See R. Barth and R. Steck (eds.), 1980, *Interdisciplinary Research Groups; Their Management and Organization*, University of Washington, Seattle, Washington; P. Diesing, 1982, *Science and Ideology in Policy Science*, Aldine, New York; S.R. Epton, R.L. Payne, and A.W. Pearson (eds.), 1983, *Managing Interdisciplinary Research*, John Wiley and Sons, Chichester.

[7] Mar, Newell and Saxberg, 1985a, *op. cit.*

[8] *Ibid.*

[9] L.D. Brown, 1983, *Managing Conflict at Organizational Interfaces*, Addison-Wesley Publishing Co., Reading, Massachusetts.

[10] Mar, 1978, *op. cit.*

[11] Lockheed Missiles & Space Company, 1983, "System Engineering Management Guide," Defense Systems Management College, Fort Belvoir, Virginia.

[12] Mar et al., 1985b, *op. cit.*

EDITORS' INTRODUCTION TO:
M.F. CASWELL
Better Resource Management Through the
Adoption of New Technologies

No matter how important a new technology or innovation is designed to be, its value in improving environmental quality or resource allocation depends on whether it is used. Dr. Caswell presents an economic framework to describe how the adoption pattern of a new technology for resource management might be predicted. Two types of remote sensing technologies, aircraft and satellite, are used to illustrate the theory. The appropriate technology for a country would be the one that would give the highest net benefits, and these benefits will depend on environmental and cultural characteristics of that country. How many countries will adopt a technology will depend on the distribution of these characteristics. One technology may not be the best for all circumstances, so some may use a new technology and some may not. The environmental impact (either positive or negative) that would result from the introduction of a new technology will therefore depend on how many countries will use it. The effectiveness of policies designed to encourage the use of a resource management innovation can be evaluated using the framework presented here.

26 BETTER RESOURCE MANAGEMENT THROUGH THE ADOPTION OF NEW TECHNOLOGIES

MARGRIET F. CASWELL
University of California
Santa Barbara, California

INTRODUCTION

New technologies have been developed in recent years that were designed to help solve major environmental problems. Many of these innovations, however, are not widely used, and the reasons behind this lack of adoption need to be understood. The decision to use a particular technology represents a complex interaction among environmental, social, and technological factors. In this chapter, I suggest that in some cases it might be best if the technology is never fully adopted despite its apparently superior characteristics.

There are few economic models that can be used to predict the likelihood that a particular technology will be used or how that likelihood will be affected by economic conditions. What is needed is a theoretical framework that encompasses large- and small-scale technologies, and includes users from regions with different physical and cultural characteristics. A starting point to develop this theory is recent work concerning the adoption of irrigation systems by individuals with farms of different land qualities; this analysis sheds new light on the general technology-adoption process.[1] An expansion of this approach will be used here to analyze adoption choices for remote sensing technologies by countries with heterogeneous topography and weather patterns.

421

In order to present briefly a theoretical framework for describing those factors that influence the adoption of new technologies, aircraft and satellite remote sensing systems will be used as examples. Finding solutions for some important global environmental issues may require that many nations act concurrently to obtain information. Many who have thought about this situation have suggested that satellite remote sensing could provide crucial information for providing an improved understanding of the area, extent, nature, and temporal dynamics of environmental parameters related to these issues, both at the global and national levels, if this system were actively employed by many nations. Few countries, however, have adopted this technology. Basic economic theory can be used to explain why adoption has not occurred and, more importantly, why there may never be full adoption of this technology.

Remote sensing is the observation and recording of energy by a system that is not in physical contact with the object or phenomena under study. Remote sensor systems on aircraft or satellites "sense" energy output (reflected, emitted, transmitted, and scattered) from features in the environment.[2] Remotely sensed data have been readily available since the advent of the airplane-mounted camera, but the modern era of remote sensing technology is generally taken to have begun with the launch of the first U.S. Earth Resources Satellite in 1972. Prior to this event, aerial observation from aircraft provided a variety of services, from basic mapping to the detection of crop damage, and remains a useful technique. The comparatively small scale and local information available from aircraft can be provided by private entrepreneurs and effectively used by individuals. Airplane-mounted sensors can be highly sophisticated, however, and there have been many developments in this technology in the last several decades.[3]

Remote sensing using satellite imagery can improve the quality and quantity of data or contribute data never before available. For either aircraft or satellite technology, data acquisition is unselective in that everything to which the sensor responds will be recorded. Therefore, part of the remote sensing information "package" also includes methods of data processing and analysis for detection, identification, categorization, measurement, error analysis, and accuracy assessment.[4] Developers of remote sensing technology often sell the "hardware" and "software" as a bundle with other complementary services.[5] This is especially true when the customer is a less developed country and may lack trained personnel.

Once the data have been processed, the information can serve as an input in user-developed models to monitor, survey, or explore the environment.[6] A Geographic Information System (GIS) may merge the remote sensing data with data generated by more conventional means and convert these data into a form that can be used by planners to review options or develop management policies. The potential for using such techniques and tools for studying the dynamics of environmental change has only begun to be understood. Recent research has focused on estimating global biomass, atmospheric carbon dioxide concentrations, deforestation/desertification, and transfrontier pollution problems, among others.

There is currently a concern, however, that the potential benefits of the satellite system are being lost due to the lack of adoption of remote sensing technology.[7] For developing countries in particular, the benefits of efficiently obtaining basic information about their resource base would seem to be high. Some experts have suggested that too many countries depend on airplane-mounted sensors for image acquisition when they should adopt the more "modern" systems available from satellites. The following analysis will show that the heterogeneity of environmental characteristics among countries may explain why there may never be full adoption of the modern technology. It will also be shown how an understanding of the adopter's environmental character-istics and social goals will help the developers of a new technology to encourage adoption and speed the process of diffusion. Such a frame-work can be used to assess the potential benefits of many technologies designed to lessen environmental degradation or to manage resources.

THE ADOPTION DECISION

The choice problem for a country can be explained in economic terms.[8] For this analysis, it will be assumed that there is one level of information that is needed by all countries.[9] For instance, each may need an accurate map of water, settlements, agriculture, forest, grasslands, and desert. Each country would benefit from this inventory and these benefits can be measured in monetary terms.[10] The benefits may differ among countries and may be associated with social goals such as deforestation detection or mineral exploitation.[11]

Two major determinants of cost and accuracy for remote sensing data processing are topography and cloud cover, and these factors differ widely among regions. Steep slopes and rugged topography increase the cost of image processing for both aircraft and satellite technologies, but proportionally more for satellite imagery than for data collected by airplane sensors. The same relationship holds for a higher average level of cloud cover because, as the probability of obtaining a clear and complete image goes down, the expected cost of obtaining a given level of information would increase. Since acquisition dates for satellite imagery must often be set in advance of short-range weather forecasts, the lower the probability of obtaining a clear image, the more images must be scheduled. It is easier to schedule an overflight by an airplane according to current conditions.

The appropriate technology for a country would be the one that would give the largest net benefits from gaining the desired level of information.[12] These net benefits are merely the total benefits less the costs (i.e., what economists would term *social profits*). The total benefits derived for a country would not depend on the way that the information was obtained, so the maximization of net benefits would entail the choice of technology that minimized the costs. Many technologies can be compared, but in the interest of simplicity, this analysis will assume that there are two primary types of remote sensing technology for image/ information acquisition: an airplane-mounted system (which will be denoted as t for traditional) and one with satellite-mounted sensors (which will be denoted as m for modern).[13] One would not, however, consider the former to be "old fashioned." Each is assumed to include the image processing appropriate for the information needed. For the following analysis, the costs of either technology are in annual terms, so there is no need to distinguish between system ownership and contracted services.[14]

The topography and the meteorological characteristics of a country can be summarized by a linear index, q, which would denote environmental quality with respect to the effectiveness of using the aircraft-mounted system.[15] The number assigned to q for any country would be derived from the long-term average cloud cover and the weighted average of the topographic features. The measure of quality could range from a low value of 0 (constant cloud cover and rugged terrain) to the highest value of 1 (no clouds and level land), and each country would be associated with a particular level of q.[16] For instance, Ethiopia might have a quality index of .45 (mountainous) while for neighboring Somalia on the coastal plain, $q = .80$ might be a reasonable index of environmental characteristics.

For each succeeding level of quality, the relative costs between the two technologies will differ. The relative advantage of the satellite system decreases as the quality of environmental conditions decreases. Although all countries with a lower value for q will have higher costs for obtaining the desired level of information, the costs of satellite-acquired data will have increased more substantially than that gathered with airplane-mounted sensors. This is because the costs of image acquisition and processing for the traditional technology are less sensitive to environmental characteristics than are the costs for the modern system. It is easier with airplane-mounted sensors to adjust for cloud cover and topography by changing timing and altitude. The differences in costs can be significant. For instance, in 1980 the costs for acquisition and processing land inventory data for a 15,000 km^2 area in Bolivia were $2.55/km^2 for aircraft and $40.12/km^2 for Landsat.[17]

Figure 26–1 illustrates the costs of obtaining the desired information for an array of environmental characteristics. It is assumed that the cost at the highest level of quality would be larger for airplane- rather than satellite-acquired images because, under perfect conditions, a single satellite image acquisition (with processing) for an area would be less costly than a series of overflights using airplane-mounted sensors.

Although satellite remote sensing (the modern technology) may be less costly for countries having the best environmental characteristics, such a cost advantage may quickly deteriorate as the quality index falls.

Because the cost functions for each technology have different characteristics with respect to the quality index, the relative net benefits that a country would receive (the total benefits from gaining the information less the technology costs) will also depend on the environmental characteristics. These properties are depicted in Figure 26–2 for a representative level of total benefits (shown as BB) and an array of qualities. If the traditional system is never more costly than the modern one and the

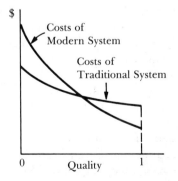

FIGURE 26–1   The costs of acquiring one level of information using modern and traditional technologies as a function of environmental quality.

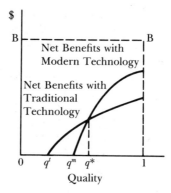

FIGURE 26–2 The net benefits that would be gained from acquiring one level of information using modern or traditional technologies for each value of environmental quality.

benefits of the information exceed those costs, then only airplane-mounted sensors would be used. Analogously, if the traditional system is always more costly, then only satellite-acquired imagery would be used. One can expect that more often there will be a trade-off between the two technologies.

There are several values of the quality index that are of particular interest (Figure 26–2). The point q^t represents the lowest level of environmental characteristics where the total benefits of information outweigh the costs of getting that information using the traditional (airplane-acquired) technology. Similarly, q^m is the lowest level of environmental characteristics where the net benefits of using the modern system is positive. It is also interesting to note that the traditional technology (although not as sophisticated as the modern system) is profitable over a wider range of environmental circumstances (from q^t to 1 rather than q^m to 1). q^* is the "switching quality"—that value of the index where the net benefits of using either technology are the same. The decision maker of a country with a quality index equal to q^* would be indifferent as to which technology was used.[18] The appropriate technology to use would depend on the characteristics of the country making the choice.

Despite the advantages of the modern technology under ideal circumstances, the traditional system is more effective when the environmental characteristics are not favorable. For countries with poor environmental characteristics (i.e., a value of q less than q^t), neither the traditional nor the modern system would be adopted because the costs would outweigh the benefits. If the information is obtained, it would not be through remote sensing techniques but through ground measurements or anecdotal evidence. For countries that have a quality index with a value between q^t and q^m, only the use of the traditional system would result in positive net benefits (Figure 26–2). For countries with an environmental index with a value greater than q^m (the quality where the total benefits of using the modern system equal the costs), the use of either of the technologies would result in positive net benefits and the decision maker would choose the system that obtained the desired information at the highest level of net benefits.

On the basis of net economic benefits, the traditional system would continue to be used by countries with a quality index having a value between q^t and q^*, while countries with environmental characteristics greater than q^* would switch from traditional to modern technology since the net benefits of obtaining the information with a satellite system are greater than if airplane-mounted sensors were used.

Changes over time in costs and technological characteristics would affect the relative net benefits of the two technologies and hence the adoption pattern. Higher technology costs for satellite systems will reduce the number of countries that will adopt that technology, while improvements in the effectiveness of that system in obtaining data under adverse environmental conditions will lead to increased adoption. An increase in the cost of the traditional system (which would occur, for example, if the price of aviation fuel increased) would be equivalent to lowering the net benefits curve for airplane technology in Figure 26–2. The lowest quality that would result in positive net benefits for that system, q^t, would be higher than before, and the switching quality would be lower. Thus, the number of countries that use airplane-mounted sensors to obtain the desired level of information would decrease while the number that would adopt the modern technology would increase.[19]

The net benefits of using either technology will increase if there is an increase in the value of the desired information. For example, the benefit of having a basic resource inventory would increase if the market value of the resource (e.g., timber or minerals) went up or if the social value of wildlife habitat increased. Since total benefits are neither a function of the technology chosen nor the value of the quality index, an increase in benefits would be equivalent to a parallel shift upward of the two net benefit curves in Figure 26–2. The switching quality, q^*, would not change, so the number of countries that would adopt the satellite system would remain the same. The number of users of the airplane-acquired imagery would increase, however, because q^t would be lower.

AGGREGATE IMPLICATIONS

It has been suggested that environmental characteristics are an important factor in determining the likelihood of a country adopting a particular remote sensing technology. The next step in the analysis is to determine the overall use of the two technologies. The framework presented here can be applied to innovations with either positive or negative environmental effects. The total number of countries that adopt a modern resource management technology will determine the global impact of that technology. For instance, if a system that radically reduced the emissions causing acid rain were invented, the resulting reduction in acid rain damage would depend on how many power plants installed the technology or how many countries mandated installation.

The total number of countries or regions that eventually adopt a technology will affect global resource management. It might not be in a country's self-interest to adopt the modern (satellite) technology, and it may use neither technology. The adoption decision will depend, in part, on the environmental characteristics of the country. If the distribution of these characteristics is known, then the total number of users can be determined and cost-effective policies could be developed to encourage adoption.

Figure 26–3 depicts a hypothetical frequency distribution of the number of countries that have a particular index of environmental characteristics.[20] One can look at the adoption process in two parts: Figure 26–3a represents the case when only the traditional system is

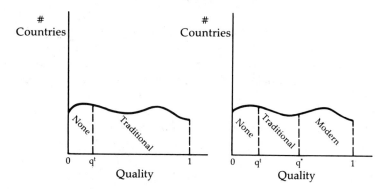

FIGURE 26–3   Using a hypothetical frequency distribution for the number of countries that have environmental characteristics q, the total number of countries using modern, traditional, or neither technology is shown for the cases when a) the modern technology is not available and b) both technologies are available.

available; Figure 26–3b represents the period after the introduction of the modern technology. Although it is clear that there have been many technological advances made over the years to improve the traditional technology, it is assumed that the characteristics of that technology remain the same over the two periods under discussion. Before the availability of satellite imagery when only airplane-mounted sensors were available, the potential adopters would fall into two groups: those likely to adopt and those who would not find it profitable to use the traditional remote sensing technology.

The size of these two groups are represented by the areas under the curve in Figure 26–3a. The total number of countries with an environmental quality index less than q^t is the area to the left of the graph designated by *None,* and the number that adopt the traditional technology is represented by the area to the right.

After the introduction of the modern technology, the adoption pattern would change. Countries would have another choice to analyze, so there would be three groups within the set of potential adopters: those that still adopt neither remote sensing technology; those that retain the traditional technology; and those that switch from the traditional to the modern satellite system (Figure 26–3b). The number of nonusers remains the same as in the preceding case (the area to the left of q^t). The area under the frequency curve between q^t and q^* represents the number of countries that do not switch; the area to the right of the switching quality, q^*, shows the level of adoption for satellite technology.

The example shown in Figure 26–3 depicts one particular pattern of frequencies (uniform) and the resulting number of users for each technology. The number of users might change substantially if the frequency distribution of countries with each value of environmental quality were different. If, for instance, there were few countries with near-perfect conditions, there may be very few countries in total that would switch.

Figure 26–4 illustrates how the distribution of quality would affect the extent of adoption. In Figure 26–4a, there are many countries with good environmental characteristics (with respect to the use of remote sensing technologies) and none with the lowest levels of quality. Figure 26–4b shows a frequency distribution of quality for the case where there are few countries with excellent environmental characteristics and the majority have qualities that range from middle to poor. In both cases, the critical levels of the quality parameters, q^t, and q^*, are the same (and they are the same as those depicted in Figure 26–2). To emphasize the

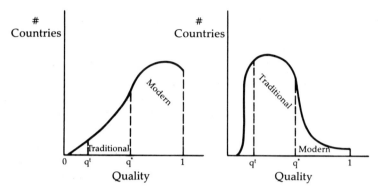

FIGURE 26–4   The importance of the real nature of the frequency distribution for the number of countries that have environmental characteristics q is shown by comparing the relative number using modern, traditional, or neither technology a) when most countries have excellent environmental characteristics and b) when most countries have only fair characteristics.

importance of the distribution, Figure 26–4 is drawn so the total number of potential adopters (the total area under the curve) is the same for each case. The level of adoption of the modern technology for the case shown in Figure 26–4a is far greater than that in Figure 26–4b. In the former case, the area under the distribution curve to the right of q^* is significantly larger than in the unimodal case of Figure 26–4b. The number of countries retaining the traditional (airplane-mounted) sensors is much smaller for the example in Figure 26–4a than that shown in Figure 26–4b.

Since there may be a global benefit from encouraging the adoption of a modern technology, it would be important to assess the effect of different policies on the total number of countries that use the satellite system. The theoretical framework described in the preceding section can be used for policy analysis. For instance, the United States has underwritten some of the costs for countries that use Landsat. Such a subsidy or investment credit serves to lower the costs of the new system and thus increase the net benefits of that technology. This is equivalent to raising the satellite net benefit curve in Figure 26–2, which results in a lower switching quality, q^*. Countries with environmental characteristics between the old and new values for q^* would be encouraged to change from traditional to modern technologies. Whether that change would be significant, however, depends on the number of countries with those characteristics. Using Figure 26–4 as an illustration, it can be seen that an equal reduction in the switching quality, q^*, would result in fewer new adopters for the case shown in Figure 26–4a than for that in Figure 26–4b.

Other policies (and other resource management technologies) can be analyzed in the same way. As another example, a technology developed to reduce emissions that cause acid rain may be much more effective when installed at new electricity-generating plants than at older ones. In this case, the measure of quality, q, would be related to the age of the equipment, with the newest designated by $q = 1$. The plant manager would choose the least costly technology given that a certain level of emission reduction must be maintained. Plants with equipment older than q^* would not install the new technology. The amount of damaging emissions that would be reduced with the modern technology would depend on the distribution of the ages of plants. The United States might have a unimodal distribution (similar to Figure 26–4b) while Japan, which has many new plants, could be characterized by the distribution in Figure 26–4a. The appropriate technology for emission reduction may be different for the two countries. More plants might find it profitable to switch to the modern system in Japan than in the United States. This illustration was presented to emphasize the importance of knowing the relative advantages of the new technology and the distribution of environmental (or physical) characteristics. One technology may not be best under all circumstances.

SUMMARY

The framework presented here can be used to: 1) direct research and development efforts toward resource management technologies with a wide range of effectiveness; 2) determine the extent of adoption for existing technologies; and 3) assess the impacts of policies designed to encourage the spread of the technology. Although the primary example used in this chapter addressed a hypothetical case involving the adoption of remote sensing technologies, the theory can be applied to a broad class of technologies that could have a global environmental impact. The minimum quality at which the traditional system will be used (q^l) and the quality beyond which the modern technology will be adopted (q^*) are functions of both economic and technological parameters; hence, the effects of technological adoption may be studied with respect to changes in these characteristics.

Providers of space systems could use the results of an empirical analysis using this framework to determine the benefits of improving the modern remote sensing technology, such as lowering the baseline cost or decreasing the sensitivity of the costs to declining environmental characteristics. The relative merits will depend on the distribution of countries in the pool of possible adopters.

New technologies may determine our success as global resource managers. These innovations may reduce the level of toxic residuals that are produced; they may reduce the amount of natural resources needed for production; they may permit us to monitor conditions more accurately; or they may process information more quickly. The contribution of any new invention or technique, however, depends on the extent to which it is used. The effectiveness of a policy designed to encourage the adoption of any innovation will depend on technological characteristics, economic costs, and social goals. None of these components should be studied in isolation. Despite the apparent superiority of a technological innovation, it may not be in society's best interest to urge complete adoption (i.e., it may not be the appropriate technology for all cases). As we seek new ways to ensure environmental quality and improve our ability to manage the resources of the biosphere, it will be important to consider all impacts of our decisions.

APPENDIX

The following are the basic assumptions necessary to derive the technology choice model. There is only one level of information that is sought by each country, and the benefits (B) that will be gained from that information do not depend on the technology used. There are two remote sensing technologies: airplane-mounted sensors ($i = t$ for traditional) and satellite systems ($i = m$ for modern). q is the index of environmental characteristics and is normalized to be equal to the effectiveness of the traditional system. $h(q)$ is the effectiveness of the modern technology when environmental characteristics are q

$$h_t(q) = q \text{ and } h_m(q) \leq q \text{ for } 0 \leq q \leq 1$$

$h'(q) > 0$ and $h''(q) > 0$, where primes denote derivatives with respect to q.

It can be seen that the disadvantages of satellite systems increase as the quality of environmental conditions decrease:

$$[h(q)/q]' = 1/q \ [h'(q) - h(q)/q] > 0$$

Figure 26–1 was derived from the following assumptions about the cost function:

$C_i(q)$ is the cost of obtaining the desired level of information with technology i when the environmental characteristics are indexed as q. Therefore, $C_i(q) = C_t(q)$ for the traditional system and $C_i(q) = C_m(h(q))$ for the modern system.

$$C'_i < 0, \; \partial C_m/\partial h(q) < 0, \; C''_i > 0, \; |C'_m| > |C'_t|, \text{ and } C''_m > C''_t$$

Figure 26–2 was derived from the following assumptions about the net benefits ($NB_i(q)$) of obtaining the desired level of information using technology i.
$NB_i(q) = B - C_i(q)$, therefore, $\partial NB_i/\partial B > 0$—net benefits will increase as total benefits increase.

$$NB'_i = - C'_i > 0 \text{ and } NB''_i = - C''_i < 0$$

so

$$NB'_t < NB'_m, \; NB''_t > NB''_m, \text{ and } NB_m(1) > NB_t(1)$$

Also, $\partial NB_m/\partial h(q) = - \partial C_m/\partial h(q) > 0$, so that any improvement in the effectiveness of the modern technology will result in increased net benefits. Comparative static results were used to analyze the adoption pattern effects that would occur from changes in economic or technological conditions.

The equilibrium condition $NB_t(q^t) = 0$ is totally differentiated to yield

$$C'_t \, dq^t = dB - dC_t$$

Therefore, $dq^t/dB = 1/C'_t < 0$. When benefits increase, the range for use of the traditional system will increase.

Also, $dq^t/dC_t = - 1/C'_t > 0$. When the cost of the traditional system increases, fewer countries will use that system. The same analysis holds for q^m.

To analyze changes in the switching quality, the equilibrium condition $NB_t(q^*) = NB_m(q^*)$ is differentiated totally to yield

$$(C'_t - C'_m) \, dq^* = dC_m - dC_t$$

Note that with the above assumptions about total benefits that $dq^*/dB = 0$. A change in benefits will not affect the switching quality, so the number of users of the modern technology will remain the same. However, $dq^*/dC_t < 0$ and $dq^*/dC_m > 0$. If the costs of the traditional system increase (all else equal), more countries would switch to the modern system. The opposite would happen if only the costs of the modern system would increase. For an equal increase in both costs (e.g., as the result of a tax), the switching quality would remain the same.

Hence, the number of users of the modern technology would not change. Since, however, $dq^t/dC_t > 0$, there would be fewer users of the traditional system.

The aggregate conditions shown in Figure 26–3 were derived with the following: Let $G(q)$ be the distribution of the quality parameters (i.e., the number of countries with characteristics q).

D = total number of potential users of the technologies

$$= \int_0^1 G(q) \, dq$$

When only the traditional technology is available

$$D = \underbrace{\int_0^{q^t} G(q) \, dq}_{\text{nonusers}} + \underbrace{\int_{q^t}^1 G(q) \, dq}_{\substack{\text{traditional} \\ \text{users}}}$$

When both technologies are available,

$$D = \underbrace{\int_0^{q^t} G(q) \, dq}_{\text{nonusers}} + \underbrace{\int_{q^t}^{q^*} G(q) \, dq}_{\substack{\text{traditional} \\ \text{users}}} + \underbrace{\int_{q^*}^1 G(q) \, dq}_{\substack{\text{modern} \\ \text{users}}}$$

Changes in the limits of the integrals (q^* and q^t) will affect the number of users in each category.

NOTES

[1] M. F. Caswell and D. Zilberman, 1985, "The Choices of Irrigation Technologies in California," *American Journal of Agricultural Economics*, Vol. 67, pp. 224–234; M. F. Caswell and D. Zilberman, 1986, "The Effects of Well Depth and Land Quality on the Choice of Irrigation Technology," *American Journal of Agricultural Economics*, Vol. 68, pp. 798–811.

[2] J. E. Estes, 1984, "Improved Information Systems: A Critical Need," *Machine Processing of Remotely Sensed Data Symposium.*

[3] R. N. Colwell, editor-in-chief, 1983, *Reference Manual of Remote Sensing*, 2nd ed., American Society of Photogrammetry, Falls Church, Virginia.

[4] *Ibid.*

[5] M. L. Burstein, 1984, "Diffusion of Knowledge-Based Products: Applications to Developing Economies," *Economic Inquiry*, Vol. 22, pp. 612–633.

[6] Estes, 1984, *op. cit.*

[7] R. L. Shelton and J. E. Estes, 1981, "Remote Sensing and Geographic Information Systems: An Unrealized Potential," *Geo-Processing*, Vol. 1, pp. 395–420.

[8] Although the term "country" is used throughout the chapter as the decision-making unit, the theory can be applied to individuals, firms, or regions.

[9] Economists (as well as those in other disciplines) are sometimes accused of making so many simplifying assumptions that they "assume away the problem." Often, however, including many details only adds to the complexity of the analysis and contributes nothing to the qualitative results.

[10] An excellent discussion of such monetary values can be found in Tayler H. Bingham, 1988, "Social Values and Environmental Quality," in *Changing the Global Environment*, D. B. Botkin, M. F. Caswell, J. E. Estes, and A. Orio (eds.), Academic Press.

[11] A single level of information in the hierarchical scheme is assumed here in order to abstract away from the choice of information level so that the analysis can concentrate on the differences in costs that obtain for each information set.

[12] The concept of "appropriate technology" as introduced by E. F. Schumacher, 1973, *Small Is Beautiful*, New York, Harper & Row is used in this chapter to describe the technology that gives the highest net benefits to a country (or region) under the circumstances that prevail in that country.

[13] This is obviously a simplification since there are many hardware/software combinations for each technology that would produce the desired level of information. In accordance with the basic economic theory of the firm, it is assumed that there is one package of inputs for each system that is most efficient, and that combination defines the technology.

[14] Although this formulation implies that there is only one provider of each technology, the analysis can be used when there is competition among the sellers. To include such a consideration would add to the notational complexity without offering further insight into the problem.

[15] Linear quality indices are often used to summarize complex characteristics. For example, the Storie soil classification system combines information on soil type, permeability, and slope into a single number, which is used to designate land quality.

[16] Although the following framework will be presented in a way that is not mathematically rigorous, the arguments can be demonstrated formally. The basic assumptions used to derive the theory are presented in the Appendix, and a complete discussion of the mathematical model can be found in M. F. Caswell, 1987, "The Adoption of Remote Sensing Technology," paper presented at the Western Economic Association Conference, Vancouver, Canada. For this chapter, however, I would like to stress the intuitive appeal of the theory.

[17] William G. Brooner, 1988, "The Application of Remote Sensing in South America: Perspectives for the Future Based on Recent Experiences," in *Changing the Global Environment*, D. B. Botkin, M. F. Caswell, J. E. Estes, and A. Orio (eds.), Academic Press.

[18] The assertion that the decision maker would be indifferent at q^* is based on the assumption that the benefits of adoption come only from gaining the information and are not a function of which technology was chosen (as would be the case if national prestige was a social goal). It also implies a static framework in that there are no expectations that relative values will change within the decision-making time frame, nor are there any other forms of uncertainty.

[19] This is a static analysis, so the dynamic adjustment process is not described.

[20] In order to use two-dimensional graphs to help to illustrate the theory, the following discussion will be based on the assumption that all countries gain the same benefits for obtaining the desired level of information. Also, the frequency distribution depicted in Figure 26-3 is drawn approximately uniform only for convenience. This assumption is dropped for the later discussion.

EDITORS' INTRODUCTION TO:
A. ENDRES
The Search for Effective Pollution Control Policies

Dr. Endres is an environmental economist who has studied the most recent developments in industrial pollution control policies, especially those being used in the USA and in West Germany. This chapter deals with the important issues of economic incentives, political reality, and technological advance. Dr. Endres uses a theoretical economic framework to contrast the main types of pollution control policies with respect to their effectiveness and economic efficiency. Several empirical studies are briefly surveyed to give an idea of the magnitude of cost savings that the market-based approaches might provide as compared to emissions standards and other command-and-control policies. Dr. Endres concludes that a stronger integration of market incentives into environmental policy would be beneficial for economic growth and environmental quality.

437

27 THE SEARCH FOR EFFECTIVE POLLUTION CONTROL POLICIES

ALFRED ENDRES
Technische Universitaet
Berlin, West Germany

INTRODUCTION

Since the environment's capacity to neutralize or assimilate our wastes has been exceeded in many areas, even seemingly local pollution episodes can have a global impact. The cumulative effects of many coastline cities dumping inadequately treated waste into their local waters threaten to destroy the viability of large areas (e.g., the Mediterranean Sea, Lake Erie). Industrial growth in the northern hemisphere may impose a global warming trend on the entire planet.

For the last twenty years, there has been an attempt by many governments to reduce the levels of pollution. In the United States, national and state environmental agencies have been established and billions of dollars have been spent to reduce air and water pollution and to rid the environment of toxic wastes.[1] Other countries have made large commitments of resources as well. Despite the enormous amount of money that has been spent, there is a concern that global pollution levels have not significantly decreased. Why have pollution control policies fallen short of their mark? The purpose of this chapter is to compare the most commonly used control policies with policies that are based on market incentives and to assess whether the benefits of changing policies would outweigh the costs.

Economists have been concerned during the last decade with characterizing the optimal level of pollution (from the standpoint of society as a whole) and designing the most efficient policies for attaining that level.[2]

The most commonly used approach to environmental policy has been the direct regulation of polluters. This policy type can take many forms, such as emission standards, abatement technology requirements, or restrictions on the quality of inputs that are used. Economists have criticized this command-and-control approach on various grounds and have proposed policies based on market incentives as alternatives or supplements to direct regulations. The main variants of these economic incentive concepts are effluent charges and transferable discharge permits. Under an effluent charge approach, each firm emitting a certain kind of pollutants has to pay a constant fee per unit of emission. The charge for different kinds of pollutants (e.g., NO_x or SO_2) would vary depending on the damages that type of emission would cause. Under the transferable discharge permit approach, the right to emit a certain amount of a particular pollutant is granted using a system of vouchers (permits). To be allowed to pollute, a firm is required to have the appropriate number of permits. These permits may be auctioned off among the potential polluters or may be given away free according to an agreed-upon criterion. In both cases, it is important to note that permits are transferable, that is, they may be traded among the polluters.

For the remainder of this chapter, emission standards will represent the command-and-control policies that are currently being used. Both effluent charges and transferable discharge permits will be used to illustrate the characteristics of market-based control policies. The most important criteria for this analysis are the efficiency, ecological accuracy, innovative push, and political feasibility of an environmental policy. The *efficiency* question relates to the ability of a policy to decrease pollution to a predetermined level at minimum abatement cost. *Ecological accuracy* denotes the ability of a policy to reach the predetermined pollution target level without trial-and-error adjustment processes. The *innovative push* relates to what extent the policy offers the pollution producer an incentive to seek technological improvements in pollution abatement. *Political feasibility* relates to a policy's appeal to interest groups influencing the political decision process.

To arrive at a fair comparison of the different environmental policies, one assumes that all three have the same pollution reduction target (i.e., the reduction of a certain pollutant's emission in a certain region by a certain amount). After a theoretical discussion, several quantitative studies are briefly surveyed to assess the abatement costs of arriving at a predetermined environmental policy target using alternative environmental policies. Finally, two examples of practical market-based environmental policies are discussed.

THEORETICAL CONSIDERATIONS

EFFICIENCY

Consider the environmental polity target of reducing pollution of a certain type X from stationary sources in a certain region to 50 percent of the prepolicy level. An environmental policy is efficient if it brings about this level of pollution reduction at minimum cost.

What properties characterize an efficient pollution control mechanism when the predetermined level of regional pollution reduction can be achieved by many different combinations of pollution reductions by the individual pollution producers? The same level of environmental quality might be reached if all polluters reduced their emissions by 50 percent or if one-half of those producers shut down. Therefore, the total regional abatement cost of achieving the predetermined pollution level will depend on the way the pollution reduction is distributed among the producers. According to economic theory, the least costly way of achieving the target would be to have each pollution producer reduce emissions until the additional (i.e., marginal) abatement costs are the same for each.[3] This is a property that characterizes an efficient pollution control policy. Those producers that have the most cost-effective technologies for reducing pollution would emit less than a producer that has higher costs for reducing emissions. Therefore, it would not be optimal (i.e., achieve the target at minimum cost) if each producer reduced emissions by the same amount (unless, of course, the group of pollution producers was identical).[4]

Consider the efficiency property of a simple *command-and-control policy* to achieve a 50 percent reduction target by making each firm halve its X emissions. Since efficiency requires that the emission reduction shares of the firms be differentiated according to the costs of reducing the last unit of pollution, a command-and-control approach assigning equal amounts of emission reduction to each firm obviously fails to meet the efficiency condition.

Policymakers might try to overcome this efficiency deficit of the command-and-control approach by setting differentiated emission standards, taking into account the abatement cost differences among different producers. Designing such emission standards would be difficult, however, since policy makers do not know the cost of reducing the last unit of pollution (i.e., the marginal abatement cost). Neither the average costs of pollution control nor the size of the producer serve as reliable proxies for the marginal cost. Moreover, the attempt to tailor efficient emission standards according to the marginal abatement costs of individual producers would result in a strong disincentive for polluters to reveal any information on these costs (and to try to keep these costs down). An effective policy must offer the incentive for producers to cooperate. Otherwise, huge enforcement costs would dissipate the benefits that might be gained by reducing the level of pollution.

An *effluent charge* that is set at the proper level, on the other hand, will induce individual producers to reduce pollution in the most cost-effective manner. A producer must decide whether the benefits of emitting more pollutants outweigh the effluent charges that they will have to pay. These benefits are merely the additional abatement cost that they would not incur. Therefore, each producer will continue to reduce its emissions up to the point where its marginal abatement cost equals the effluent charge. For units up to this point, abatement is cheaper than paying this tax. For units beyond that point, paying the tax is cheaper than abatement. With this approach, the effluent charge per unit of emission for a certain pollutant is identical for all producers emitting that pollutant. Therefore, in the effluent-charge equilibrium, the marginal abatement costs for each firm are equal to the effluent charge and thus equal to each other, as required by the efficiency (cost-minimization) conditon.

The second market-based pollution control policy is of auctioning *transferable discharge permits*. For the units of effluent emitted, permits must be purchased. Each pollution producer must decide whether to reduce pollution (and bear the costs of doing so) or to bid for discharge permits in order to continue polluting. Since marginal abatement costs rise with the level of pollution reduction, it is reasonable for each producer to carry abatement up to the point where the marginal abatement cost has increased to the level of the permit price. If the producers reduced fewer emissions, they would be paying more in permit costs than they would have for pollution reduction. Reducing emissions beyond that price would be unwise for the producer since, for each additional unit, abatement would be costlier than the permit to emit that unit. For units still emitted at a level where marginal abatement cost exceeds the permit price, permits are bought. Thus, the rationale for the efficiency of a transferable discharge permit auction is analogous to the argument in the effluent charge case: Since all producers adjust by equating their marginal abatement cost to the permit price and this price is identical for all firms, their marginal abatement costs are identical to each other when the permit market equilibrium is achieved. The transferable discharge permits auction meets the efficiency condition just as the effluent charge does.

These efficiency properties of the transferable discharge permit system are not disturbed if permits are given away instead of being auctioned off. Consider the case of permits being given away to producers in proportion to their prepolicy emission levels. The permits are devalued to 50 percent in order to meet the environmental policy target. Then, there will be an incentive for the high abatement cost polluters to buy permits from their low-cost fellows in order to avoid pollution reduction to a full 50 percent. There will be an incentive for low-cost polluters to sell these permits, thereby reducing more than 50 percent of their emissions. In a competitive permit market equilibrium, producers will end up in an efficient situation. They will abate pollutants up to the point where marginal abatement costs equal the permit price and equal each other, just as in the auction model.

ECOLOGICAL ACCURACY

Regarding the ability to attain precisely a predetermined pollution target, emission standards perform well because the emission level is under direct control of the environmental agency.[5] The goal of reducing regional emissions by 50 percent is achieved if each producer reduces its emissions by 50 percent.

In contrast, the effluent-charge approach suffers from the fact that the charge acts as a "price" or "tax" for discharging to which each producer is free to adjust its discharge quantity. Therefore, the only thing the administrating agency can do is guess the polluters' marginal abatement costs and thereby try to foresee the polluters' abatement reactions to alternative tax rates. The agency then sets the tax rate at a level estimated to induce pollution reduction by the appropriate amount. It is likely that the agency will misjudge the marginal abatement costs and therefore will miss the pollution target. If the resulting equilibrium pollution level is higher than the target, the effluent charge would have to be increased. If the pollution level is lower, the charge could be decreased. Of course, such a process of trial and error is itself costly and politically hard to justify. In addition, effluent charges would have to be adjusted if the aggregate marginal abatement costs change, due, for example, to economic growth or inflation.

In terms of ecological accuracy, it is the advantage of a transferable discharge permit policy that the level of pollution is under direct control of the environmental agency through the number of permits auctioned off. In a transferable discharge permit system, inflation and economic growth would not cause an increase of emissions since the number of permits is fixed (although the price of the permits would increase). Therefore, of the three policy approaches under discussion, both transferable discharge permits and emission standards would lead to a high level of ecological accuracy. The effluent charge approach would probably result in a trial-and-error adjustment process before reaching the targeted pollution level.

INNOVATIVE PUSH

The innovative push of an environmental policy is the extent to which that policy will encourage improvements in the technologies used to reduce pollution.

When a command-and-control policy such as an emission standard is used, there is an incentive for producers to develop and apply techniques able to meet the required standards at lower cost. There is no incentive, however, to look for new methods to abate more pollutants than are called for by the standard.

On the other hand, an effluent charge produces a twofold incentive for polluters to invent and apply new methods of pollution control: Whenever the new method achieves the producers' given abatement level at a lower cost than the old one, producers benefit directly. Moreover, whenever the new method abates more than the old equilibrium abatement level at an additional cost lower than the old method, producers save on the charges they no longer have to pay (net of the added abatement cost) by pushing abatement beyond the old equilibrium level. The new equilibrium level is defined by the quantity of emissions reduction that equalizes the new (lower) marginal abatement cost with the effluent charge.

For transferable discharge permits, the issue of innovative push is less clear. The chance to achieve given abatement levels at lower cost gives some incentive to innovate, as with the effluent charge. But the incentive to introduce new methods of reducing pollution beyond the pretechnical-progress level turns out to be lower than in the case of the effluent charge. The proof of this can be found in the Appendix.

By considering only the criterion of innovative push when evaluating the three policies, it can be seen that the market-based approach of effluent charges is superior to either standards or permits.

POLITICAL FEASIBILITY

Transferable discharge permits and effluent charges are by far the most popular pollution control policies among environmental economists. In practical environmental policy, however, command-and-control approaches are the ones that are most often used. The reasons for this divergence are many: there is a concern by those in the political process of shaping environmental policies that market-based approaches give producers the "right to pollute." There is a belief that effluent charges and transferable discharge permits explicitly give polluters the right to use and degrade our environmental resources. There are three important points often overlooked with this critique, however.

1. In a command-and-control policy such as emission standards, firms are allowed to pollute up to a limit. This also implies that property rights have been implicitly allocated to the polluters.
2. Effluent charges and transferable discharge permit auctions are "tough" on polluters in that they make them pay for their rights to pollute. Emission standards are "soft" on polluters in that producers get the right to use the environment free of charge.
3. Even though the quantity of emissions from the individual source is dependent on economic considerations with the market-based approaches, the total amount of pollution in the region would be just as restricted as with the command-and-control approach. Therefore, any level of environmental quality that can be attained by an emission standard can also be attained using effluent charges or transferable discharge permits if the charge or the number of permits, respectively, is specified correctly. Policymakers who are concerned about increasing the level of environmental quality need not depend on command-and-control policies to achieve their goals.

When some policies are known to be more effective than others, they would be more likely to be accepted through the political process. It is even likely that more ambitious environmental quality goals are feasible in the political process if efficient policies are used. Industry opposition to environmental policies decreases when control costs decrease. This would be particularly true if transferable discharge permits were initially distributed at no cost. In this case, the efficiency gains could be partially appropriated by the polluting firms.[6]

One reason that command-and-control policies have proven to be more politically feasible than market-based control measures is that direct regulations usually apply less stringently to established industries (or plants) than to new facilities. Therefore, direct regulations have been favored by established industries so they could retain an advantage. For example, the 1970 U.S. Clean Air Act helped to preserve aging industries in the Midwest and Northeast at the expense of the developing areas in the Southwest.[7]

Possibly the most important reason for polluters consenting to the command-and-control approach in environmental policy is that, with a policy such as emission standards, the generation of emissions is allowed and costless up to the limit defined by the regulatory standards. The only regulatory cost for polluters with this policy is that of holding down emissions to the maximum level compatible with the regulation. The burden may even be lightened by subsidies for pollution control investment. With effluent charges, polluters pay twice—once for reducing emissions and once for paying the charges. If transferable discharge permits are allocated through an auction, polluters would also pay twice—for purchasing the permits and for reducing emissions. Thus, even though they may be more efficient in the sense of using fewer resources for pollution control than command-and-control approaches, market-based control policies may be much more expensive to polluters than standards would be. Of course, if the transferable discharge permits are allocated free of charge instead of being auctioned, polluting firms would be less likely to oppose this policy. Support for auctioned permits might come from environmental organizations since they could also purchase the permits and thus lower the level of pollution. Thus, when considering political feasibility, transferable discharge permits appear superior to the other approaches in that they would potentially gain support by both polluters and environmentalists.

MARKET-BASED INCENTIVES: EMPIRICAL EXAMPLES

There is an increasing number of interesting simulation studies comparing market-based incentive policies to the command-and-control approach. The criterion for the comparison is economic efficiency as defined earlier—the least costly method of reaching the environmental quality goal. The market-based incentive policy that is most frequently analyzed is that of transferable discharge permits.

In a 1984 study conducted in St. Louis, Missouri, Atkinson and Tietenberg[8] compared an emission standard, an emission permit system, and an ambient permit system with respect to the control costs to achieve the ambient standard designated by a state implementation plan. The emission permit system is the same as the transferable discharge permit approach defined earlier—the permits to pollute in different locations within the region are traded on a one-to-one basis. With the ambient permit system, however, the emissions from different points in the region are traded according to ratios that take into account the differences in the impact the emissions would have on ambient air quality. This system has the potential of meeting the ambient air quality standard at each receptor point with a higher level of total emissions than either of the other two policies because the transmission properties of the emissions from different points in the region are incorporated. Although this feature will result in cost savings, it should be realized that receptor points that had a higher level of air quality than required by the standard would deteriorate in quality if the ambient permit system was used.

The results of the Atkinson and Tietenberg study showed that the primary ambient standard would be attained at

- a control cost of $7525 per day when the emission standard is used
- a control cost of $882 per day when the emission permit system is used
- a control cost of $83 per day when the ambient permit system is used.

The cost saving that could be obtained from using market-based incentive approaches such as the permit system could be substantial.

Another study that confirms the qualitative results described earlier was conducted by O'Neil et al.[9] They compared the costs of increasing the levels of dissolved oxygen in the Fox River in Wisconsin using either the emission standard approach or a transferable discharge permit policy. Their results show a control cost of $11.1 million per year when the emission standard is used and a control cost of $5.4 million per year when the transferable discharge permit system is used.

Studies briefly reviewed here and others[10] suggest that the efficiency of environmental policies can be increased considerably by switching from command-and-control policies to approaches using market-based incentives for pollution control. If these gains in efficiency are distributed between business and the environment (as explained in note 6 for the case of the transferable discharge permit policy with free initial distribution), better environmental quality may be achieved at lower resource cost.

In the case of a transferable discharge permit auction, the efficient result is brought about because the polluters with low costs of abatement will demand fewer permits than the high-cost ones. The permit auction system succeeds in having the high levels of abatement for the low abatement cost firms compensate for the low abatement levels of the high abatement cost polluters.[11]

MARKET-BASED INCENTIVES: "PRAGMATIC" EXAMPLES

Practical environmental policy is a blend of many different approaches. Even though market-based incentives are not often used at this time, there are some interesting elements of it in the environmental policy of some countries. Two examples are briefly reviewed here.

AN EFFLUENT CHARGE IN WEST GERMANY

In the Federal Republic of Germany, all firms and municipalities discharging effluents directly into water courses are liable for effluent charges according to the wastewater charges act (*Abwasserabgabengesetz*) of 1976.[12] The charge is assessed per damage unit, and these units are calculated on the basis of the discharged quantity of oxydizable matter, suspended solids, and toxicity. The West German wastewater charge is not the same as the pure effluent charge described in the theoretical discussion earlier. The wastewater charge is designed to supplement the command-and-control approach in an overall water pollution policy. Nontradable permits are required for the discharge of effluents, and polluters are required to install reasonably available control technology (*allgemein anerkannte Regeln der Technik*) according to the Federal Water Act (*Wasserhaushaltsgesetz*) or to meet more restrictive standards imposed by regional authorities. The need to comply with these standards reduces the ability of firms to choose between the reduction of emissions (saving the charge) and the generation of emissions (saving abatement cost). This freedom of choice, however, is the effluent charge's key to efficiency. Therefore, it cannot be expected that the West German wastewater charge would attain the same level of cost savings as the effluent charge as recommended by economists.

In spite of these and other reservations, the German wastewater charges seem to have produced a significant ecological impact. According to a recent report of the Federal Ministry of the Interior, many (particularly large) polluters increased their pollution abatement investments to evade the charge. Also, an additional effect of the charge was to speed up compliance to the regulations of the Federal Water Act.

A TRANSFERABLE DISCHARGE PERMIT SYSTEM IN THE UNITED STATES

In the United States, a "pragmatic" transferable discharge permit system has been developed—the Controlled Emission Trading Policy.[13] Its most important elements are the bubble and the offset concepts. With the *bubble concept,* an emission standard does not have to be met by each individual source of pollution but by a group of sources within a proscribed region. Thus, polluters operating the sources may agree to restrict pollution of sources with low abatement costs more than a command-and-control emission standard would demand and to compensate this extra abatement by a lower reduction in high-cost sources. Of course, the bubble arrangement will be approved by the regulatory agency only if it is considered to be ecologically equivalent to the traditional emission standard approach. With the *offset concept*, a new source of pollution is allowed to begin operation if its additional pollution is more than compensated for by a reduction in the emissions of sources already operating in the area.

The idea that both practical concepts have in common with the theoretical transferable discharge permit approach is that emission rights are not fixed to individual sources and may be traded among sources. There are, however, differences between theory and practice. The transferable permit idea assumes a perfectly competitive permit market while bubble and offset transactions are mostly bilateral trades with the considerable involvement by a regulatory agency. Moreover, the offset policy requires that new sources meet New Source Performance Standards,[14] a restriction not implied by the transferable discharge permit approach.

There is also a problem in defining the geographical area in which emissions are tradeable. On efficiency grounds, the area should be as large as possible: If the goal is to reduce regional emissions by a certain percentage, this will be achieved at least cost if emissions are transferable throughout the whole region. However, there are distributional problems with this approach. Trading throughout the entire region may lead to an unacceptable accumulation of emissions in some areas. To avoid "hot spot" problems, an asymmetric transferability of emission rights may be warranted: The region may be split into subregions, which would be defined by geographic and social characteristics. In addition to the emission standard for the region as a whole, a "hot spot" standard may be specified for each region. As soon as this limit is reached in one subregion, emissions may be allowed to be traded only out of this subregion (or within it). As these examples illustrate, there may be many compromises made between economic efficiency and political feasibility that would result in improved environmental quality at a reduced cost.

SUMMARY

Theoretical reasoning and empirical evidence suggest that the economy and the environment could gain by moving from an environmental policy that depends on command-and-control approaches to a system of market-based incentives. In practice, however, resistance in the political process is strong. Perhaps environmental economists would be more effective if they urged a step-by-step integration of market-based incentives into the current command-and-control framework rather than recommending a completely new environmental policy based solely on economic incentives. Of course, these market-based incentive approaches cannot be taken directly out of environmental economics textbooks but have to be modified in the light of particular circumstances. The examples given earlier have shown that this can be done.

Presently, transferable discharge permits with free initial distribution of rights seem to be the most promising road to this end. The problems with this approach do not appear to be insurmountable. As for effluent charges, there are some effective policies used in practice, particularly in the area of water pollution control. These policies could be improved by tying tax liability closer to the amount and noxiousness of the pollution generated. Progress in measurement techniques will substantially facilitate this task.

The solution to global environmental problems will require the cooperation of many countries throughout the world. Government budgets are limited, however, so the most cost-effective pollution control policies will need to be used to obtain the desired levels of water and air quality. Currently, most governments use command-and-control policies, such as emission standards. As this chapter has suggested, the use of market-based incentive approaches has the potential to reduce control costs substantially. Even modified types of transferable discharge permits or effluent charge systems would reduce the costs of pollution control. The most advanced technologies will not be effective in improving environmental quality if they are not used. Only market-based approaches will give the polluter the incentive to choose the technology that will reduce emissions in the least costly way.

APPENDIX

The incentive to introduce new methods of reducing pollution is less for transferable discharge permits than in the case of an effluent charge. To show very briefly that this is true, a simple case will be illustrated. Consider a competitive industry that produces pollution as a by-product of the manufacturing process. For simplicity, it will be assumed that all firms are similar enough that we can use the "representative firm" to typify the situation. Figure 27–1 will be used to show how firm i will

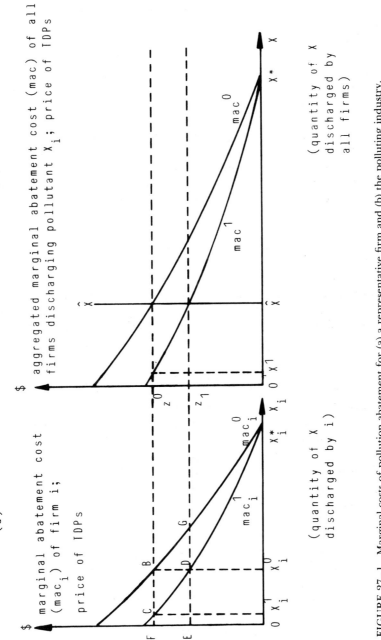

FIGURE 27–1 Marginal costs of pollution abatement for (a) a representative firm and (b) the polluting industry.

react to industry conditions. Figure 27–1a shows the marginal costs of pollution abatement for firm i, and Figure 27–1b shows the marginal abatement costs for the industry. The point $X_i = X_i^*$ represents a situation when emissions are not controlled and $X_i = 0$ would exist if there were total control (no emissions allowed). The costs of reducing each additional unit of emissions from the benchmark value of X_i^* using the technology available in the initial period are shown in the marginal abatement cost curve (mac_i^0). The equivalent curve for the industry is represented in Figure 27—1b by mac^0. The efficient permit price (where mac^0 meets $\hat{X}$, the permit supply curve) would then be set at z^0. The firm would respond to such a price by initially demanding X_i^0 permits (that level of emissions where $mac_i^0 = z_0$). Each firm in the industry would reduce emissions up to the point where the marginal cost of abatement would equal the price of the discharge permit.

Suppose that there is a technological innovation that can reduce emissions at a lower additional cost per unit. This case is represented by the lower marginal abatement curve in Figure 27–1a, mac_i^1. If firm i was the only polluter that used the new technology (i.e., the permit price remained at z^0), then that firm would continue to reduce emissions until the level associated with point C was reached—where $mac_i^1 = z^0$. The use of the new technology would result in a savings for firm i, which is the area X_i^*CB in Figure 27–1a. If all firms have access to the new technology, however, the industry marginal abatement curve becomes mac_1 and the new equilibrium permit price would be z^1. When the new permit price is considered, firm i would only have the incentive to reduce emissions up to the point where $mac_i^1 = z^1$ (point D in Figure 27—1a). The equilibrium level of pollution (X_i^0) of the average firm i does not change by the introduction of the new technique and neither does the equilibrium total level of pollution $\hat{X}$.

Since the permit price will be reduced when the cost effects of the new innovation are realized, the savings in control costs for each firm will be less. If the lower level of savings is not sufficient to cover the costs of introducing the new technique, then the firm will hesitate to innovate. If this is true for all firms, the polluters face a prisoner's dilemma. Each firm would gain if all firms would innovate. However, it is attractive for each individual firm to take a free rider attitude—a firm that waits while others innovate will benefit from the lower permit price without incurring the costs of innovation.

In the long run, the decreasing permit price, induced by the technical progress, reduces the incentive for further innovation.

An environmental agency might be able to avoid this problem by stabilizing the price of the permits. Also, there are automatic tendencies countervailing the price-decreasing effect of innovation in the abatement technique. Growth in the number and size of the polluting firms of the region as well as inflation make the aggregate marginal abatement cost increase. Thereby, the demand curve in the permit market shifts to the right, pushing the permit price up.

NOTES

[1] B. Commoner, 1987, "The Environment," *New Yorker*, June 15, pp. 46–71.

[2] For an overview of these issues, see M.A. Freeman, 1982, *Air and Water Pollution Control: A Benefit Cost Assessment*, New York; E.J. Mishan, 1974, "What Is the Optimal Level of Pollution?" *Journal of Political Economy*, pp. 1287–1299; P.B. Downing, 1984, *Environmental Economics and Policy*, Boston-Toronto; A. Endres, 1985, "Umwelt- und Ressourcenökonomie," Darmstadt, West Germany.

[3] The marginal abatement cost is the cost of eliminating an additional unit of pollution. It is plausible that for each firm marginal abatement costs rise with the level of pollution abatement. For instance, to abate the first unit of a pollutant "a wet towel covering the smoke stack" will do; to abate an additional unit in a situation where 99 percent of the emissions have been already abated requires sophisticated installations.

[4] To see this, for example, consider a situation where the overall pollution reduction is split in a way not equalizing the marginal abatement costs of two firms i and j. Then, if the marginal abatement cost of i is higher than that of j, total abatement cost can be reduced without changing the overall level of pollution when i's share of pollution abatement is reduced by one unit and this reduction is compensated by a one-unit increase in j's share. Given this possibility of a reduction in total abatement cost, the initial situation cannot have been efficient (cost minimizing). However, this shift of one unit of pollution reduction from i to j makes i's marginal abatement cost decrease and j's marginal abatement cost increase. So the process of shifting pollution abatement shares reduces the difference between the marginal abatement costs of the two firms. If, after the shift of one unit, the marginal abatement cost of i is still higher than that of j, another unit shifted further reduces total abatement cost. Shifting the load of pollution reduction from i to j continues to reduce the abatement cost in the region (i.e., to make the situation more efficient) until the marginal abatement costs of i and j have equalized. In this situation, no more cost-reducing shifts are possible, and efficient pollution abatement is achieved.

[5] The ecological accuracy of the command-and-control policy suffers if the emission standard is given in terms of emission *concentrations*. With increasing production, total emissions may increase even though the standard is met.

[6] Consider a command-and-control policy and a more efficient transferable permit policy. The efficiency gains of the latter policy can be distributed in several ways between polluters and the government (which we assume is representing the best interests of society). Suppose a certain regional pollution reduction is aimed at with transferable discharge permits which are initially distributed without cost, instead of using the command-and-control approach. If the permit policy is used to arrive at the same level of pollution as aimed at by the command-and-control approach, the efficiency gains of the

former policy are all appropriated by the polluters. They reduce the same amount of pollution as they would have done under command-and-control at lower cost. If, on the contrary, the more efficient policy is used to arrive at more pollution reduction up to the point where the cost to polluters is equal to their cost under command-and-control, efficiency gains all go to the government. Here, equilibrium pollution under the permit policy is lower than under command-and-control but the cost situation of polluters is the same under either policy. In the two polar cases of efficiency gains distribution just mentioned, either polluters or the government gain. There are intermediate cases, where both sides gain, sharing the benefits from the more efficient policy. Here, the transferable discharge policy achieves higher environmental quality at lower cost.

[7] R.W. Crandell, 1983, *Controlling Industrial Pollution*, Washington, DC, Ch. 7.

[8] S.E. Atkinson and T.H. Tietenberg, 1984, "Approaches for Reaching Ambient Standards in Non-Attainment Areas," *Land Economics*, Vol. 60, pp. 148–159.

[9] W. O'Neil et al., 1983, "Transferable Discharge Permits and Economic Efficiency: The Fox River," *Journal of Environmental Economics and Management*, Vol. 10, pp. 346–355.

[10] R.M. Lyon, 1982, "Auctions and Alternative Procedures for Allocating Pollution Rights," *Land Economics*, Vol. 58, pp. 16–32; M.T. Maloney and B. Yandle, 1984, "Estimating the Cost of Air Pollution Control Regulation," *Journal of Environmental Economics and Management*, Vol. 11, pp. 244–263; J.J. Opaluch and R.M. Kashmanian, 1985, "Assessing the Viability of Marketable Permit Systems: An Applicaton to Hazardous Waste Management," *Land Economics*, Vol. 61, pp. 263–271; A.M. McGartland and W.E. Oates, 1985, "Marketable Permits for the Prevention of Environmental Deterioration," *Journal of Environmental Economics and Management*, Vol. 12, pp. 207–228.

[11] Analogous reasoning applies to the transferable discharge permit system with initially free distribution and for effluent charges.

[12] G.M. Brown and R.W. Johnson, 1984, "Pollution Control by Effluent Charges: It Works in the FRG, Why Not in the U.S.?" *Natural Resources Journal*, Vol. 24, pp. 929–966.

[13] H. Bonus, 1984, "Marktwirtschaftliche Konzepte im Umweltschutz," Stuttgart, West Germany; R.W. Crandall, 1983, *op. cit.;* T.H. Tietenberg, 1985, *Emissions Trading*, Washington, DC.

[14] The New Source Performance Standards are nationwide emissions limits for new sources specified by the U.S. Environmental Protection Agency. In low air quality regions, local authorities may even require the more restrictive Lowest Achievable Emission Rate Standards. In clean air regions, local authorities may require that the Best Available Control Technology Standards are met. In practice, these standards define emissions limits that lie between the limits of the New Source Performance Standards and the Lowest Achievable Emission Rate Standards. The idea behind all three kinds of standards is that new sources should install the pollution control equipment consistent with the actual state of technology.

INDEX

Acid rain, 22, 97, 122, 170, 188, 428, 431
see also Chapter 7
Advanced Very High Resolution Radiometer (AVHRR), 224, 273, 277–278
Advanced Visible and Infrared Imaging Spectrometer (AVIRIS), 195, 293
Africa, 6, 120, 128–130, 137–138, 148, 209–210, 247–248, 277, 348–350, 355
Agriculture, 106–108, 112, 223–227, 321–322, 396
alley cropping, 123, 129
arable land, 137, 141–147
shifting, 19, 129, 139
see also Dust Bowl; Irrigation; Chapter 8; Chapter 9

Air quality control regions, 104–105
Albedo, 25, 45, 150
Allelopathy, 126
Amazon, 94, 149, 223–227, 344–345
Appropriate technology, 126, 151, 424, 426, 431
Argentina, 92, 216–221
Arid lands, 21, 196, 247
Atomic Absorption Spectrometry (AAS), 171
Atmospheric chemistry, 74
see also Chapter 3; Chapter 4; Chapter 7
Australia, 148, 193
Bacteria, 12, 40, 126
see also Chapter 4
Bangladesh, 207–209

Bioeconomic equilibrium, 317
Biogeochemistry, 12, 30, 39–40
Biogeography, 76
Biological diversity, 21, 25, 27, 31–32, 61, 75–78, 102, 124–125
  see also Chapter 6
Biomass, 43, 139, 150, 277
Biome, 8, 12, 223, 279
BIOSOMA, 385–389
Biosphere, 4, 12–13, 24, 27, 29–30, 42, 45, 73, 78, 81, 110–113, 121, 286, 342
  see also Chapter 4
Biotechnology, 126, 390
Biotic provinces, 76, 79
Bolivia, 380, 425
Boundary Waters Canoe Area, 10
Brazil, 92–97, 139, 223–227, 354
Calcium, 12, 65
Canada, 9–10, 56, 104, 235
Carbon dioxide, 22, 44–45, 51–52, 65, 74, 80, 97, 111, 122, 126, 150, 170, 181
  see also Greenhouse effect; Chapter 21
Cartography, 195–196, 198–200, 221–223, 234–239, 254, 261–262, 294, 423
Chipko movement, 359
Climate change, 10, 20, 22, 25, 74, 97, 122, 147–150
  see also Greenhouse effect; Chapter 10; Chapter 21
Coastal zone, 12, 25, 60, 363
  see also Chapter 5
Common property resources, 319, 343, 351, 354, 357
  see also Chapter 21
Deforestation, 19, 122, 207–209, 223–229, 321, 358, 362
  see also Chapter 6; Chapter 9

Desalinization, 394
Desert, see Arid lands
Desertification, 21, 123
Developing nations, 20, 22, 26, 93–94, 320, 399, 423
  see also specific countries; Chapter 13; Chapter 14; Chapter 22
Diatoms, 11
Dose response, 106, 379
Drought, 144, 209–210, 322
Dust Bowl, 24
Earth Observing System (EOS), Chapter 19
Ecological accuracy, 440, 443–444
Ecological Monitoring Unit (EMU), 247
Ecosystem, 6, 8–9, 15, 78, 92, 94, 147, 279
Effluent charge, 442–445
El Nino, 7, 209
Energy, 170, 180, 188, 191–194
Endangered species, 9, 21, 92, 125, 237–238, 251
  see also Extinction
Erosion, 12, 24, 43–44, 123, 129, 143–147, 207, 237
Ethics
  egalitarian, 331, 335
  elitist, 331, 335
  utilitarian, 331–332, 335
Evolution, 10–12, 19, 31, 39, 56
Exhaustible resources, see Nonrenewable resources
Expressed preference approach, 375
Externalities, 322, 330, 371–372
External effects, see Externalities
Extinction, 19, 23, 25, 65, 124–125, 360, 364

see also Biological diversity; Chapter 6

Federal Republic of Germany, 110, 448

Fire, 9, 18–19, 139
see also Chapter 10

Fisheries, 30, 72, 81, 351–357

Flood, 63, 207–208, 217, 220, 322, 347

Forest, 5, 9, 19, 23, 25, 108–110, 207–209, 216–219, 238, 274–277
management, 357–359
tropical, 19, 21, 223–227, 254
see also Chapter 9
see also Deforestation; Chapter 6

Fossil fuels, 22, 25, 102, 188, 333

Fuel wood, 139, 321, 357

Gaia hypothesis, 44–45, 65–66, 73

Gas chromatograph mass spectrometry, 172

Genetic engineering, *see* Biotechnology

Genetic resources, *see* Biological diversity

Geographic Information System (GIS), 206–207, 210, 233–234, 252–253, 261, 290–291, 300–301, 423

Germany, *see* Federal Republic of Germany

Global, 7, 27

Global Environment Monitoring System (GEMS), *see* Chapter 16

Global Resource Information Data Base (GRID), *see* Chapter 16

Grazing, *see* Overgrazing

Greenhouse effect, 22, 25, 122, 149

Groundwater resources, 124, 176

Hazardous waste, 177–178, 259

Herbicide, 126, 173

Hydrocarbon, 105, 177–178, 182, 188, 191

Indonesia, 139, 194, 348, 353–354

Information system, *see* Geographic Information System

Innovative push, 440, 444

Integrated pest management, 126

Intergenerational factors, 328

International Geosphere–Biosphere Programme (IGBP), 271, 273, 280, 295

International Satellite Land Surface Climatology Program (ISLSCP), 279

Irreversible effects, 143, 149, 314, 380

Irrigation, 124, 345–351, 396

Kenya, 139, 314, 358

Kirkland's Warbler, 9

Landsat satellite, 190–194, 218–219, 224–227, 238, 272–273, 293, 425, 430

Limited entry, 353

Livestock, 19–20, 127

Long-term Ecological Research Program (LTER), 273

Maasai, 343, 361

Man and the Biosphere Program, 273

Man, early, 19

Madagascar, 92

Mangrove, 207–209

Mapping, *see* Cartography

Market-based incentives, *see* Effluent charge; Transferable discharge permits

Marsh, George Perkins, 8, 29, 341, 369–372, 380–381

Maximum sustainable yield, *see* Sustainable use

Mexico, 60, 64–65

Microbes
  ecology, 63–65
  mats, 54–56
  see also Bacteria

Monitoring, 106, 151, 171, 174–176, 205, 209–212, 219, 223–227, 246, 248–250, 294–295, 355, 363, 414–416
  see also Chapter 17

Natural resource management, 9, 121–122, 206–207, 246–248
  see also Chapter 22

Natural selection, 39

Nature reserves, 94–97, 237, 273–277, 315, 353, 358–362, 399

Nepal, 139, 358

Net benefits, *see* Net returns

Net returns, 317, 331, 424–427, 430

New Guinea, 355–356

Nigeria, 145, 345

Nitrogen, *see* Chapter 4; Chapter 7

Nitrogen dioxide, 122, 170
  see also Oxidants

Nonrenewable resources, 121, 320
  see also Chapter 12

Norway, 110

Nuclear, 319, 333
  see also Chapter 10; Chapter 17

Oceans, 248–250
  see also Chapter 5

Open access, 319, 343

Overgrazing, 20–21

Oxidants, photochemical, 106, 122
  see also Ozone

Ozone, 12, 104–111, 122

Panama, 361

Pareto criterion, 329–330, 334–335, 377

Peru, 26, 206–207, 222

Pesticides, 21, 126–127, 169, 181–182, 396

Philippines, 347–348, 354–355

Phosphorus, 6, 52

Plate tectonics, 41

Political feasibility, 440, 445

Population growth, 8–9, 20–21, 78, 85, 321

Remote sensing, 4, 29,
  see also Chapter 12; Chapter 14; Chapter 15; Chapter 17; Chapter 18; Chapter 19; Chapter 26

Renewable resources, 130, 246–248, 316–320, 328
  see also Fisheries; Forest; Wildlife

Revealed preference approach, 374

Reverse osmosis, 180

Savanna, 18, 140

SED principle, 312–313

SEDSURE principle, 312, 319–322

Sequestering capacity, 111

Shadow price, 327–329

Shuttle Imaging Radar (SIR), 194, 196, 208

Social prices, 312, 315

Social profits, *see* Net returns

Soils, 18, 39, 55, 102, 123, 129, 173, 175, 182, 206–207, 237, 254, 275–276
  see also Erosion; Chapter 9

Species–area relationship, 92
Sri Lanka, 139, 361
Stability, ecological, 8, 42, 45–46, 54, 319
Sulfur oxides, 39, 55, 65, 170
  see also Acid rain; Chapter 7
Supercritical fluids, 181
SURE principle, 312, 316–320
Sustainable development, see Sustainable use
Sustainable use, 6, 26, 82–85, 121, 357
  see also Chapter 20
Sweden, 110
Synthetic Aperture Radar (SAR), 298
Systeme Probatoir d'Observation de la Terre (SPOT), 197–198, 200, 222
System engineering, 412–416
Tanzania, 361
Technology adoption, 215–216, 423–428
Thalassogaia, 73
Thematic Mapper (TM), 190–192, 198–200, 222, 272–273, 279
Thermal destruction, hazardous wastes, 177–178
Thermal effluents, 170, 261, 292

Thermal Infrared Imaging Spectrometer (TIMS), 196, 292
Tillage, conservation, 123, 129
Toxic substances, 39, 123, 181
  see also Chapter 7
Tragedy of the Commons, 319, 343
Transferable discharge permits, 442–446
United Nations Environment Programme (UNEP), 24, 83, 212, 243
United Nations Food and Agriculture Organization (FAO), 94, 245–248, 347, 352
United Nations Stockholm Conference, 24, 243, 246, 329–330
Upwellings, 6, 12
Vinal (Prosopis ruscifolia), 216–219
Waste reclamation, 178–181
Water treatment, 180
Wildlife, 20, 123, 314, 359–362
World Health Organization (WHO), 245–246
World Meteorological Organization (WMO), 244–245
Zambia, 139, 360
Zimbabwe, 361